PENGUIN BOOKS

MASTERS OF BATTLE

Terry Brighton is the author of *Hell Riders: The Truth about the Charge
of the Light Brigade*. He has worked on the curatorial staff of The
Queen's Royal Lancers Regimental Museum for many years and is an
Associate of the Museums Association. He lives in Lincoln.

Masters of Battle

Monty, Patton and Rommel at War

TERRY BRIGHTON

PENGUIN BOOKS

PENGUIN BOOKS

Published by the Penguin Group
Penguin Books Ltd, 80 Strand, London WC2R ORL, England
Penguin Group (USA) Inc., 375 Hudson Street, New York, New York 10014, USA
Penguin Group (Canada), 90 Eglinton Avenue East, Suite 700, Toronto, Ontario, Canada M4P 2Y3
(a division of Pearson Penguin Canada Inc.)
Penguin Ireland, 25 St Stephen's Green, Dublin 2, Ireland
(a division of Penguin Books Ltd)
Penguin Group (Australia), 250 Camberwell Road, Camberwell, Victoria 3124, Australia
(a division of Pearson Australia Group Pty Ltd)
Penguin Books India Pvt Ltd, 11 Community Centre, Panchsheel Park, New Delhi – 110 017, India
Penguin Group (NZ), 67 Apollo Drive, Rosedale, North Shore 0632, New Zealand
(a division of Pearson New Zealand Ltd)
Penguin Books (South Africa) (Pty) Ltd, 24 Sturdee Avenue, Rosebank, Johannesburg 2196, South Africa

Penguin Books Ltd, Registered Offices: 80 Strand, London WC2R ORL, England

www.penguin.com

First published by Viking 2008
Published in Penguin Books 2009
1

Copyright © Terry Brighton, 2008
All rights reserved

Typeset by Rowland Phototypesetting Ltd, Bury St Edmunds, Suffolk
Printed in Great Britain by Clays Ltd, St Ives plc

A CIP catalogue record for this book is available from the British Library

ISBN: 978–0–141–02985–6

www.greenpenguin.co.uk

To Linda, with all of my love

In any specific action we always have the choice between the most audacious and the most careful solution. Some people think that the theory of war always advises the latter. That assumption is false. If the theory does advise anything, it is the nature of war to advise the most decisive, that is, the most audacious. Theory leaves it to the military leader, however, to act according to his own courage, according to the spirit of the enterprise and his self-confidence. Make your choice, therefore, according to this inner force; but never forget that no military leader has ever become great without audacity.

Carl von Clausewitz, *Die Grundsätze des Kriegführens*
(Principles of War), 1812

Montgomery . . . the greatest living soldier
(General Dwight D. Eisenhower)
Montgomery is Britain's greatest general
(General Gerd von Rundstedt)

★

Patton is America's best (General Gerd von Rundstedt)
General Patton, the rootin', tootin', hip-shootin' commander of
American Forces (NBC)

★

Rommel: the boldest *Panzerwaffen* general we have in the
German army (Adolf Hitler)
Rommel, Rommel, Rommel – what else matters but beating him?
(Winston Churchill)

★

There was a general consensus that they should put Montgomery
and Patton and Rommel in the same ring and take off the gloves
and let 'em go at it.
(Bill Mauldin, in *An American Oral History of WWII*, Studs Terkel)

Contents

List of Illustrations

Center in California (US National Archives, Washington, DC)

List of Maps

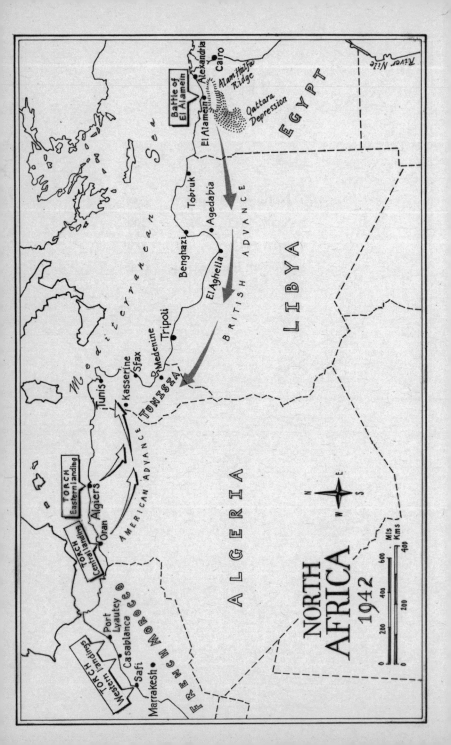

NORTH AFRICA 1942

Cairo
Alexandria
River Nile
EGYPT
Battle of El Alamein
El Alamein
Alam Halfa Ridge
Qattara Depression
Tobruk
Agedabia
Benghazi
El Agheila
BRITISH ADVANCE
LIBYA
Tripoli
Medenine
Sfax
Kasserine
Tunis
TUNISIA
AMERICAN ADVANCE
ALGERIA
TORCH Eastern landing
Algiers
TORCH Central landing
Oran
Port Lyautey
Casablanca
Safi
Marrakesh
FRENCH MOROCCO
TORCH Western landings
Mediterranean Sea

Mls
Kms
0   200   400   600
0   200   400

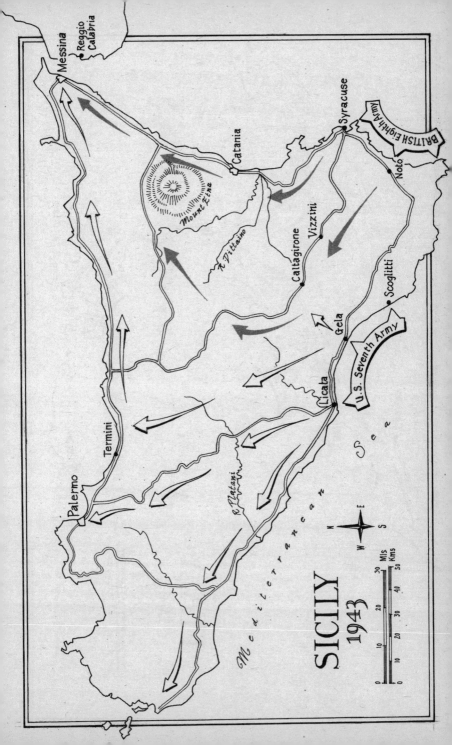

SICILY
1943

Messina

Reggio
Calabria

Catania

Monte Etna

R. Dittaino

Syracuse

BRITISH EIGHTH ARMY

Noto

Vizzini

Caltagirone

Scoglitti

Palermo

Termini

Gela

Licata

U.S. Seventh Army

R. Platani

Mediterranean Sea

N
W E S

Mls
Kms

30   20   10   0
50 40 30 20 10 0

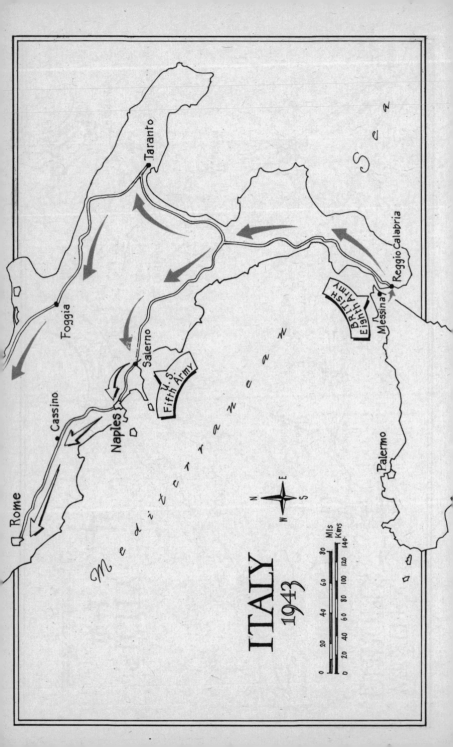

ITALY
1943

Rome

Cassino.

Naples

Salerno

U.S. Fifth Army

Foggia

Taranto

BRITISH Eighth Army

Messina

Reggio Calabria

Palermo

Mediterranean

N Sea

Mls
0   20   40   60   80
0   20   40   60   80 100 120 140
Kms

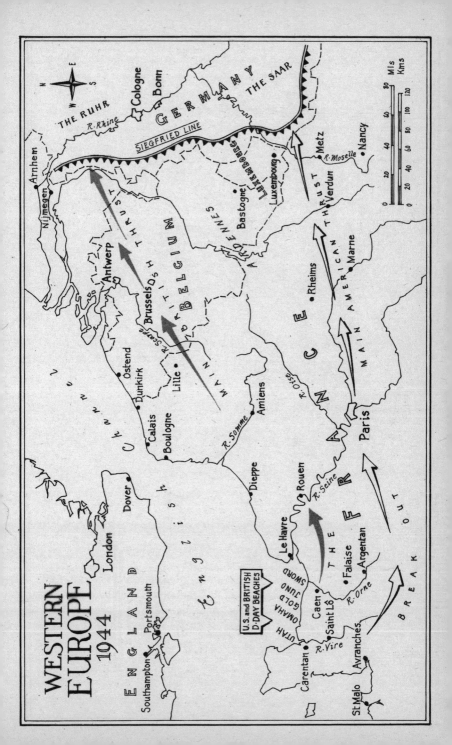

Prologue: The Ego at War

In the Second World War, Great Britain, the US and Germany each produced one land-force commander who stood out from the rest: Bernard Montgomery, George Patton and Erwin Rommel. These three armour-plated egos were, in their own opinion but also in the judgement of their contemporaries, the greatest generals of the war.

All three were arrogant, publicity-seeking and personally flawed, but with a genius for the command of men and an unrivalled enthusiasm for combat. All had spectacular success on the battlefield. Each understood the war in terms of his own ambitions and the attempts of the other two to thwart them. Rommel became the only German general known by name in Britain and America before most had even heard of Montgomery and Patton. They had to compete with him as larger-than-life personalities in whom their armies could believe before they could beat him on the battlefield. Yet as they fought for the headlines the hostility expressed by the two allies was directed not at their mutual enemy but at one another. Rommel, aware that the men and armour under their command outnumbered his own, remained confident that his superior tactical skills could defeat both of them.

It was a very personal contest: the clash of mighty armies perceived as a bout between three men. In *Masters of Battle*, for the first time in the literature of the Second World War, all three are 'put in the same ring' and allowed to 'go at it' against a backdrop of the great tank battles of North Africa, the invasions of Sicily and Italy, the Normandy landings and the push through France and Belgium into Germany.

★

Montgomery, Patton and Rommel were all born in November, but in different years, between 1885 and 1891, and under the same astrological sign: Scorpio, from the scorpion, known for its venomous sting. Each of the three was to live up to that.

Montgomery was a small man with a shrill voice, but his appearance belied the size of his ego. Convinced that only he knew how to conduct the war he treated his superiors with contempt and snubbed even Churchill. His victory at El Alamein against the previously invincible Rommel inspired the British press to compare him with Wellington, a sentiment he heartily endorsed. King George VI, visiting him in North Africa, said he was delighted to discover that Monty was not after *his* job. Montgomery led British forces in the invasion of Sicily, and rewrote the plan for the D-Day invasion, during which he commanded all Allied ground troops and attempted once more to outsmart Rommel, who commanded the coastal defences.

Patton was nicknamed 'Old Blood and Guts' because of his enthusiasm for battle, and General Eisenhower joked that he probably wore his combat helmet in bed. He certainly wore an ivory-handled Colt revolver everywhere and put on what he called his 'warrior face' to deliver obscene and profane speeches to the troops. He led American troops to their first victory in North Africa and commanded US forces in the invasion of Sicily. After D-Day he led the breakout from Normandy, the only Allied commander to emulate Rommel's blitzkrieg (lightning war). As his armoured columns raced towards the Rhine he boasted that he would be first into Berlin and 'personally shoot that son-of-a-bitch Adolf Hitler'.

Rommel's firm-set face and goggled cap became an icon of the desert war after Hitler personally gave him command of the Deutsches Afrika Korps. He pressed the British back to El Alamein, defeated the Americans at Kasserine, and was nicknamed Wüsten-fuchs (Desert Fox) for the uncanny brilliance of his battle tactics. General Auchinleck found it necessary to tell his beaten British army that 'Rommel is not superhuman . . . it would be undesirable to attribute supernatural powers to him.' After his defeat and pursuit across North Africa by Montgomery, Rommel was put in charge of defending the French coast. There, he planned to beat back the Allied invasion and win the war for Germany.

Both Montgomery and Patton described their battle with Rommel as a personal contest. Monty chose a metaphor from the tennis court: 'I feel that I have won the first game when it was Rommel's service. Next time it will be my service, the score being one—love.' Patton likened it to a medieval joust mounted on tanks: 'The two armies could watch. I would shoot at Rommel. He would shoot at me. If I killed him, I'd be the champ. America would win the war.' Both men had the greatest respect for their enemy. Monty kept a portrait of the German in his command caravan while Patton studied Rommel's book on tactics. Rommel returned the compliment: 'Montgomery never made a serious strategic mistake . . . [and] in the Patton Army we saw the most astonishing achievement in mobile warfare.'

In a surreal counterpoint to their respect for Rommel, the allies Montgomery and Patton loathed each other with a rare intensity. Monty told his staff officers that Patton was 'a foul-mouthed lover of war' who lacked his own military insight. Patton called Monty 'a cocky little Limey' and claimed he could 'outfight that little fart anytime'. When they were thrown together for the invasion of Sicily, each commanding their nation's forces, the island proved too small for two such egos and the campaign was determined more by the fight between them than by their fight with the enemy. When they clashed again in Normandy, competing to break through Rommel's defences, the very outcome of the war was at stake. Monty's advance faltered and Patton, leading the breakout, said American troops would 'save the face of the little monkey'. Monty planned his own strike inland and demanded that men and fuel be transferred from Patton's army to his own as they raced to be first across the Rhine.

Masters of Battle brings together not only the mutual respect of the foes and the furious animosity of the allies, but also the volcanic relationships of the three generals with their chiefs. Monty attempted to keep 'Winston's podgy finger' out of *his* battles. Patton believed that Eisenhower had his eyes on the White House rather than the war. Rommel realized, too late to save himself, the truth about the Führer he had once idolized.

Montgomery, Patton and Rommel were students of war before they were warriors, and all three were familiar with Carl von Clausewitz's *Principles of War*, first published in 1812 and still the primary text for would-be military leaders when Monty was at the Royal Military College, Sandhurst, Patton at the US Military Academy at West Point, and Rommel at the Königliche Kriegsschule (Imperial War School) in Danzig. Patton bought his copy of the book while honeymooning in London in 1910 and may have ignored his new wife to read it, incurring her suggestion that he preferred Clausewitz to her own charms.

Clausewitz argued that in every battle situation the military leader must choose between 'the most audacious' and 'the most careful' action and concluded that 'no military leader has ever become great without audacity'. He might in evidence, had he been able to observe the Second World War, have pointed out Rommel, noting his *Fingerspitzengefühl* (the instinctive and immediate response to battle situations) and talent for blitzkrieg, and also Patton, the only Allied commander to match Rommel at his own game and whose motto was borrowed from Frederick the Great: '*L'audace, l'audace, l'audace − tout jour* [*sic*] *l'audace.*'

Montgomery made 'carefulness' his primary battle-plan and his victories depended on it. Only his genius for *Materialschlacht* (the slow build-up of superior manpower and supplies before engagement) could have defeated Rommel at El Alamein and on the Normandy coast. The qualities that make a great military leader are then perhaps more complex than Clausewitz allowed, and in Rommel and Patton (well matched in audaciousness) and Monty (the master of carefulness) we can observe these two command styles brought face to face in the most crucial campaigns of the war.

★

4

Masters of Battle tells the story of three extraordinary men, each central to the war effort of Great Britain, the United States and Germany respectively. The explosive passions of their relationships with each other and with their political masters rival the pyrotechnics of their tank battles in determining the conduct and outcome of the war. Through the mutual respect of the arch-enemies Monty and Rommel, and the mutual animosity of the allies Monty and Patton, this book presents the Second World War as it was seen and experienced by three of its most flamboyant, controversial and influential commanders.

Terry Brighton
Lincoln, February 2008

PART ONE

1. Introductions – Three Portraits

In December 1940 the Lord Mayor of Heidenheim, a small town fifty miles east of Stuttgart, sent a Christmas gift parcel to *Wehrmacht* troops born in the town and now serving abroad. It contained a fir branch as a token of the trees that decorated their homes, *Magenbrot* (locally made biscuits), cigars, and a colour postcard of Major-General Erwin Rommel, Heidenheim's most famous son.

Rommel was forty-nine years old and had commanded 7th Panzer Division during the invasion of France. The spectacular success of his armoured blitzkrieg made him the first divisional commander to reach the English Channel coast and his name was celebrated throughout Germany. Jealous voices in the *Oberkommando der Wehrmacht* (High Command of the Armed Forces) already whispered that he was Hitler's favourite general, but the troops applauded him and none more than those also born in Heidenheim. One soldier replied to the Mayor:

My greatest thanks for the Rommel card. This picture catches our general exactly as he is in real life. Hard and relentless on himself and his men. It was with this face that he himself fired a round from a flare gun into the vision slit of a French tank, forcing it to retire. This is 'our Rommel'. Can I ask you to send me more cards for my comrades?

Although sent ostensibly by the Mayor of Heidenheim the parcel was funded by the local branch of the National Socialist Party. The postcard had been reproduced from a portrait painted by war artist Wolfgang Willrich. Sergeant Willrich believed that art should portray the 'heroic ideal' defined in racial as well as military terms and had produced a book of drawings, *Des edlen ewiges Reich* (The Everlasting Nobility of the Reich). He was recruited by Propaganda Minister Joseph Goebbels and attached to the Propaganda Company that travelled with 7th Division in its thrust across France.

The portrait of Rommel showed him in uniform and greatcoat with a cap and goggles, wearing the Knight's Cross of the Iron Cross and the Pour le Mérite. War correspondent Hanns Gert von Esebeck described in print the Rommel caught on canvas by Willrich: 'He has a high forehead, a strong, forceful nose, prominent cheekbones, a narrow mouth with tight lips, and a chin of great determination. The strong lines around his nostrils and the corners of his mouth, relax only when he smiles. His clear, blue eyes, penetrating and focused, reveal the cunning which marks the man.'

At the end of the war the victorious Allies portrayed Rommel for reasons of their own as 'the good German' and 'conspirator against Hitler'. Much was made of the fact that he had never joined the Nazi Party. But neither did Willrich, who considered his work to be a record of racial purity and who in his portrait of the general caught a quite different Rommel. Here was the original Rommel formed by the nationalist bias of the Second Reich in which he grew up and the war academy he attended as a young man. He was a nationalist and a devout believer in the Führer, and co-operated happily with both the portrait painter and the propaganda minister in presenting an image, not only of the victory of the *Wehrmacht* over its enemies, but also of the German superiority that *they* defined in racial terms. Rommel appears to have had no problem with that and accepted Willrich as happily as he did Lieutenant Karl-August Hanke, the 'Party man' attached to 7th Division: 'I won't have to watch my tongue, but some of the others will be on guard.' When Goebbels wrote in his diary that Rommel was 'not just sympathetic to the National Socialists; he is a National Socialist', he was aware of the general's non-membership but implying a deeper identity. Willrich saw this too and caught it brilliantly on canvas: an innate sense of superiority that fed the darkest roots of Nazi doctrine.

Thousands of copies of the Willrich portrait were printed by the Volksbund für das Deutschtum im Ausland (National League for Germans Abroad) and distributed among the troops. So popular had the general become that poster-sized reproductions were printed by the propaganda ministry and several of these appeared in the windows of Heidenheim. Its citizens were proud that this hero of the Reich was one of them.

It was this portrait, rather than any photograph of his foe, that Montgomery chose to hang above the desk in his command caravan, perhaps because *this* was the Rommel – fiercely nationalist, devoted to the Führer, convinced of ultimate German victory – that he had to defeat before the course of the war, and the deepest beliefs of the man himself, could be changed.

Erwin Johannes Eugen Rommel was born in Heidenheim in the German state of Württemberg on Sunday, 15 November 1891. According to his sister Helena, the boy who was to become the much feared 'desert fox' began life as 'a very gentle and docile child who took after his mother. He had a white skin and hair so pale that we called him the "white bear".'

The Germans knew Württemberg as 'the home of common sense', a backhanded compliment meant to indicate the population's lack of intellect and sophistication, and at first Erwin appeared true to type. Helena noted that he spoke slowly and spent all his leisure time 'in the fields and woods'. His father (also called Erwin) was a schoolteacher in Heidenheim and this was not the bookish son he might have expected. When Erwin Senior was appointed headmaster of the *Realgymnasium* (which prepared pupils for university) in nearby Aalen the difference between them widened. School life was difficult for the 'headmaster's son', who lagged behind his classmates in every academic subject.

The father–son relationship was saved by a glider. Erwin Junior was obsessed by aeroplanes and airships, and when he was fourteen he and his friend Keitel decided to build a full-size glider in a field near Aalen. They had hopes that it might really fly but to get the aerodynamics right required a number of complex mathematical calculations. His father's obsession was mathematics and in this project the two found common ground. The two boys built the glider, and although they had no way of getting it off the ground, Erwin was so convinced of its aerodynamic qualities that he persuaded anyone who would listen (and some of his later biographers) that it flew for thirty yards.

From this came a decision as the end of his schooldays approached, to become an engineer in the Zeppelin works at Friedrichshafen. Erwin Senior, unhappy with any child of his working in a factory but aware that Erwin was not destined for university, came to a 'common-sense' conclusion: the best career for a young man who excelled in practical and outdoor pursuits was the army. The officer class was still largely dominated by the Prussian aristocracy (the 'vons') but had opened up to the bourgeoisie and a headmaster's son stood a fair chance. The idea of a military career appealed to him and particularly the thought of joining the engineers. The army, however, did not see Erwin Rommel as prime officer material and his father's letter of support hardly helped by describing his highest quality as 'good at gymnastics'. His application was rejected by the engineers and next by the artillery. If the infantry had not accepted him he might have ended up in the Zeppelin works where his best friend Keitel had already found work.

On 19 July 1910 at Weingarten, a small garrison town near Stuttgart, the eighteen-year-old Rommel enlisted in 124th Württemberg Infantry Regiment as a *Fahnenjunker* (officer cadet). Württemberg had been one of the several states brought together with Prussia in 1871 to form the German Reich. In this new country, only twenty years older than Rommel himself, the one institution that immediately took on a national identity was the Imperial German Army, with the Kaiser as its Supreme Commander. Rommel earnestly adopted this nationalist ethos, becoming for the first time aware of himself as a 'German' rather than a 'Württemberger', fiercely loyal to Kaiser Wilhelm II and the Reich.

While attending the Königliche Kriegsschule in Danzig for officer training he met Lucie Maria Mollin at a ball held in the officers' mess. She was a slim, dark-haired language student and despite their differences – she wished to attend every dance held in the city; he was eager to save every *pfennig* for the motorcycle he wanted – they became lovers. Lucie's mother was not happy about the relationship: the Mollins were Catholic and the Rommels were Protestant. Erwin and Lucie agreed secretly that they would eventually marry but told no one.

Rommel graduated from the *Kriegsschule* without distinguishing himself. The training officer who filled in his final report described him as 'of medium height, thin and physically rather awkward and delicate . . . firm in character with immense willpower and a keen enthusiasm . . . with a strong sense of duty'. He was considered to be merely 'average' in all marking categories except one: '*Führung – Gut*' (Leadership – Good).

He and Lucie celebrated his graduation by having their photograph taken. She wore a flared suit and wide-brimmed hat, he the high-collared tunic and *pickelhaube* (spiked helmet) of an infantry officer, his hands stuck firmly in the pockets of a greatcoat that bulked out his wiry frame. The middle-class provincial boy had made something of himself; in the new Reich civilians were expected to step from the pavement into the road to give precedence to officers in uniform. He was short-sighted in one eye and sometimes sported a monocle, Prussian style, rather than the *pince-nez* spectacles his father wore. But he was not one of 'the vons' and, as he would discover more than thirty years later when 'they' still predominated in the General Staff, the greatest of victories on the battlefield could not alter that.

In January 1912 he was commissioned as a second lieutenant and rejoined his regiment at its barracks in the old monastery in Weingarten. The Kaiser was building up his army and navy in preparation for the great battle that most believed must eventually come. Germany felt itself hemmed in by Russia to the east and France to the west, and when France agreed to assist Russia in any future European war this hardened into a sense of 'being surrounded'. The *Entente Cordiale* between France and Britain made a potential enemy of the latter too. The nationalist fervour of the Reich included a deeply felt suspicion that these nations would use any means to prevent Germany taking its rightful place among the European nations.

Rommel's main tasks at Weingarten were the drilling and training of new recruits. Now that they were apart he wrote to Lucie every day, but to keep their continuing relationship secret from her mother he sent his letters care of the Danzig post office instead of to her home. Each one began, '*Meine liebste Lu*', and was signed, '*Dein Erwin*'. They were a lover's letters and on 28 March he told her, 'I'm looking forward hugely to your long letter. I hope you're going to make it really intimate.'

If this relationship was kept from Lucie's mother, a matter just as crucial was kept from Lucie herself. In Weingarten he was seeing a teenage fruit-seller, Walburga Stemmer, and in the spring of 1913 Walburga gave birth to his illegitimate daughter, Gertrud. There is no evidence that he considered 'doing the right thing'. By then Lieutenant Rommel had made the army his 'first family' and, in the unwritten ethos of the officer corps, a fruit-seller was an impossible match if he hoped for further advancement. There was even a word for it: *Kavaliersdelikt* (gentleman's mistake). Rommel knew the child would be seen as an unfortunate slip that need not harm his career; marrying the mother would abruptly end it. He informed Walburga that he would support the child financially. At the same time he confessed all to Lucie and their 'understanding' continued.

When war came to Europe in 1914 no one was quite sure why. It was as if the European nations had prepared for it and now that its time had come any cause would do, and the assassination of Archduke Ferdinand was suitably bloody. In the first week of August Germany declared war on Russia and France, and Britain declared war on Germany. Rommel, like most Germans, saw the war as a defensive act against the hostile alliance of Britain, France and Russia. He described the scene at the regimental barracks in Weingarten on 6 August, the day that 124th Infantry expected the order to mobilize:

I greeted the men of 7th Company whom I would lead into battle. Their young faces beamed with joy and anticipation. Nothing could be better than leading such soldiers against an enemy. At 1800 hours Colonel Haas inspected his regiment of riflemen – all in field grey – and just then the mobilization order came through. The cheering of young Germans eager for battle rang through the ancient cloister.

The young man whose first career choice had been the airship factory was going to fight for the Kaiser and the Reich. His only concern was that he might reach the front too late for the first battle.

In London on 26 February 1944 George Bernard Shaw, then eighty-eight years old, visited the Chelsea studio of his friend Augustus John. Britain's leading portrait painter, John had painted Thomas Hardy, W. B. Yeats, Dylan Thomas and Shaw himself. Now his latest subject had expressed a wish to meet the playwright.

John's studio was located on Tite Street. Outside, a chauffeur sat waiting in a Rolls-Royce. As Shaw reached the top of the stairs he found John with brush and palette in hand. Sitting stiffly upright on the dais, in beret, battledress and medal ribbons, was General Bernard Montgomery. Shaw was struck first of all by his diminutive stature – only five feet seven inches, with a head that seemed too small for even his slender frame – compared with the hugeness of his reputation, lauded by the Prime Minister, Winston Churchill, and fêted by the British public for his victory over Rommel's Deutsches Afrika Korps. Then he noted the details: 'What a nose! What eyes!'

Surprisingly there was an immediate rapport between the

clean-shaven soldier and the exuberantly bearded playwright. Shaw attempted, without success, to persuade Monty of the importance of the beard as an attribute of greatness. They did agree that only five per cent of generals were good at their jobs. Shaw wrote later that he was surprised to discover a soldier intelligent enough to want to talk to him at all.

After the sitting Montgomery returned by Rolls-Royce to his 21st Army Group headquarters established in St Paul's School in Hammersmith, which he had attended as a boy. The school had been evacuated and from there he was planning the D-Day landings and the invasion of occupied France. He wrote wryly: 'My office was located in the room of the High Master. Although I had been a school prefect, I had never entered that room before. I had to become a Commander-in-Chief to do so.'

Shaw returned to his home in Whitehall Court to write a critical note to John. He felt that, as it stood, the portrait did not properly catch its subject and in explaining why, the playwright left an impression of Montgomery at the height of his fame:

I had to talk all over the shop to amuse your sitter and keep his mind off the fighting. And I noted the extreme unlikeness between you. You were massive in contrast with that intensely compacted hank of steel wire, who looked as if you might have taken him out of your pocket. Your portrait of B. M. fills the canvas, suggesting a large tall man. It does not look at you, and Monty always does this with intense effect. He concentrates all space into a small spot like a burning glass.

Shaw suggested that John take a petrol rag, rub out the portrait and start again: 'Paint a small figure looking at you as he looked at me from the dais . . . his expression one of piercing scrutiny, his eyes unforgettable. The background: the vast totality of desert Africa.'

As Monty's fame grew he became obsessed with commissioning portraits of himself and always by highly acclaimed artists. Previously there had been Neville Lewis; later there would be Frank

Salisbury and James Gunn. All depicted a calm, confident and alert man. Only Augustus John caught the arrogance and contempt, and the incongruity between his small, bird-like head, emphasized by shallow cheeks and a sharp nose, and the general's uniform. Shaw advised the artist to paint instead the Monty that an adoring British public saw, a small man who conquered something huge: the desert, the Deutsches Afrika Korps and Rommel. But John, with an artist's insight, painted the general seen only by those who knew him best, primarily his army staff: a far more complex and less appealing man.

When Montgomery saw the finished portrait he agreed with Shaw. The artist had misunderstood the task he had been given, best expressed by the general's ADC, Christopher 'Kit' Dawnay, of representing 'the nation's conquering hero'. Monty told Dawnay: 'John's picture of me is no good. I am not going to buy it,' even though he had agreed to pay five hundred pounds for it.

Bernard Law Montgomery was born in St Mark's Vicarage, Kennington Oval, London, on 17 November 1887, the fourth child of the Reverend Henry Montgomery and his wife Maud. Two years later the family moved to Tasmania where his father was consecrated bishop. According to Monty both he and his father were bullied by his mother, whom he called 'the enemy':

Certainly I can say that my own childhood was unhappy. This was due to a clash of wills between my mother and myself. My early life was a series of fierce battles, from which my mother invariably emerged the victor. But the constant defeats and beatings with a cane, and these were frequent, in no way deterred me. I learnt early to stand or fall on my own.

Monty's description of his early years carries his view that the childhood determined the boy, i.e. that the mother he portrays as a domestic tyrant was the cause of his difficult, often arrogant nature. Biographers have generally accepted this – for which virtually all the evidence comes from Monty himself – far too uncritically. There is sufficient independent evidence to suggest that the boy determined the childhood.

His brother Harold remembered him as 'the bad boy of the family, mischievous by nature'. Another brother, Brian, contradicted Monty's account of his upbringing. He described their mother as a young woman struggling, often on her own, to cope with a large family and only able to do so by insisting on routine and obedience. Bernard was the cause of continual friction: 'He defied his mother's authority when his wishes ran contrary to her own ideas. Bernard did not grow up in an atmosphere of fear – of mother – or develop any characteristic because of her. The simple truth is that he inherited from her an indomitable will which frequently clashed with her own inflexible purpose.'

Andrew Holden, a friend of the Montgomery children in Hobart, Tasmania, described a children's party at which the hostess called for quiet while she explained a game they were to play. All fell silent except Bernard, who called out: 'Silence in the pig market, the old sow speaks.'

In later life Monty became very aware that others found him awkward and a loner, and in his *Memoirs* he dangled such tempting pseudo-Freudian bait in the face of future biographers that many have taken at least a bite at it. At the peak of his fame, as he sat for Augustus John, he had so totally ostracized his mother that he refused even to meet her. But it is likely that the aloofness and conceit John caught in the face of the architect of the D-Day landings were there from the beginning. The portrait was rejected because it revealed Monty as he had always been and not the public image painted by his victories, an ecstatic prime minister and the press. His mother's real failing, perhaps, was to stand up to him.

In 1901 the family returned to London where his father became secretary of the Society for the Propagation of the Gospel and Monty, aged fourteen, entered St Paul's School. He chose to join the Army Class, which followed a more practical course than the academic classes. His mother was furious as it had been hoped he might go into the Church, but Bernard would not change his mind. This was the first battle with her that he won.

Although not for centuries a military family the Montgomerys originally came to fame by killing Englishmen. Roger de Montgomery, from Falaise in Normandy, was a cousin of William the Conqueror and served as his second-in-command at Hastings in 1066, where he commanded on the right flank. After the battle William granted him large estates on the Welsh border in the area now known as Montgomery.

The Norman blood had thinned somewhat since. Brian recalled that at St Paul's Bernard was nicknamed 'Monkey' because he was 'small, tough, wiry and extremely quick in all his movements'. By Monty's own admission he did no work there. Even his knowledge of the Bible was not put to best use. During a scripture lesson on King David's intrigue against Uriah the Hittite to 'acquire' the latter's wife for himself, the form master caught Bernard passing a ditty of his own composing around the class:

> David on his palace roof in his night attire,
> Saw a lady in her bath, her name it was Uriah.
> He sent a message to the battle, if you would be savèd,
> Put Uriah in the line of fire, Yours sincerely, David.

Bernard was caned in front of the class. The room in which this corporal punishment took place later became one of those he commandeered as commander-in-chief.

His final school report described him as 'rather backward for his age' and suggested that 'to have a serious chance for Sandhurst he must give more time for work'. Despite that he passed the entry examination and entered the Royal Military College, Sandhurst, on 30 January 1907 when he was nineteen. According to his entry records he was five feet seven inches tall, weighed a little under ten stone and had a chest measurement of thirty-four inches.

A young man who neither smoked nor drank and showed no

interest in girls did not fit easily into an all-male environment. Yet from the start he proved himself a leader, both in the manner the college expected, recognized by his promotion to cadet lance corporal after only six weeks, and in a way it deplored, as prime mover in a gang of bullies whose rowdy conduct culminated in the 'ragging' of a cadet they disliked. While this man was held fast, Monty set fire to his shirt tails. The cadet's backside was badly burned and he had to be admitted to hospital. Monty commented: 'He was unable to sit down with any comfort for some time.' At first the college commandant ruled that Montgomery was not fit to receive a commission, but after an appeal from his mother this was commuted to the loss of his lance corporal's stripes and six months added to his course.

He passed out of Sandhurst in the summer of 1908. Those with money or a high academic showing at the college could virtually take their choice of the most select regiments. Monty had neither. He joined the lowly Royal Warwickshire Regiment because he 'quite liked the cap badge'. That December he was posted to the regiment's 1st Battalion stationed at Peshawar on the North West Frontier of India. There was no fighting to be done. Lacking the private financial resources of his fellow officers, he was unable to buy horses and polo ponies and was thereby excluded from two of their favourite pastimes; he excluded himself by choice from the third. He was an outsider and conscious of it, writing, as a dedicated abstainer: 'An expression heard frequently was that so-and-so was a "good mixer". A good mixer was a man who had never been known to refuse a drink.'

Nevertheless his unruly nature broke through. In 1910 the battalion moved to Bombay where he bought a motorcycle, a form of transport disapproved of by the equestrian gentlemen of the officers' mess. Along with others who at least had the excuse of being drunk he wrecked the Bombay Yacht Club in an after-dinner mêlée. Monty was sober and claimed he took part just for the hell of it. He was disciplined for 'uproarious proceedings'.

In 1911 when he was the battalion sports officer, the German armoured cruiser *Gneisenau* visited Bombay and the men aboard challenged the British to a football match. Adjutant Tomes told him not to field their best players so that they would not embarrass the visitors by too great a winning margin. Montgomery fielded the full First XI and they won forty–nil. He told Tomes: 'I was taking no risks with those bastards.' The obscenity was a rare slip. The tactic was to become his trademark.

The regiment returned to England in January 1913 and was based at Napier Barracks in Shorncliffe near Dover. Monty was promoted to lieutenant, purchased a second-hand Ford car and became known as a reckless driver; he gained a place in the army hockey team. It was here that he made the first real friend of his career, Captain Lefroy, with whom he had 'long talks about the Army and what was wrong with it; he helped me with advice about what books to read'.

On the evening of 29 July 1914 Monty was playing tennis in Folkestone when an officer came running to summon him back to barracks because the order to mobilize had been received. The regiment's immediate task was to guard the mouth of the Thames near Sheerness lest the Germans attempt a pre-emptive landing. On 4 August Britain declared war on Germany. He wrote to his mother: 'No one knows when we shall go as everything is being kept very secret. We are working night and day to get ready.'

On 24 September 1945 Robert Murphy, General Eisenhower's diplomatic adviser, drove to General Patton's military headquarters in Bad Tölz beside Lake Tegernsee in Bavaria. He arrived early for their lunch date and was shown into a room where the general sat in a relaxed pose, his right hand resting on his helmet on the table beside him, and the artist at work to one side almost hidden by his huge canvas.

The two Americans spoke while the artist worked. Boleslaw Czedekowski was a Polish portrait painter who counted among his subjects many of Europe's aristocratic élite. He had previously lived in New York where only the wealthiest could afford to sit for him. Now based in Vienna, he was currently staying at Patton's HQ while he completed this latest commission.

Patton, in full military dress, wore seven rows of medal ribbons, the gold-buckled belt and ivory-handled pistol that were his trademark, and breeches; he held a riding crop in his left hand. This four-star general displayed twenty stars in all, counting those on his helmet, his epaulettes and the twin points of his shirt collar. At fifty-nine he was recognized as the most brilliant and unpredictable American field commander of the war. He had recently returned from a visit home, where cheering crowds had lined the streets of Boston and Los Angeles as 'the conquering hero' drove through in an open-topped car. He had dressed for those appearances exactly as he did now for Czedekowski.

After the surrender of all German forces in Europe, Patton had been appointed military governor of Bavaria. His task was to rebuild the civil administration and he did so by reinstating the previous holders of civil-service posts regardless of whether they had been Nazi Party members or not. This clashed with the policy of denazification declared by the Allied Military Government, and at a press conference on 22 September, two days before his sitting for Czedekowski, war correspondents pressed Patton to explain himself. In doing so he likened membership of the Nazi Party to being a Democrat or Republican, because it had been a 'requirement' for civil servants in wartime Germany and did not identify them as Nazi activists. The following morning in the US press it was the simile, not the explanation, that made the headlines, and Patton was in trouble.

Now as he kept his pose for Czedekowski and chatted to Murphy, Patton took a call from General Walter Bedell Smith, Eisenhower's chief of staff. He was ordered to retract his comments of two days before and his continued command of Third Army was dependent on his doing so. On finishing the call Patton told Murphy that Bedell Smith was 'a son-of-a-bitch'.

None of this showed in the face of the man on the canvas. In fact, for Patton, this portrait was not so much about the face as its context. The artist's task was to paint 'the warrior' in a classical stance as seen in the family portraits that graced the stately homes and castles of Europe: army commanders with their shako or chapka on a table beside them and the pastoral scene behind marked by an obligatory cannon, now become for Patton a burnished helmet and a single artillery gun. Ironically neither Montgomery nor Rommel was painted in this way, and it was the general from the New World who presented himself as the latest in an unbroken line of warriors pictured through the centuries.

Two of those warriors stood much closer to him than the rest. Their portraits formed one of his earliest memories as a small boy: his father reading aloud stories of his Civil War ancestors to him in the parlour of his family home, while he stared in awe at the portraits on the wall behind his mother's chair. His grandfather had been a Confederate colonel and these two men, Robert E. Lee and Stonewall Jackson, were 'family'.

George Smith Patton Junior was born on Wednesday, 11 November 1885, in his family's ranch house in San Gabriel, near Pasadena, California. His father had given up a law practice in order to manage the family's two-thousand-acre Lake Vineyard estate, but the roots that mattered most to the Pattons were in Virginia and in the Confederate South of the Civil War.

His grandfather had studied under Stonewall Jackson at the Virginia Military Institute and commanded 22nd Virginia Infantry in the war. His grandfather's brother, Waller Tazewell Patton, had been wounded at Bull Run and killed at Gettysburg. Georgie, as he was known to the family, grew up on stories of those men and, by his own account, 'I got all excited.' His main interest was soldiers and their weapons: 'When I was a little boy at home I used to wear a wooden sword and say to myself, "George S. Patton, General".'

He was a little 'slow' and was at first educated at home. Passages from Homer, Shakespeare and Kipling, read to him by his father, fired his imagination. His brother Robert was present when he attempted to re-enact an ancient feat of armoured mobile warfare with the help of several young cousins and a farm wagon: 'The boys hauled the wagon to the top of a small hill overlooking Lake Vineyard's turkey shed. At Georgie's order, the cousins crouched in the wagon preparing to hurl sticks as spears. Georgie pushed the wagon down the hill. It ripped through the flock of turkeys at peak speed and killed and mangled quite a few.'

Although living in California George Patton Senior maintained the traditional prejudices of a Virginian gentleman. He was intolerant of Catholics and Jews although the family's Mexican servants were treated well; not so many generations before, the Pattons had kept slaves. George began to judge others on the basis of their background and 'breeding'. Robert believed that 'He learned, through observing Papa's genteel expressions of ingrained bigotry, that the Pattons and their Anglo-Saxon Protestant kind were better than other people.'

He was eleven when he began attending Stephen Cutter Clark's Classical School for Boys in Pasadena in September 1897 and could neither read nor write properly. In the years ahead he experienced particular difficulty with spelling, leading some biographers retrospectively to diagnose dyslexia, a condition unknown at the time. By the time he left school five years later he still wanted to be a soldier. In fact he had decided to become a brigadier and the direct route – the illustrious US Military Academy at West Point – was ruled out by his poor academic showing. In any case attendance at the Virginia Military Institute was a family tradition. He enrolled there in September 1903. According to his entry sheet he stood six feet one inch tall and weighed 167 pounds. His blond hair and good looks were not recorded, but the tailor who measured him for his first uniform recognized him as a Patton. Although he felt at home among the 'Southern gentlemen' who were his fellow cadets, this was too provincial an institution for a young man who wanted to be a brigadier. He sat a competitive entrance examination for West Point and came first. Dyslexia or not, whenever high-quality work could get him what he wanted he produced it.

Arriving at West Point in June 1904 he found that the Patton name and his Confederate heritage counted for nothing. Writing home to his father on 24 July he described his classmates as 'nice fellows but very few indeed are born gentlemen . . . the only ones of that type are Southerners'. This superior attitude hardly endeared him to them and they took their revenge, picking as their weapon of choice his nickname 'Georgie'. This tall, handsome cadet impressed all he met with his military bearing, until he spoke. He had a particularly high-pitched voice and in an all-male environment the epithet 'Georgie' (and the inflexion put upon it) gained a quite different connotation.

Patton failed his first year, which had to be repeated. Most biographers blame his supposed dyslexia but it was as likely his new obsession with fencing, particularly fighting with the broadsword. He became one of the best swordsmen in his class at the expense of other work. The possibility of failing West Point altogether seems to have shaken him and once more he got down to work. He noted orders to himself in a black-leather notebook: 'You must do your damndest and win. By perseverance and eternal desire any man can be great.' At the end of his second year he passed the exams with grades that put him in the top third of his class.

In February 1908 he was promoted to cadet adjutant. Now he had not only the position of command he sought but also a stage. At morning parade he read the orders of the day. He led the cadets when they marched. He was determined to make an impression and claimed with exquisite hyperbole that he changed his uniform 'fifteen times a day' in order always to be meticulously dressed. For the first time he had to command others and there could be no men more difficult to command than his own classmates. His problem – how to establish a 'command presence' while giving orders in a squeaky voice – was solved by deliberately cultivating a macho personality: a *loud* voice, a malicious look that he called his 'warrior face' and practised in front of a mirror, and the habitual use of obscene and profane language.

It worked – perhaps too well, for his assurance of his own authority was caricatured in the *Howitzer* (the cadet yearbook), which imagined his response on the parade square to the barracks being hit by a violent earthquake:

Men came tumbling out in all stages of dishabille. Suddenly the Cadet Adjutant appeared, faultlessly attired as usual. Walking with firm step across the area, he halted, executed a proper about-face, and the stentorian tones rang out: 'Battalion Attention-n-n-n! Cadets will refrain from being unduly shaken up. There will be no yelling. The earthquake will cease immediately.'

During his final year at West Point he fell in love with Beatrice Ayer. He had known her for almost six years but their courtship had been low-key because her family lived some distance away in Boston. On 6 April 1909 he wrote to tell her that 'I would rather have your love than the world and all.' Later he added a telling condition: 'Before everything else I am a soldier.'

When he graduated from West Point in 1909 Patton was commissioned as second lieutenant in the 15th Cavalry and assigned to the detachment based at Fort Sheridan north of Chicago. It was not the station he had hoped for. A squadron of the 15th was based at Fort Myer near Washington, DC, which had once been part of Robert E. Lee's Virginia estate and was now the residence of the Army's chief of staff. *That* was a prime stopover on the route to high military rank. Fort Sheridan was a backwater.

He set about getting himself noticed and excelled both in sword drill and on the polo field. He married Beatrice in Boston on 26 May 1910, and they honeymooned for a month in Britain and France. He took his black notebook with him and in London he recorded purchasing a copy of Carl von Clausewitz's *Principles of War*. He may have ignored her to read it because later when they were apart for the first time he wrote in reply to a letter from her: 'What in H do you mean by "Doing what I like best to do – reading"? I would rather a damned sight look at something

26

else than a book and you know what it is too. It looks like a skunk.'

The 15th Cavalry squadron at Fort Myer traditionally showcased its horsemanship by excelling on the polo field, yet for many years the engineers based at Washington Barracks had trounced them. Captain Frank McCoy, who had seen Patton's skill at polo demonstrated at Fort Sheridan, suggested: 'Let's get George brought to Washington. He'll turn the scales against the Engineers.' Patton moved to Fort Myer in December 1911. The 15th Cavalry provided mounted units for state ceremonials and visiting dignitaries, and this offered the perfect stage for a man whose bearing and meticulous appearance, even in the saddle, set him apart. He and Beatrice figured prominently in Washington society. He mixed with politicians and senior officers of the War Department, and before long he was taking regular early-morning rides along an equestrian trail with Secretary of War Henry L. Stimson.

In 1912, aged twenty-six, he represented the US at the Stockholm Olympics, competing in the multi-event pentathlon and coming fifth overall. He claimed he should have won, arguing that in the pistol shooting when the judges rated his second shot a 'miss', it had in fact passed straight through the hole made by his first. He won the fencing category by defeating the French champion. Instead of returning immediately to the US he spent two weeks at the French Cavalry School at Saumur studying French sabre drills.

On 23 September 1913 he began a course at the Mounted Service School at Fort Riley, Kansas, although he was no ordinary student: he was appointed Master of the Sword, a new post created to denote the cavalry's top instructor in swordsmanship. While there he designed a new cavalry sword, adopted by the Ordnance Department as the US Sabre Mark 1913 but commonly known as 'the Patton sabre'. The design said much about the man. Previous swords had been designed for the optimum blend of offensive and defensive qualities. The Patton sabre was primarily a weapon for 'thrusting forward'.

When the Great War began in August 1914 President Woodrow Wilson announced that the United States would remain neutral and said that the conflict could best be resolved by applying international law. Patton was furious. In his opinion the only international law was 'the best army' and his should be put to the test.

On 5 August 1914 Rommel left Weingarten on one of the trains that were departing at regular intervals for the western front: 'At Kornwestheim I saw my mother and two brothers and sister for a few moments, then the locomotive whistled – a last clasp of hands – and we were off. It was dark when we crossed the Rhine and searchlights crossed the sky seeking enemy aeroplanes and airships.'

In Shorncliffe, Montgomery was busy with attack drills and other, less predictable preparations: 'All officers' swords were to go to the armourers' shop to be sharpened. It was not clear to me why, since I had never used my sword except for saluting. I had my hair cut by a barber in Folkestone. Being totally ignorant about war, I asked the Commanding Officer if it was necessary to take any money with me.'

Patton was desperate to join them. Recalling that he had set himself the target of becoming a brigadier by the age of twenty-seven, he complained in a letter to his father that he had now passed that benchmark and 'I am not a First Lieutenant.' Distinguishing himself in the European war offered a short route to promotion, and although the President declared that the United States would remain neutral, Patton believed his country must eventually be drawn in.

Patton hoped for a long war to maximize his chances of taking part, the consensus among British officers at Shorncliffe was that the enemy could be beaten in three weeks, while German officers leaving Weingarten for the front expected an *Entscheidungsschlacht* (decisive battle) to resolve the matter quickly in their favour.

2. First Blood

The twenty-two-year-old Lieutenant Erwin Rommel arrived at Diedenhofen on 20 August 1914 and marched towards Longwy in the thickly forested Ardennes. His war began the very next day. The Imperial German Army, working to a plan developed by Count von Schlieffen, had first attacked Anglo-French forces to the west, and expected a quick victory. This would allow the transfer of troops to confront Russian forces in the east. The 124th Württemberg Infantry Regiment, part of the German Fifth Army, was tasked with engaging French troops near Longwy. Rommel was ordered to lead a five-man patrol and reconnoitre a town on the front line; as it was eight miles away the patrol travelled by horse and cart. The man given the rein lost control of the two Belgian bays as they approached the front, the horses raced past the last German outpost at the gallop and it seemed likely that Rommel would be delivered without a shot to the enemy. He wrote: 'No amount of yanking or yelling helped, but finally we landed, unhurt, in a manure pile.' It was later discovered that the town was unoccupied.

Two days later, twenty-six-year-old Lieutenant Bernard Montgomery disembarked from the SS *Caledonia* at Boulogne and continued by train towards Le Cateau. Monty's war began five days later and just as badly. After the first major battle of the war at Mons on the Franco-Belgian border, the British Expeditionary Force under Sir John French was retreating. To slow the pursuing German forces the Royal Warwickshire Regiment was ordered to counter-attack uphill at Le Cateau. Montgomery led his platoon through a hail of bullets from rifles and Maxim guns, and what he called a 'perfect storm of shrapnel fire' in which 'men fell like nine-pins'. It was for him a forlorn charge with an ignominious end that in all probability saved his life: 'Waving my sword I ran forward in front of my platoon, but I had only gone six paces when I tripped over my scabbard, the sword fell from my hand, and I fell flat on my face. By the time I picked myself up and rushed after my men I found that most of them had been killed.'

Meanwhile at Fort Riley, Kansas, twenty-eight-year-old Second Lieutenant George Patton attempted to circumnavigate President Wilson's declaration of US neutrality. He wrote to tell Major General Leonard Wood that he wanted a year's leave 'on some pretext' so that he could serve in the French Army and that he would 'never apply to the United States for help if I get in trouble or captured'. Wood replied that nothing could be done. 'We don't want to waste youngsters of your sort in the service of foreign nations. Stick to the present job.' Patton took out his frustration by writing a poem, 'To Wilson', deriding the President for 'making soldiers play the part of dogs who will not fight; of dogs who, fearful not to growl, fear far too much to bite'.

On the southern flank of the main German thrust through Belgium into France, Fifth Army crossed the French border below Luxembourg. Early on the morning of 22 August, 124th Infantry Regiment approached the village of Bleid and it was here that Rommel made his 'first kill'. Handing his horse, Rappen, to his orderly, Haenle, he led the men of his platoon in loose formation through a potato field covered with a thick ground fog. When a salvo of shots rang out he led a charge towards an enemy who fled before they could be seen. Buildings and hedges loomed out of the fog and Rommel went forward with Sergeant Ostertag and two men to inspect them, uncertain whether they had reached Bleid or not. The French had lost their sense of direction too: 'Frenchmen came running back down the road on our left, beyond the bushes where we stood. It was easy to shoot them down at a distance of about ten paces through a break in the bushes. Our four rifles put dozens of Frenchmen out of the fight.'

By 14 September Rommel's regiment had reached the hills around Varennes on the edge of the Argonne forest. When an advance faltered under relentless fire from French infantry he took a rifle from a wounded man and dashed forward.

I saw five Frenchmen twenty paces ahead, standing and firing. I lifted the rifle to my cheek and fired twice, and two men fell. I fired again and – nothing. The magazine was empty! There was no time to reload and my only hope was the bayonet! It was one against three and the French fired as I rushed at them. I was knocked head over heels by a bullet and landed in front of them. The shot had shattered my left thigh and blood was spurting from a wound as big as my fist. I expected another shot to finish me off, but my men came rushing up, shouting, and the enemy fell back. I pressed against the wound with my right hand, while rolling to take cover behind an oak tree.

Corporal Rauch used a belt as a tourniquet and two men carried him three miles to the field hospital at Montblainville, by which time he was unconscious from loss of blood. A few days later he was decorated with the Iron Cross, Second Class, and sent home to recover.

For Montgomery the 'first kill' came unexpectedly. The German advance had been halted, and by mid-September the British had pushed forward to the Aisne. The two sides entrenched, leaving a six-hundred-yard strip of land between them, the first indication that the mobile warfare of the early months was coming to an end. Monty was now commanding his own company (250 men): 'So I ride a horse, as all company commanders are mounted,' he wrote to tell his mother. 'I have a big beard. I have not washed my face or hands for ten days. Don't forget the cigarettes, will you?' He was smoking more than twenty Capstan Navy Cut Medium cigarettes each day as the two armies sat watching each other. Then on 27 September: 'I can see the Germans in their trenches and occasionally we exchange shots; I bagged one man this morning and one horse. The shrapnel is screaming overhead and altogether it is rather different from a Sunday at home.'

On the morning of 13 October in an attempt to outflank the enemy line to the north near Ypres, the Warwickshires were ordered to evict German troops from the village of Méteren. Because the enemy had entrenched immediately in front of the houses and had use of the church tower as an observation post, surprise was impossible and there was no artillery support. It was raining heavily as Monty drew his sword, ordered the men of his platoon to fix bayonets and led a charge:

I saw in front of me a trench full of Germans, one of whom was aiming his rifle at me. No one had taught me how to kill a German with a sword. The only sword exercise I knew was saluting drill. I hurled myself through the air and kicked him as hard as I could in the lower part of the stomach; the blow was well aimed at a tender spot. He fell to the ground in great pain and I took my first prisoner!

The trench had been taken but the enemy still occupied the village. In the afternoon another assault was ordered. Monty walked ahead of his platoon in the pouring rain 'to see what the positions looked like' and as he turned to signal to them he was shot by a sniper firing from the upper window of a house. The bullet passed through his chest from back to front – he had his back to Méteren – and punctured his right lung:

I collapsed, bleeding profusely. A soldier from my platoon ran forward and plugged the wound with my field dressing; the sniper shot him through the head and he collapsed on top of me. I managed to shout that no more men were to come to me until it was dark. It was then 3 p.m. and raining. The sniper kept firing at me and I received one more bullet in the left knee. The man lying on me took all the bullets and saved my life.

It was three hours before night fell and two men went forward to carry him in using a greatcoat as a stretcher. At the dressing station his wound was considered fatal and orderlies dug a grave for him. When he failed to die he was transferred by ambulance to the French military hospital in St Omer. By the time he recovered consciousness the next day, he was in the Royal Herbert Hospital in London. He had been promoted to captain and awarded the Distinguished Service Order for 'conspicuous gallantry'.

Three months after his thigh wound Lieutenant Rommel returned to the fight with a vengeance. By the beginning of 1915 both sides had given up hope of a quick victory and begun digging-in along a vast front. The 'war of manoeuvre' had stalled against hundreds of miles of barbed wire and trenches, and the necessity to transfer troops to the eastern front meant that German forces were less able to attempt concentrated thrusts in the west. But they were not yet ready to sit in the trenches and an offensive was ordered in the Argonne sector.

It was here that Rommel, commanding 9th Company, first showed his genius for forward action. On 29 January 124th Infantry Regiment attacked a French position protected by deep bands of barbed wire backed by blockhouses at sixty-yard intervals. As Rommel and his men climbed out of a forward trench and raced forward they came under small-arms fire from the French: 'We ran into deep barbed-wire entanglements. As I followed a small track which ran through the wire, a burst of enemy fire forced me to take cover. Bullets ricocheted all around as I crawled on all fours.' He shouted and waved for his men to follow but no one moved. He crawled back to them with his pistol drawn: 'Obey my orders *now* or I will shoot you.' This time the whole company followed him through the wire. On the far side they found a blockhouse unoccupied and took shelter there.

The rest of the regiment had failed to breach the wire, leaving them isolated in the midst of the enemy. Rommel sent back an urgent message: 'Nine Company has penetrated to a strong French position south of start line. Request urgent support. Also machine-gun ammunition and grenades.' When the French counter-attacked in force he ordered his men to throw their last grenades and evacuate the blockhouse as enemy infantrymen stormed in. His company was now trapped between the blockhouse and the wire: 'From their position, the French could fire on us with rifles and machine-guns. Then a battalion runner shouted across the wire: "Rommel to withdraw, support not possible."'

He estimated that a retreat back through the wire under constant fire would result in 'a minimum loss of fifty per cent and perhaps the whole company'. Typical of the man, he chose to ignore his orders and turned his company to attack and retake the blockhouse:

It might then be possible to pass back through the wire, with only a more distant enemy fire to hinder us. Speed was crucial. We had to be clear before the French recovered from our shock attack. We charged and seized the blockhouse, and saw the red trousers of the Frenchmen dashing away through the bushes. This was our chance – as the French ran west, we ran east, and scrambled through the wire. We were almost through before the enemy returned to the blockhouse and fired on us.

Five men were wounded and had to be carried; the rest escaped unhurt. By his own force of will (and his pistol) Rommel had pressed his company forward while the advance to either side stalled, and when ordered to retreat he instead attacked, wrong-footed the enemy and brought his men safely back to the German line. He was immediately awarded the Iron Cross, First Class, the first lieutenant in his regiment to receive that honour. For the first time the men of 124th were heard to say, 'Where Rommel is, there is the front.'

The 'front', however, was becoming a static line of trenches manned by Maxim guns and from which the only advance was a suicidal dash 'over the top'. Rommel was to see little of that. He was transferred to the newly formed Württemberg Gebirgs-bataillon (Württemberg Mountain Battalion) at Arlberg in the

Austrian alps for training in mountain warfare. While there he took leave to return to Danzig and marry Lucie Mollin. Back with his battalion he wrote to '*Meine liebste Lu*' every day.

Captain Montgomery spent the first weeks of 1915 at the Royal Herbert Hospital recovering from the damage to his right lung. From his hospital bed he analysed his regiment's first action of the war and found all the lessons of Sandhurst jettisoned in one go:

The Commanding Officer galloped up and shouted to us to attack the enemy on the forward hill at once. This was the only order; there was no reconnaissance, no plan, no covering fire. We rushed up the hill, came under heavy fire, and there were many casualties. Nobody knew what to do, so we returned to the original position. If this was real war it did not seem to make any sense.

In February, discharged from hospital as 'medically unfit for service overseas', he was appointed chief of staff of 91st Infantry Brigade, formed to train the mass of new recruits responding to Kitchener's appeal for volunteers. He found General Mackenzie, the commanding officer, to be 'a very nice person, but quite useless, and I really ran the Brigade and they all knew it'. He would remain a staff officer for the rest of the war. Unlike Rommel, whose sole wish on recovery was to return to the front, Monty believed he could make a greater contribution to the war effort by attempting to reform the British Army where it was failing: far behind the line.

Across the Atlantic Patton followed the war jealously in press and radio reports. His hopes were raised on 7 May when a German submarine torpedoed and sank the Cunard liner *Lusitania* off the Irish coast with the loss of 1198 lives. Among the drowned were 128 Americans. President Wilson sent a diplomatic protest to Berlin but still saw no reason for the US to enter the war, declaring that a country could be 'too proud to fight'. Patton answered that in a letter to his father: 'In any other country or age that pride has been called by another name. We ought to declare war. If Wilson had as much blood in him as the liver of a louse is thought to contain he would do this.'

He graduated from the Mounted Service School at Fort Riley in June, and in September was posted to 8th Cavalry at Fort Bliss, located outside El Paso in Texas. From there he led regular mounted patrols along the border with Mexico where a civil war had broken out between the Mexican president, Venustiano Carranza, and the bandit leader Francisco ('Pancho') Villa.

Patton had fallen behind Monty and Rommel in rank and, more crucially, in an appreciation of the reality of war in the age of the Maxim gun. On 30 September he wrote to tell Beatrice about the first regimental parade at Fort Bliss: 'It was a fine sight all with sabres drawn and all my [the Patton] sabre. It gives you a thrill and my eyes filled with tears . . . it is the call of one's ancestors and the glory of combat. It seems to me that at the head of a regiment of cavalry anything would be possible.'

If that experience seemed to belong to the Civil War period, so did El Paso. The men wore Colts and spurs, cowgirls flirted in rowdy saloons with soldiers and outlaws alike, and cattle-rustling was the career of choice for some. Patton made himself at home in the saloons, and found a friend in Dave Allison, a town marshal who claimed that he 'killed several Mexicans a month'.

When the US announced its support for Carranza, making an enemy of Pancho Villa, the latter vowed that he and his band of five hundred *pistoleros* would take revenge on Americans wherever he found them. In January 1916 he took sixteen Americans from a train at Santa Ysabel and shot them. In March he crossed the border and attacked the garrison town of Columbus, killing six American soldiers and twelve civilians. In response President Wilson sent 10,000 troops under General John ('Black Jack') Pershing, the commanding officer at Fort Bliss, on a punitive expedition into Mexico. Patton went as the general's ADC.

Although Pershing established a headquarters at Colonia Dublán, he used his open-topped Dodge car as a mobile command post in which he could roam the desert. It was Patton's first experience of a general who, instead of commanding from the rear, moved relentlessly between his front-line units and gave orders from the saddle – or in this case from the front passenger seat of his Dodge. Pershing was autocratic, pressed his staff hard and tolerated no failure. Sending Patton through bandit territory with a message for 11th Cavalry, whose precise location was unknown, Pershing told him to 'deliver the message or don't come back'. The aide was smitten and Pershing became his role-model for the commander *he* hoped to become. His only regret was that Pershing allowed his cavalrymen to carry only firearms and had ordered their Patton sabres to be left behind at Fort Bliss.

On 14 May Pershing sent Patton and ten men into the town of Rubio in three Dodge cars to buy supplies for the troops and grain for the horses. He had no other orders, but decided to investigate the nearby San Miguelito ranch house. Thought to be the family home of Julio Cárdenas, one of the bandit leaders, it had previously been searched and no trace of him found. Patton deployed his men around the house and walked towards an arched gate leading into the courtyard. When he was fifteen yards away three bandits rode out and made straight for him. He drew a Colt revolver from his holster. 'All three shot at me, one bullet threw gravel on me. I fired back, five times.' Only two of his five shots found their target, hitting a bandit in the arm and wounding the same man's horse.

Patton's men opened fire, allowing him to take cover against the wall while he reloaded. The wounded bandit made for the safety of the house while the other two spurred their mounts on. Patton fired at the nearest horse as they passed in front of him, bringing it down. He and his men waited until the rider was back on his feet and then fired together. 'It was only ten yards, we all hit him, he crumpled up.' They turned their weapons on the third bandit as he raced away and brought him down too. The first wounded man now jumped from a window and made a run for it. They fired and he raised a hand in surrender, but when one of Patton's men approached him, he fired back with a pistol in his other hand. The shot missed and the soldier 'blew his brains out'. This man was Julio Cárdenas.

The reporters accompanying Pershing's expedition had previously found little to report. Now they had a western-style shoot-out at a Mexican ranch led by a pistol-toting American officer. The first account appeared in the *New York Times* under the headline 'DRAMATIC FIGHT AT RANCH – LIEUT. PATTON AND TEN MEN KILLED THREE BANDITS'. The story was taken up in headlines across the country, and a Boston paper gave Patton the name by which he first became known to Americans: 'Bandit-killer'. Several reports noted that this was the first time the US cavalry had *driven* into combat.

Patton had made his first kills using an ivory-handled Colt .45 purchased in El Paso. He cut two notches in the grip and told Beatrice: 'You are probably wondering if my conscience hurts me for killing a man. It does not.' He had become a national hero, but feared this might be his only moment of glory if the US did not soon enter the European war.

Only weeks after the shoot-out in Mexico, Field Marshal Douglas Haig's July 1916 offensive on the Somme gave the lie to any lingering association of glory with the war in France. Regiment by regiment, infantrymen were ordered out of forward trenches into an iron mêlée of machine-gun bullets and exploding shells. British losses on the first day alone were 19,000 killed and 38,000 wounded. Montgomery's 91st Infantry Brigade, renumbered 104th and posted to the Somme, suffered thirty per cent casualties in a single attack. The Allies gained six miles of ground.

The western front had become a stand-off with both sides dug-in and their parallel trenches running three hundred miles down from the northern coast of France. As a staff officer, Monty was struck by how easily generals ordered whole divisions 'over the top' with only a slim chance of taking the enemy position and the certainty of mass casualties. The 'cannon-fodder' mentality of nineteenth-century British commanders lived on in Haig and his peers, so that orders to attack routinely ended with the words 'to be pressed home regardless of loss'. Monty wrote: 'The frightful casualties appalled me. The so-called "good fighting generals" of the war appeared to me to be those who had a complete disregard for human life.' Nevertheless his reaction was not that of the war poets, who wrote about the waste of a whole generation of young men hurled pointlessly at the guns, but that of the career soldier: such a loss, if it gained little or nothing, was an *inefficient* way of making war.

Not all of the failings of the British Army could be blamed on a 'cannon-fodder' mentality. Monty admitted that 'to an ambitious young officer with an enquiring mind [himself] many things seemed wrong'. He identified the most pressing problem as poor communication between a command post and the front line, and was astonished to find one infantry brigade using carrier pigeons to relay messages. He ridiculed the anxiety of the brigade commander when no bird had returned to Headquarters for some time: 'At last the cry went up: "The pigeon." It alighted in its loft. Soldiers rushed to get the news and the Brigade Commander roared out: "Give me the message." It was handed to him, and this is what he read: "I am absolutely fed up with carrying this bloody bird about France."'

That brigade commanders were unaware of the location and situation of their forward troops and were unable to adapt their battle plan in response to enemy actions was bad enough. Worse, Monty thought, was that he had never yet set eyes on his own commander-in-chief, Field Marshal Haig. It would be some time before he did and he was beaten to it by the most unlikely dinner guest of all to visit Haig's French headquarters: the American bandit-killer, George Patton.

The United States declared war on Germany on 6 April 1917 after German U-boats were ordered to sink all Atlantic shipping, regardless of neutrality. General Pershing was appointed to command an American Expeditionary Force to fight in France and selected George Patton to accompany him as his ADC with the new rank of captain. In May Pershing and his headquarters staff travelled on HMS *Baltic* from New York to Liverpool. In London they were introduced to King George V, his prime minister Lloyd George, and the Minister for Munitions, Winston S. Churchill.

In Paris, where Pershing established the AEF headquarters, Patton took an apartment off the Champs-Élysées. He accompanied Pershing to the British headquarters and dined with Haig. The field marshal, a cavalryman of the old school, enquired about the Patton sabre, which went down well with the American. Afterwards Haig noted that 'The ADC is a fire-eater and longs for the fray.' But that was the problem. Patton's duties were administrative and he longed for 'a real job'. While Monty was revelling in his work as a staff officer, Patton admitted to his diary that 'I have always talked blood and murder and am looked on as an advocate of close up fighting. I could never look myself in the face if I was a staff officer and comparatively safe.'

He bought a twelve-cylinder Packard car and took out his frustration along the wide Parisian avenues. He knew that the US Army was as yet in no condition to challenge the Imperial German Army. The first American troops to reach Paris were welcomed with shouts of '*les hommes au chapeau de cowboy*', and in September Pershing moved his headquarters to Chaumont, closer to the front, but only to establish a training camp. Patton looked for a faster route to the cutting edge of the war and found it in a new British invention: the tank.

The British 'landship', nicknamed 'Big Willie', was a thirty-ton tank powered by a Daimler engine and armed with a six-pounder gun, able to advance at four miles per hour and cross trenches up to fourteen feet in width. This new weapon appealed to Patton: 'Tanks are only used in attacks. There is a fifty percent chance they won't work at all, but if they do they will work like Hell.' This, then, would be his route to front-line combat, glory and high command. He put in a request to be considered for command if a US tank unit was formed, stressing that after the action at Rubio in Mexico he had become 'the only American officer who has ever made an attack in a motor vehicle'.

From a German perspective the US had joined the war but not the fighting. Italy had joined the Allies too and presented a more immediate threat. Rommel's Württemberg Gebirgsbataillon was ordered to the eastern front in preparation for an attack on Italian forces massed along the Isonzo river. The offensive began on 24 October with an artillery bombardment from a thousand guns. The following day the Gebirgsbataillon was tasked to take the enemy strongpoint of Mount Matajur, a seven-thousand-feet-high peak south-west of Caporetto. Unhappy with his assigned role in support of the main assault, Rommel persuaded his senior officer to allow him to mount an independent attack on the right flank with three rifle companies and a machine-gun company, a total of three hundred men.

Crossing the front line before first light, Rommel's detachment made a bayonet charge on an Italian artillery battery and 'caught the gunners out at ablutions', taking the position without either side firing a shot. He left one company to hold the battery while he led the rest of his men uphill, surprising and taking more Italian positions until the number of prisoners reached a thousand. Finally the enemy realized what was happening. A full division fired on the single company Rommel had left behind to hold the battery:

If the Italians attacked, the whole company would be destroyed . . . the enemy would then recover his position on the hill, cutting off the rest of my detachment . . . we could only be saved by making an attack of our own. Our heavy machine-guns opened up a steady fire on the enemy, while we charged the enemy flank with furious determination.

He deployed his three remaining companies to attack the Italians from the rear and right flank, and the whole division – twelve officers and five hundred men – surrendered. His prisoner count now totalled fifteen hundred men.

Rommel continued his advance two miles along the uphill road towards the heavily fortified position atop Mount Matajur. Because he had to leave men behind to hold the positions already taken and to guard the prisoners, he now had only a hundred men with him and estimated that the Salerno Brigade holding the summit must number more than a thousand. He ignored an order to withdraw and pushed his men forward under cover of darkness. At dawn the next day, three hundred yards from the summit, he positioned his machine-guns and riflemen at intervals around the perimeter of the enemy camp and had them open fire for three minutes. He then walked in holding a white handkerchief. The twelve hundred officers and men of the Salerno Brigade, convinced they were surrounded by a superior force, surrendered without a shot. Rommel had taken the Mount Matajur, not by force of numbers but by force of will.

He was promoted to captain and posted as a junior staff officer to the headquarters of the Württemberg Army Corps. This assignment, intended as a further distinction, disgusted him because it meant that he no longer commanded men in action. He saw no possibility of making a contribution to the war effort from the rear, as Monty did. Like Patton he recognized that the only way he could fight was from the front.

In November 1917 the US Army decided to establish a tank school in Langres and Patton was appointed to command it with the new rank of major. The 1st Tank Brigade would be equipped with the French tank, powered by a Renault engine and with a smaller gun than the British 'land-ship' but having the advantage of a fully revolving turret.

First Patton visited the French Tank School at Chamlieu to study the Renault tank. He arrived in the town after dark. Struck by a feeling that he had been there before he asked his driver if the French camp was over the hill ahead. The driver told him that the camp was further on but that there was an ancient Roman camp there. The next day Patton visited the site. He was a firm believer in reincarnation and his own previous lives as a soldier, and knew that 'this place had once been home'. He wrote a poem about it that ended with the lines:

> And now again I am here for war
> Where as Roman and knight I have been;
> Again I practice to fight the Hun
> And attack him by machine.

He was undoubtedly sincere, although he did own an 1852 edition of *Caesar's Commentaries*, which describes the Roman occupation of that area.

On 20 November the British mounted the first massed tank attack of the war at Cambrai. This shock assault with five hundred tanks broke through German defences and advanced an astonishing seven miles in four hours. Patton visited the British and talked to their tank commanders. He later spent two weeks with a French front-line tank unit. While comparing the two types of tank and the tactics used, he could not help comparing the two nationalities: 'I like the French much better than the British, possibly because they do not drink tea.' He was convinced that the tank was the future of land warfare.

The German 'spring offensive', launched on 21 March 1918 against the British sector of the front, was an attempt to win the war before US troops were ready to fight. Berlin calculated that by the summer Pershing would have a million American troops in France and be ready to unleash them. Aiming to defeat the British and French before then, the most battle-hardened troops of the Imperial Army crossed at three points – the Somme, Aisne and Marne sectors of the front – and took forty miles in two weeks, inflicting 175,000 British casualties. But by April, with German losses exceeding 300,000 killed or wounded, the operation came to a halt.

The Allies now planned an offensive of their own to commence in August and to end the war within three months. On 18 July Montgomery was appointed chief of staff of 47th Infantry Division with the rank of lieutenant colonel. The division played a crucial role in the initial attack and he wrote to tell his mother that 'We really have got the Bosche on the hop, he is thoroughly disorganized.' But so was 47th Infantry Division. Monty had to keep his senior commander, General Goringe, informed on the situation at the front and issue orders on the general's behalf to forward commanders. He found again that there was no adequate system of communication between the front line and the command post in the rear. He was not going to use carrier pigeons: 'We devised a system of sending officers with wireless sets up to the leading battalions and they sent messages back by wireless.'

The American Expeditionary Force took part in this same offensive, and Patton commanded 1st Tank Brigade equipped with 140 Renault tanks. Before their first battle on 12 September – the first massed tank attack by the US Army – he issued Special Instructions to the brigade:

No tank is to be surrendered or abandoned to the enemy. If you are left alone in the midst of the enemy, keep shooting. If your gun is disabled use your pistols and squash the enemy with your tracks. Remember that you are the first American tanks. You must establish the fact that AMERICAN TANKS DO NOT SURRENDER.

Two American infantry divisions attacked the St Mihiel salient where the German line bulged out as far as the Meuse river, supported by Patton's tanks. Although he was expected to command his brigade from the rear, he had the same communication problem as Monty and came up with a very different solution. With no radio link between the command post and his tank commanders, he walked after the tanks taking a telephone wire with him, and when that ran out he carried on with four runners. His presence at the front paid off when tanks approaching Essey were ordered back by a French officer, who considered enemy shelling too intense for the town to be taken. Patton ordered them in and risked his life by walking behind the tanks to encourage the men.

It took the AEF only two days to take the salient. The first use of American tanks in battle had proven the cavalry role Patton advocated for them: to dash ahead and punch a hole through the enemy line, which could then be exploited by infantry coming up behind. Despite that, he was reprimanded for leaving his command post and accompanying the tanks into combat. In his defence he quoted the communication problem. Only to Beatrice did he admit his personal need to experience the 'blood and guts' of combat, saying that he wanted to duck as shells exploded but refused to do so. He found the danger and the admiration of 'the men lying down' to be 'a great stimulus'.

Less than two weeks later the AEF, including Patton's 1st Tank Brigade, faced a greater test. They were to make a frontal attack on a twenty-mile section of the German line between the Meuse river and the Argonne forest, protected by trenches, deep bands of barbed wire and concrete machine-gun posts. Patton thought that 'If the Bosch fights he will give us Hell.'

The Meuse–Argonne offensive was launched at 0530 hours on 26 September, with Patton's 140 tanks moving through a morning mist that quickly hid them from his command post. Ignoring the former reprimand he walked forward with ten runners, maintaining visual contact. Soon after 0900 and five miles from their starting point the tanks reached Cheppy and were met by enemy shells. The accompanying infantrymen took cover below the brow of a hill. Five of his tanks had come to a stop where two wide trenches could not be crossed and the troops who were detailed to break down the sides of the trenches took cover at every exploding shell. Patton ordered them to continue digging under fire and dealt severely with a man who refused: 'I think I killed one man here. He would not work so I hit him over the head with a shovel.'

When the sides of the trenches had been collapsed and the tanks continued towards Cheppy, Patton followed on foot, yelling to the infantrymen, 'Let's go get them.' At first they followed him, but on the brow of the hill they were raked by machine-gun fire and fell back: 'I called for volunteers and went forward . . . Six of us started and soon there were only three, but we could see the

45

machine-guns right ahead so we yelled to keep up our courage and went on.'

Another man fell and now only his orderly, Joe Angelo, was with him. Then Patton was hit by a machine-gun bullet that passed through his right hip: 'The only man left helped me to a shell hole and the Bosch shot over the top and was very close. The bullet came out just at the crack of my bottom about two inches to the left of my rectum. It was fired at about fifty yards so made a hole about the size of a dollar where it came out.'

Angelo stopped the bleeding with a field dressing, but it was more than an hour before enemy fire subsided and Patton could be rescued by a stretcher party and carried two miles to an ambulance. He was later transported by train to the base hospital south of Dijon, whence he wrote to tell Beatrice that when they dressed the wound he 'looked as if I just had a baby'.

Patton was promoted to the rank of colonel and the US press called him 'The Hero of the Tanks'. He was more concerned that he had put in only two days' actual fighting: 'Peace looks possible but I rather hope not for I would like to have a few more fights.' The war ended on 11 November 1918, his thirty-third birthday.

Colonel Montgomery was the only one of the three to have witnessed the slaughter at the Somme and at Passchendaele, where both sides were dug in and the static war of attrition wasted a generation to win and lose again a furlong of shell-pocked ground. It has almost become *de rigueur* to draw a causal connection between this experience and the nature of his generalship in the Second World War, but in truth he had no problem with a high casualty rate if it was necessary for victory; at El Alamein those of his commanders who complained about the high rate of casualties incurred in carrying out his battle plan were told to get on with it or be replaced. His concern in the Great War was that men were sent to their deaths for no gain and that this was an *inefficient* way to win a war. He blamed poor planning by inept generals, and poor communication between command posts and the front line.

He felt he could do more as a staff officer by improving planning and communication procedures than he could by leading troops at the front. Monty was from the outset 'thinking beyond the front', convinced that battles were won or lost by decisions made at HQ, both before they commenced and as they unfolded.

The Great War did not fashion Monty's future military career, but it allowed him to discover how his own needs and talents could be applied to the art of making war. His difficult relationship with his mother had been the result of 'a clash of wills' in which neither would give way, and it was to her in a letter written on 3 September 1916 that he revealed where his deepest victory in the present war lay: 'As Chief of Staff I have to work out plans for the operations, and see that all branches of the Staff work with my plans . . . they all do what I tell them like lambs.' He had discovered his forte in war – to plan what others will do and to ensure that they do it.

Captain Rommel experienced a very different war from Monty. He fought on the western front in its early, mobile phase, when a rapid advance with maximum firepower concentrated at the point of attack could throw the enemy off-balance and achieve astonishing results. He was moved to the eastern front where conditions were similar, before the two sides fully dug in and static trench warfare meant that the forward movement of small units had no chance of success. While Monty believed that a battle was won by Headquarters staff, Rommel concluded that tactical victory depended on the on-the-spot decisions and rapid forward movement of front-line commanders. He felt that he had to be at the front to command effectively and was disgusted when assigned to staff duties. But we cannot draw a causal connection between the predominantly mobile warfare he experienced in the Great War and the nature of his generalship in the Second World War. If Rommel had served at the Somme and at Passchendaele he would not have drawn different lessons: he would probably have been killed while leading from the front.

Rommel's daring had won him the admiration of his men but

not the recognition he sought from his superiors. In November 1917 General Otto von Below promised the coveted Pour le Mérite (the 'Blue Max') to the first officer to take the highest Italian position, Mount Matajur. Rommel captured the peak but the Blue Max was awarded to Lieutenant Walther Schnieber. He complained to his battalion commander that Schnieber had taken a quite different summit, and was advised to forget the matter. Instead he made a formal complaint to the commander of the Alpine Corps; he received no reply. In December 1917 after he had taken the Italian village of Longarone the Kaiser awarded him the Blue Max, but the lack of recognition for his earlier achievement still rankled. After the war he pestered the official army historian to make changes to the account of that action and demanded that an appendix be added to detail his own part in it. He later wrote his own account of his Great War actions, ostensibly to pass on to a new generation the tactical lessons learned, but perhaps his first-person narrative was as much an attempt to put the record straight. Its most famous reader would be Adolf Hitler.

Colonel Patton, like Rommel, missed the static front of the trenches. The cavalryman devoted to saddle and sabre had pioneered tank warfare for the US Army, and his tank brigade entered the war at its final stage when the German Army was giving ground and rapid movement forward was likely to succeed. He developed a doctrine of combat that he applied in the Great War and which he would follow in the war to come:

Move swiftly, strike vigorously.

War means fighting. Fighting means killing, not digging trenches.

When the enemy wavers, throw caution to the winds. A violent pursuit will finish the show. Caution leads to a new battle.

Patton felt that he had to be at the front to command effectively, and that he could achieve nothing as a staff officer. He unwittingly defined what he shared with Rommel, and the crux of their difference with Monty, when he wrote: 'In the next war, victory will depend on EXECUTION, not PLANS.'

The tank had yet to come into its own. Almost all of Monty's experience of higher command, as chief of staff of 47th Infantry Division, had been while the British were 'holding the line' in a war of attrition. When Haig ordered an offensive on 18 August 1918, telling his commanders that 'risks ought now to be incurred', Monty did not have sufficient faith in the ten tanks under his control to push them forward, and used them for 'mopping-up' behind the front line as it moved forward, an early manifestation of the caution that later became his trademark. Rommel had no experience whatsoever of these new machines. Only Patton had commanded tanks in an offensive operation. He believed that instead of spreading them between units to be used in a support role, they should be concentrated and used as a spearhead to break the enemy line, creating a gap that the infantry could exploit. Even then the tank was not done, but should 'assume the role of pursuit cavalry and ride the enemy to death'. The first shipment of American-built tanks arrived in France as the Great War ended. They were faster than the British and French tanks but the armistice left them no role. Patton wrote wistfully about what he could have done with 'a few hundred of them . . . if there was only a war on'.

He had also discovered a darker truth about himself: that he enjoyed combat at its most extreme, both in the near certainty of being killed and in killing. On 11 November he wrote in his diary: 'Peace was signed. Many flags. Wrote a poem.' His verse longed for the return of 'the horns of mighty Mars' when he could know again 'the white-hot joy of taking human life'. He had become addicted to the 'adrenaline rush' of war. He wrote to tell Pershing that 'war is the only place where a man really lives' and that he already felt homesick for 'the noise of the shells and the machine-guns'.

The different experiences of Monty, Patton and Rommel in the Great War helped to fashion their style of command as generals in the Second World War but cannot be considered its cause. These three men had intrinsically different mind-sets before the war began, and learned what they did because they were open to those particular lessons. Monty, removed from the Somme and given continuous experience of mobile war, would not have committed himself to command from the front. Neither Rommel nor Patton, placed on the Somme, would have dedicated their military careers to meticulous planning at Headquarters. All three had found their true role in combat and celebrated the armistice with mixed feelings, for peace is no anodyne for men born to fight.

3. How to Fight in Armour

'And so it had all been in vain. We did our duty, two million died, and all of it was for nothing.' That was how Captain Rommel felt as he returned to the barracks at Weingarten in November 1918, but the words were written by another young soldier travelling home from the western front: Corporal Adolf Hitler.

Hitler, like Rommel, had won the Iron Cross for bravery. Now they had lost everything they believed in: the Kaiser had abdicated, the Reich had failed and the army had been defeated. Except that for both men the army had not been defeated on the battlefield but in the armistice negotiations. They and many other veterans believed in the *Dolchstoss* ('dagger in the back'): the Imperial Army had been betrayed by the politicians.

At least Rommel had his new wife, Lucie. He bought a motor-cycle and worked on his stamp collection, and was chosen to join the *Reichswehr*, the new skeleton army Germany was permitted by the terms of the armistice. As its functions were restricted to internal security and it was allowed no tanks or armoured vehicles, it was little more than an armed police force. Nevertheless it was sorely needed. Hunger and unemployment recruited willing revolutionaries for the 'red councils' that sprang up across Germany in hope of a Russian-style revolution and the country was in chaos.

In the summer of 1919 Rommel was ordered to Friedrichshafen on Lake Constance to command 32nd Internal Security Company. To his credit he recognized that much of the unrest had more to do with empty stomachs than Bolshevism, and when faced with an angry crowd threatening to storm Gmund town hall he ordered his men to use fire hoses in place of their firearms and dispersed the mob without bloodshed.

Adolf Hitler saw things differently. It seemed to him that the Bolsheviks, supported by 'international Jewry', would soon make of their country a Communist republic and that the sooner their blood was spilled the better. Several right-wing groups, formed mostly of disillusioned soldiers with their trust in political solutions gone, agreed with that.

Beatrice was waiting on the quayside when Colonel George S. Patton arrived in New York on the SS *Patria* on 18 March 1919. Press reporters and photographers were waiting for him too and the next day his picture appeared in the *Herald Tribune* under the headline: 'TANK FIGHTERS BACK HOME. COLONEL PATTON TELLS HOW BIG MACHINES BY HUNDREDS ATTACKED GERMANS.'

The end of the war allowed a smaller army, and on 11 July Congress reduced the strength of the Tank Corps from 12,000 officers and men to 2650. The corps was renamed 304th Brigade and posted to Fort Meade, Maryland. Patton and Beatrice moved into the wooden barracks and found themselves next door to Lieutenant Colonel Dwight D. ('Ike') Eisenhower and his wife Mamie. The two men quickly struck up an unlikely friendship. The Pattons had a housekeeper, an English cook, six Mexican servants, two cars and a stable of six horses. Eisenhower was the son of an impoverished railroad employee from Kansas; all he had was Mamie. Their friendship was built on a mutual belief in the tank. Patton commanded the French-built Renault tanks of 304th while Ike commanded a battalion of American-built Liberty tanks. The upper echelons of the US Army still considered the correct role of the tank to be in support of the infantry. Ike disagreed: 'George and I thought this was wrong. Tanks could have a more valuable and spectacular role. We believed . . . that they should attack by surprise and mass.'

No one was listening. The National Defense Act of 1920 reduced the strength of the army and abolished the Tank Corps, attaching tanks to the infantry. In October 1920 Patton transferred at his own request to 3rd Cavalry Regiment at Fort Myer, Virginia, where he took command of the 3rd Squadron. The tank had been a great idea – he still believed in it – but with no support from the top the 'big machines' seemed unlikely to play any major role in future deployments of US forces.

His quarters by the Potomac overlooked Washington, DC. From there he could make the social connections with politicians and the 'top brass' that might ensure a role for himself, although his habit of asking ladies if they wished to view his battle scars – a positive response involved him dropping his trousers – proved a little *risqué* for the city élite. He wrote to tell Pershing: 'A lot of the officers and men are already talking of the next war. Of course they are my sentiments.'

The first seeds of that next war were sown in July 1921 when Hitler became leader of the right-wing German Workers Party. He changed its name to the Nationalsozialistische Deutsche Arbeiterpartei (National Socialist German Workers Party), which some shortened to 'NSDAP' but most preferred 'Nazipartei'. The new Nazi Party was stridently anti-Communist, anti-Semitic, and its leaders (with a good proportion of its members) believed that their ends might most quickly be achieved by violent means.

A tennis match in Cologne against Field Marshal Sir William Robertson, at which Colonel Bernard Montgomery's lobbying on behalf of his own future impressed more than his use of the racket, ensured that on returning to England he attended the Staff College at Camberley. He knew that to be the best route to further advancement, but his fellow students found him conceited and all too ready to give his opinion on any military topic in full and without conceding any interruption. Oral tradition has it that one of the students, found guilty of some offence, was punished by being ordered to sit next to Monty at breakfast for a week.

He survived radical cuts in the British Army and in January 1921 was appointed chief of staff of 17th Infantry Brigade, based in Cork in southern Ireland and operating against the nationalists of the Irish Republican Army. The brigade of nine thousand men was the largest in Ireland. Monty was responsible for organizing patrols and house-to-house searches, and soon discovered that no quarter was given on either side. A Sinn Féin bomb exploded as a regimental band marched by, killing and wounding many of the men. When a detachment of the same regiment later cornered twenty Sinn Féin activists in a cottage, and those men refused to surrender, the soldiers set fire to the thatched roof and shot the survivors as they fled the flames. There is no reason to believe that Monty disapproved. He wrote that 'It developed into a murder campaign in which the soldiers became very skilful and more than held their own.' In a letter to Major A. E. Percival he said that the way to win was to be ruthless: 'Oliver Cromwell, or the Germans, would have settled it in a very short time.' He told his father: 'We have had two officers murdered in the last fortnight . . . Any Republican soldier who interferes with any officer or soldier is shot at once.'

On 29 March 1922 the IRA's Cork City Brigade boarded and took over the *Upnor*, a British ship carrying machine-guns, rifles, pistols, grenades and half a million rounds of ammunition. For a time it was feared that, armed as never before, the IRA would attempt a *coup d'état*.

In Germany on the evening of 8 November 1923, Hitler's Nazis did exactly that. Party support was strongest in the country's second city, Munich, and it was there that he and six hundred supporters took over a beer hall in which local politicians were addressing a meeting. He announced that he intended to take control of the city as a first step in replacing the Berlin administration. The politicians, presented with his arguments and the even more persuasive guns of his Stormtroopers, gave their support.

The following morning he led three thousand men into the centre, intending to occupy all government buildings. He knew the army and the police would be ordered to stop him but believed they and the whole city would rise up in his support. They did not. Soldiers on the Residenzstrasse shot sixteen Nazis dead, Hitler was arrested and the attempted *putsch* was over. He had learned that before he could seize power he had to seize the minds and aspirations of the German people. During his eight months in prison he set about doing exactly that, dictating *Mein Kampf* (My Struggle) to his deputy, Rudolf Hess. The book argued that Germans were again hemmed in on all sides by their enemies and must fight once more for their rightful *Lebensraum* (living space).

In January 1921 Rommel was appointed to command of 13th Infantry Regiment based at Stuttgart. He spent nine years there, training young soldiers and attending courses to further his own military career. His old 124th Regiment was one of the many that had to be disbanded and he formed an Old Comrades Association. He took Lucie on a motorcycle tour of Italy to revisit the battlefield of the Isonzo. They went to Longarone. A few years before, Rommel's unit had been first into the town and taken many Italian prisoners there; now, when the locals discovered that he was a German officer, he felt it better to leave earlier than he and Lucie had planned. Their son, Manfred, was born on Christmas Eve 1928. This news was too much for Walburga Stemmer, who killed herself. Rommel continued to write to Gertrud, now aged fifteen.

In October 1929 he became an instructor at Dresden Infantry School. His lectures on tactics were illustrated by the Great War infantry actions in which he had taken part and his narrative style went down well with the young soldiers. During his first year he noted that among his students' favourite reading was a book by Adolf Hitler. Sales of *Mein Kampf*, which had been modest since its publication four years earlier, were rising abruptly. After the Wall Street crash of 1929 the US called in its loans to Germany, the German economy collapsed, unemployment soared and the nation was ready to believe that only an extreme solution could help. Many were attracted to Communism. Others joined the Nazi Party. Hitler's appeal spread far beyond his power-base of disillusioned veterans. Bankers and industrialists financed him because they believed his anti-Communist stance would benefit big business. The army supported him because he promised to tear up the Treaty of Versailles and remilitarize the country. He took a stand against the humiliating conditions imposed on Germany after the Great War, but bread came before pride and it was his promise to tackle the country's rampant unemployment that won him most support.

In the 603-seat Reichstag, the number of Nazi seats leaped from a mere twelve in 1928 to 230 in 1932, making the Nazis the largest party. Hitler was appointed Chancellor of Germany on 30 January 1933. That evening Nazi Party members gathered outside the Chancellery brandishing burning torches. When he appeared they chanted '*Sieg Heil! Sieg Heil!*' (Hail Victory! Hail Victory!) Joseph Goebbels wrote in his diary that 'The new Reich has been born.' On 27 February when the Reichstag burned down and a Dutch Communist Marinus van der Lubbe was arrested at the scene, Hermann Göring declared that to save the country from revolution 'every Communist must be shot where he is found'. When documents were 'discovered' that outlined a Communist terror campaign Hitler asked the Reichstag to delegate all its powers to him, and he got the two-thirds majority he needed to suspend the constitution. On 23 March the 'Enabling Act' made him dictator of Germany, not by *putsch* but by democratic process.

Rommel approved of a leader who placed the army at the centre of his plans for Germany. Restrictions imposed by the armistice were now discarded. The army was to be doubled in size by the end of the year. Tanks were soon rolling off the assembly lines of Krupp and Daimler-Benz. There was talk of taking back the 'lost territories'. Rommel had no hesitation in August 1934 when he was required (along with all members of the armed forces) to swear a personal oath of allegiance to his Supreme Commander:

I swear by God this sacred oath, that I will render unconditional obedience to Adolf Hitler, the Führer of the Reich, supreme commander of the armed forces, and that I will at all times be prepared, as a brave soldier, to give my life for him.

Soon after that he met the Führer for the first time. Rommel had taken command of 3rd Jäger (Rifle) Battalion of 17th Infantry Regiment stationed at Goslar in the Harz mountains. On 30 September Hitler visited Goslar and it was planned that the men of the battalion would line the streets as an honour guard, but when Rommel discovered that the SS insisted on placing a line of their own men in front of his riflemen for the Führer's protection, he protested that this was an insult and refused to turn them out. Himmler and Goebbels were both present, and agreed with him. The order for the SS was cancelled. They almost certainly brought this strong-willed colonel to Hitler's notice and later he inspected the guard of honour with Rommel at his side.

In October 1935 he was posted to the prestigious *Kriegsschule* at Potsdam, near Berlin, as an instructor. The school was full to overflowing. The army was still expanding, as was Hitler's new *Luftwaffe* (Air Force) and *Kriegsmarine* (Navy). Thousands of new army officers were in training and the school's 250-seat main hall was full for every one of Rommel's lectures. Whether or not he was influenced by *Mein Kampf*, which had by now become a bestseller, he took copies of his lectures to a publisher, Voggenreiter in Potsdam, and the manuscript was accepted for publication.

Rommel met Hitler for the second time in September 1936 when he was attached to the Führer's escort for the huge Nazi Party rally at Nuremberg. He was in charge of security, and when Hitler went out for a drive Rommel was instructed to ensure that no more than six vehicles followed him. The number of party 'worthies' wishing to do so filled more than twenty cars and they were not about to be stopped by a mere colonel. They were stopped, however, by the two tanks he ordered to block the road. It was Rommel's first use of tanks and that evening Hitler congratulated him for it.

Rommel's book, *Infanterie greift an* (The Infantry Attacks) was published by Voggenreiter in 1937. Although intended for military readers it told a 'good story'. His battle narratives were about the men involved as much as the manoeuvres. The book quickly became the most popular gift presented by admiring parents to teenage sons on joining the Hitler Jugend (Hitler Youth) or as a Christmas present to Nazi Party members. Tens of thousands of copies were sold. *Infanterie greift an* became a bestseller and Rommel's royalty payments amounted to so much that he asked Voggenreiter to pay him only fifteen thousand Reichsmarks each year and keep the rest on account. On his tax returns he declared only the fifteen thousand and not the larger amount held for him by his publisher.

His own reading that year included *Achtung! Panzer!* by Heinz Guderian. In it his own tactical preferences were transferred from the infantry to armour. Guderian proposed the greater use of tanks to 'move faster than has hitherto been done, to keep moving despite the enemy's defensive fire and to make it harder for him to build up defensive positions, and to carry the attack deep into the enemy's defences'.

At this time Hitler was on the lookout for officers with front-line experience and the tactical abilities that made them fit for promotion. Nicolas von Below, his *Luftwaffe* adjutant, had studied under Rommel at the Dresden Infantry School and brought *Infanterie greift an* to the Führer's attention. Hitler 'recognized' his own Great War experiences in the book, and from that moment Rommel was marked out for advancement. In February 1937 he was appointed liaison officer to Baldur von Schirach, head of the Hitler Youth. He did not join the Nazi Party but he attended two nine-day Nazi indoctrination courses, both of which included a presentation of 'the Jewish problem'. His sympathies were strongly pro-Hitler and, by extension, pro-Nazi.

When the IRA crisis ended in 1922 Monty was posted as chief of staff to 8th Infantry Brigade in Plymouth, and moved the following year to the headquarters of 49th Territorial Army Division in York. He trained troops, organized tactical exercises and wrote instructional booklets. In January 1926 he was appointed to the Staff College in Camberley as an instructor. His lectures were on infantry tactics, but in an article published the same month he wrote that 'The time is coming when the tanks will be the assault arm of the army, the artillery will be the arm which makes the assault possible, and the infantry the arm which occupies the conquered area.' He became close friends with the director of studies, Alan Brooke, who also advocated a wider role for the tank in any future war.

Monty appeared to be the quintessential bachelor until he fell in love for the first time during a golfing holiday to Dinard in Brittany. He was thirty-eight and the unlikely object of his affection was Betty Anderson, a blonde seventeen-year-old. His brother Brian described how he entertained her by 'drawing pictures in the sand to illustrate his ideas for the employment of armoured fighting vehicles'. Having marshalled his forces Monty made his advance. Not surprisingly she beat a hasty retreat. Undaunted, he set about winning her with the meticulous preparation he would later put into his battle plans. Hearing that Betty was to accompany her parents on a skiing holiday to Lenk in Switzerland, he discovered by some means where they would be staying. When they arrived at the Wildstrubel Hotel he was standing in the lobby. Perhaps to divert his attention Miss Anderson introduced him to Betty Carver, a widow of Monty's age, holidaying with her two sons aged twelve and thirteen. Monty wrote that he 'soon made friends with the boys and with their mother'.

The next January Monty again took a skiing holiday at Lenk, and Betty Carver and her sons were there too. 'By the time the holiday was over I had fallen in love.' They were married on 27 July 1927. A son, David, was born in their bungalow at Camberley on 18 August the following year. While Betty cared for the baby, Monty took his step-sons, John and Dick, horse-riding on army mounts.

In 1929 the War Office appointed him secretary of a committee tasked with rewriting the *Infantry Training Manual*. Monty wrote the first draft without consulting any other member of the committee; they read and commented on it. He helpfully suggested to the War Office that the committee could be disbanded: he would complete the manual incorporating their amendments. 'I produced the book, omitting all the amendments the committee had put forward.' The War Office sent the final manuscript to Basil Liddell Hart, the main contributor to the 1921 edition and now military correspondent of *The Times*, who pointed out what Monty had omitted: 'It drops out the passages which explained how reserves could exploit enemy weak points and be passed through the line.' Liddell Hart was the country's leading exponent of mobile warfare and it was his theory of the 'expanding torrent' – penetrating the enemy line and advancing rapidly without waiting to 'mop up' enemy forces – that Monty had left out of the manual.

In Liddell Hart's mobile warfare a battle had to be determined by the requirements of a successful attack – exploiting a break-through or an enemy weakness – and this could not be planned in advance. For Monty a battle had to be determined by the require-ments of a successful defence, which could be so meticulously planned and prepared that any weakness (which could be exploited by an enemy working to Liddell Hart's scenario) was eliminated. Liddell Hart proposed fighting to win; the corollary of that was the possibility of losing. Monty proposed fighting not to lose; the corollary of *that* was the greater difficulty of attaining a clear-cut victory.

Meanwhile there was no clear victor in the uneasy relationship between Monty and his mother which flared up on each visit home. On one occasion he ordered the gardener to cut back a tree that was blocking the light from a window, and when his mother protested he had no right to make such a decision, a furious row took place. When his father died, aged eighty-five, he convinced himself that the death had been hastened by his mother's inad-equate nursing (although he had not been a witness to it) and his visits ceased.

On 17 January 1931 he was made commanding officer of his old battalion, the Royal Warwickshire, just as it sailed to Port Said in Egypt *en route* to Jerusalem. He arrived there to find an army in which ceremonial parades took pride of place and realistic exercises were virtually unheard-of. He implemented a rigorous training programme and abolished the traditional Sunday church parade. More controversially, he established a regimental brothel run by the garrison adjutant. Lieutenant Burge described what the men called 'Monty's brothel':

The ladies were inspected by our own Medical Officer. The soldiers signed a chitty saying they had used the prescribed prophylactic. As a result our VD rate was extremely low. You couldn't stop the men getting a woman; you could ensure it was healthy.

The commander-in-chief in Egypt was impressed by Monty's modern training methods and turned a blind eye to the brothel, complaining to London only about his 'high-handedness' in dealing with senior officers and suggesting that he needed to 'cultivate tact'.

After a period as senior instructor at the Indian Army Staff College in Quetta, he returned to England in May 1937 to command 9th Infantry Brigade in Portsmouth, with the rank of brigadier. War with Nazi Germany was being spoken of, and if hostilities broke out his brigade would join a British Expeditionary Force. That summer while Monty trained the men on Salisbury Plain, Betty booked into a seaside hotel with David, by then nine years old. While playing with him on the beach she was stung on the leg by an insect and developed septicaemia. She died on 19 October. Monty described himself as 'utterly defeated'. He broke down at the funeral. Later he told his step-son Dick: 'It is hard to bear and I am afraid I break into tears whenever I think of her. I get desperately lonely and sad.' Monty had banned his mother from attending the funeral, still convinced that her poor nursing had hastened his father's death. He arranged for David to stay with Major Tom Reynolds (the headmaster of the boy's preparatory school) and his wife. He had no time for mourning and threw himself into his work. War with Germany grew increasingly likely and he determined to make 9th Infantry Brigade 'as good as any in England'.

In 1937, the year that Rommel's *Infanterie greift an* and Guderian's *Achtung! Panzer!* appeared, there was also a new book by Liddell Hart. In *Europe in Arms* he expounded the use of tanks as an independent striking force to make deep penetrations into enemy territory. In Germany both Rommel and Guderian were familiar with his work, and Montgomery had corresponded with him since their differences over the *Infantry Training Manual*. Liddell Hart and Guderian championed the mobile warfare they claimed could result from exploiting a gap forced in the enemy line. It was impossible to predict when and where this breakthrough might occur, and when it did there was a necessity to respond quickly before the enemy could bring in reserves to fill it. Such a battle could not be planned in advance and would be won by acting rapidly to exploit an enemy mistake or a weakness.

In response Monty wrote his own, very different treatise. 'The problem of the Encounter Battle as affected by modern British War Establishments' appeared in the September 1937 edition of the *Royal Engineers Journal*. He argued that a battle was won or lost in the headquarters of the commander *before* it began:

He must make a plan and begin early to force his will on the enemy . . . If he has no plan he will be made to conform gradually to the enemy's plan. Success will go to the commander who has a plan and does not allow his formation to drift aimlessly into battle, but puts it in to fight on a proper plan from the beginning. He must not be unduly influenced by local situations on the battle front.

Monty believed that a commander fighting to such a plan could not be beaten by the Liddell Hart/Guderian approach. If his planning and preparation was thorough, all weakness in the defence and all possibility of a mistake (that the enemy could exploit) would have been eliminated.

In Guderian's *Achtung! Panzer!* and Monty's 'Encounter Battle', two opposing methods of commanding a battle were proposed. In the coming war they would be put to the test by Rommel and Patton (following Guderian's ideas) and by Monty (following his own master plan), and culminate in the great battle at El Alamein.

'Ride the enemy to death. *L'audace, l'audace, l'audace – tout jour [sic] l'audace.*' Patton copied this quotation from Frederick the Great into his notebook while he was at Fort Myer and it became the watchword of his military career. While his fellow officers found time only for the *Cavalry Journal*, he read Plutarch, Cromwell, Napoleon and Clausewitz. He 'recognized' himself in their pages and claimed that he had fought and died as a soldier while marching with Caesar's Tenth Legion and with Napoleon on the retreat from Moscow. He took Beatrice to see every war film showing at the cinema, but almost always stormed out before the end because the uniforms or weapons were inaccurately represented.

His talent for soldiering was beyond doubt, and in September 1923 he began a course at the Command and General Service School at Fort Leavenworth, Kansas, roughly equivalent to the British Staff College at Camberley. He studied 'under a strange blue light', purchased because it was advertised as restoring hair to a balding scalp. His hard work paid off – he does not say whether the lamp did too – and he graduated from the school twenty-fifth out of 248, and secretly passed on his course notes to Dwight Eisenhower who was to attend the next year.

By now Patton had two daughters, but it was only when Beatrice gave birth to a son on Christmas Eve 1924 that he took an interest in his offspring and ensured that the child was named George. The following March he was posted to the US Army's Hawaiian Division at Schofield Barracks, Honolulu, as chief operations officer, but after criticizing his commanding officer, General Smith, he was demoted to chief intelligence officer for 'being too outspoken'. Smith reported that 'this officer would be invaluable in time of war but is a disturbing element in time of peace'. Patton took all of it as a compliment.

In 1929 he was offered the post of US military attaché in London. He turned it down and told Beatrice that that was to avoid any possibility that he might gain an English son-in-law:

We have two daughters. If we go to London it stands to reason that one or both of them will marry an Englishman. Well-bred Englishmen are the most attractive bastards in the world, and they always need all the money they can lay their hands on. They are men's men, and they are totally inconsiderate of their wives and daughters.

He returned to Fort Myer as executive officer of 3rd Cavalry in July 1932. That same month twenty thousand unemployed veterans of the Great War from across the US travelled to Washington to protest. By now the Great Depression was biting hard. The veterans felt particularly aggrieved and demanded that Congress bring forward the pension payable to them in 1945. Many of these 'bonus marchers' set up shacks and tents in the capital. On 28 July a man died while scuffling with police, the fracas got out of hand and President Herbert Hoover ordered in troops to maintain order. Patton led two hundred mounted cavalrymen up Pennsylvania Avenue towards a mob armed mostly with bricks. The infantry fired gas grenades and when the mob stood its ground the cavalry charged. Patton was in earnest: 'Bricks flew, sabres rose and fell with a comforting smack, and the mob ran. We moved on after them, occasionally meeting serious resistance. Two of us charged at a gallop and had some nice work at close range.' That evening he and his men 'cleared out' the marchers' makeshift camp, during which a number of tents and shacks were 'accidentally' burned down.

The next morning one of the marchers asked to see him. Staring into the face of Joe Angelo, who had saved his life in France, Patton said, 'I do not know this man. Take him away.' It was an unforgivable snub, although he explained later that he had given Angelo the money to set himself up in business and felt his appearance among the protesters to be a personal insult.

Early in 1935 he was returned to Hawaii as chief intelligence officer. His role was to collect and collate intelligence. In Europe the increasing ties between Germany, Italy and Japan suggested the possibility of future hostilities between the US and Japan, and one of his tasks was to plan internal security with reference to the Japanese inhabitants of the island. His 'plan' was not subtle: 'Arrest and intern persons of the Orange race . . . to retain them as hostages.' He wrote a report warning that a Japanese attack on Pearl Harbor was feasible and that measures should be taken to limit the effects of any such strike. Nothing was done.

He expected another war but feared that at fifty he was already too old to have any part in it. He took out his frustrations on the polo field and was temporarily relieved of captaincy of the army polo team for swearing at the opposing team captain and 'using foul language in front of ladies'. He began drinking too much and there were rumours of casual affairs. He wrote a poem that ended:

> Then here's to blood and blasphemy!
> And here's to whores and drink!
> In life you know you're living.
> In death we only stink.

When Beatrice's niece Jean Gordon came to visit he asked them both to accompany him on a three-day horse-buying trip, but his wife felt unwell and Jean went alone with 'Uncle George'. She was twenty-one and a vivacious brunette, and when they returned Beatrice 'knew' what had happened. At the time she said nothing. Later she told their daughter Ruth Ellen that she stayed with Patton 'because I am all that he really has, and I love him and he loves me'.

In June 1937 while he was out riding with Beatrice her horse kicked out and broke his leg. It took six months to heal but he kept busy. He read the writings of Liddell Hart, and when Guderian's and Rommel's books were translated by US Army Intelligence he was among the first to see copies. Historians make much of the fact that he read Rommel's book on infantry tactics and applied what he learned to beating him, but it was Guderian's book that gave the American the greater insight into the potential use of tanks.

The US Army had begun production of the M2 light tank (nicknamed 'Mae West' after the buxom actress because of its twin turrets). Armed only with a machine-gun, this was an infantry support vehicle rather than the attack vehicle proposed by Guderian and Liddell Hart, whose theories found no favour in Washington. Patton was already a believer. In a lecture on mechanized forces to officers at Fort Myer, he had outlined how tanks held as an offensive reserve and sent forward *en masse* at the right moment could determine the outcome of a battle. He wrote articles for the *Cavalry Journal* pressing the case for the tank, yet in that same journal the most senior officers in the country ridiculed proponents of the tank as 'obsessed with a mania for excluding the horse from war'. Patton also noted that Liddell Hart, by pressing the case for the tank too hard in the country that invented it, had become viewed as eccentric and *The Times* had dropped him as its military correspondent. Patton wanted tanks in the US Army, but more than that he wanted an appointment for himself in a fighting unit. Discreetly, he moderated his advocacy of the tank, proposing the use of armoured and mounted cavalry in combination.

As his broken leg began to heal and he could walk on crutches, he hobbled into the stables and beat Beatrice's horse with one crutch. It was the first incidence of an extreme behavioural trait that was to reappear with greater consequences in the war to come.

In March 1938 Hitler claimed more *Lebensraum* for Germany by annexing Austria, followed in October by the 'peaceful' occupation of the Sudetenland. When he toured the latter territory by train he put Rommel in charge of the Führerbegleitbataillon (Führer escort battalion) consisting of 93 NCOs and 274 men.

After this trip Rommel was promoted to colonel, and on 10 November took up his new appointment as commandant of the *Kriegsschule* at Wiener Neustadt, south of Vienna. He, Lucie and Manfred moved into a bungalow located between the war school and the Messerschmitt aircraft factory.

Hitler soon called on him again. In March 1939 the president of Czechoslovakia, Emil Hácha, was given the choice of watching his country bombarded by the *Luftwaffe* or 'inviting' German troops into Prague. Hacha signed the invitation and the Führer put Rommel in charge of his personal security for a triumphal visit to the capital. When an SS escort was delayed, Rommel advised Hitler to continue into Prague anyway. He bragged to Lucie: 'I persuaded him to drive on under my personal protection. He put himself in my hands.' Hitler mentioned that he had been inspired by reading Philipp Bouhler's biography *Napoleon – Kometenpfad eines Genies*. On the wall of his study Rommel had an engraving of Napoleon and now he felt that he was serving a man of similar stature and ambition. He told Lucie, 'Isn't it wonderful that we have this man?'

Patton ended the inter-war period where he had begun it, at Fort Myer, Virginia, commanding 3rd Cavalry. He and Beatrice moved into the commander's house on Jackson Avenue. Beatrice went riding with Eleanor Roosevelt, wife of the President, while Patton enjoyed the company of the German military attaché, General Friedrich von Boetticher, who was a military historian and Civil War enthusiast.

The two men had long been friends and now by mutual consent neither mentioned Hitler or the political situation in Germany. They visited the site of the 1864 battle of Wilderness. Patton explained the Confederate and Yankee deployments and pointed to a hill from which, he said, General Early had directed the troops. When von Boetticher told him that every historian of the battle placed Early elsewhere, they argued about it until an elderly man arrived to say, 'General Early was on that rise. I was at this battle as a boy.' Patton nodded: 'Of course. I saw him myself.' Shortly after that, von Boetticher attended a dinner party at the Pattons' home and was challenged by another guest on Hitler's anti-Semitism. He told Beatrice not to invite him again as his presence might harm her husband's career.

Patton trained his men hard. In August 1939 he took part in 3rd Cavalry manoeuvres at Fort Belvoir, Virginia. As Lieutenant James H. Polk, commanding G Troop, advanced his men towards the 'enemy', Patton arrived to demand, 'What the hell is going on here?' He wanted to know why Polk had not opened fire with live ammunition. The lieutenant replied that it was against peacetime regulations to open fire before reaching the specified range. Patton shouted that he had 'helped write those stupid regulations for damn fools like you. Now let's have a real exercise.' Polk ordered live rounds to be fired and there followed 'a hell-roaring battle that violated every regulation in the book. It was obvious to all of us that our colonel was preparing us for the real thing.'

In London, editorials in *The Times* had predicted war within a year, and a news feature dated 6 July 1938 reported a major military exercise involving 'three regiments . . . guns, tanks and lorries' planned and commanded by Brigadier Montgomery. In fact Britain was woefully unprepared for war. There would have been no exercise at all if Monty had not pressed for it, planned it and commanded it. Unfortunately his personal style did not help communicate its lessons to his peers, with one senior officer describing his written assessment as 'not a report but a hymn of self-praise'. General Wavell was Monty's superior officer but received short shrift when he asked (in all seriousness) what particular problems a gas attack might cause the kilted soldiers of the Black Watch. Monty replied: 'Scotsmen do not wear drawers under the kilt; the result, therefore, might be very unpleasant.'

Posted in November 1938 to command 8th Division in Haifa, in Palestine, to put down an Arab rebellion caused by Jewish immigration (which had increased following the rise of the Nazis and other anti-Semitic regimes in Europe), he proved himself to be ruthless. He told his troops that the insurgents 'must be hunted relentlessly; when engaged in battle with them we must shoot to kill . . . This is the surest way to end the war.' Within three months he reported to the War Office that 'The rebellion is smashed; it is very difficult to find Arabs to kill; they have had the stuffing knocked right out of them.'

He returned to England and took command of 3rd Infantry Division on 28 August 1939. The Prime Minister, Neville Chamberlain, had not yet authorized the army to mobilize, but a great many preparations for war with Germany were under way.

On 25 August Rommel reported to Hitler in the Reich Chancellery in Berlin and was appointed to command the Führerhauptquartier (Führer Headquarters) for the imminent invasion of Poland. The mobile HQ would be established on Hitler's special train. Rommel's task was to guard the Führer and his staff while they were on it or travelling out from it to visit army units. He was given the same 367-man escort battalion he had commanded on the tour of the Sudetenland, and promoted to a rank appropriate to his responsibility: that of major general.

Rommel moved the escort battalion to Bad Polzin, a railway town close to the Polish border where German forces were gathering for the invasion and where he would assume responsibility for Hitler's security. He set up an office in the waiting room. The army massed on the border included 2.4 million men, 2574 tanks and 7710 artillery and anti-tank guns. Reckoning by numbers alone, the Poles more than matched them with 3.65 million men, more than three thousand tanks and at least fifteen thousand guns. The numbers, however, took no account of the new tactics that German forces would employ, developed by General Heinz Guderian.

On 31 August he wrote to tell Lucie that 'The Führer knows what is right for us.' That evening he sat ready in the station waiting room until the call came: 'The invasion will begin at 0450 hours tomorrow.'

4. Rommel's Blitzkrieg

The Führer's train left Berlin at 2100 hours on Sunday, 3 September 1939, travelling east. It included a *Flakwagen* (armoured wagon with anti-aircraft guns mounted at each end), the *Führerwagen* (Hitler's drawing room and bathroom) and the *Befehlswagen* (command and communications centre). Among those on board were Reichsführer-SS Heinrich Himmler, Martin Bormann, chief of the High Command Wilhelm Keitel, and his chief of staff Alfred Jodl. At Bad Polzin they were joined by General Rommel and the escort battalion. In the early hours of Monday morning Hitler's mobile headquarters, now under Rommel's command, crossed the border into Poland – and into the Second World War.

The invasion codenamed Fall Weiss (Operation White) was in its fourth day. With the speed and firepower of the panzers as their spearhead, ten German divisions supported by the *Luftwaffe* had swept through Polish frontier forces and pressed inland. Hitler's train kept pace with them. He travelled out from it each day in an armour-plated Mercedes, accompanied by Rommel's escort troops in armoured cars, to visit German troops at the front.

Both Rommel and his wife were fervent supporters of the Führer. On 4 September Lucie wrote: 'May the dear Lord protect him, and you too, my beloved Erwin. All of my friends ask you to plead with him not to expose himself to unnecessary danger. Our nation cannot afford to lose him.' Like most Germans, they were particularly eager to see the recapture of Danzig, the 'German' city taken from them by the Treaty of Versailles. But the Rommels had a personal interest too: Danzig, where they had met and fallen in love, was 'their' city.

Following the rapid advance of the panzer divisions through Poland, the train reached Plietnitz on 5 September. Rommel replied to Lucie: 'I have big problems with the Führer. He always wants to be right up with the forward troops. He seems to enjoy being under fire.' Because Rommel always accompanied Hitler to the front he gained a first-hand insight into the operation of blitzkrieg and its effect on the enemy in this first major military operation to be fronted by tanks. The roads were littered with burned-out Polish Army vehicles, which he at first credited to the *Luftwaffe*, until reports described how they had been overtaken and destroyed by the speed and firepower of the panzers. This he saw as his own infantry tactics – using the shock effect of a rapidly moving and concentrated attacking force – applied to armoured warfare.

On 10 September the train reached Kielce south of Warsaw and by now a surprising rapport had developed between the Supreme Commander and the newly appointed general. The Führer treated him as more than a bodyguard. Rommel wrote to tell Lucie that he had been invited to attend the daily situation conference chaired by Hitler and was 'occasionally allowed to speak'. He 'sometimes sat next to him at lunch'. While Hitler was attracted by Rommel's aggressive tactics displayed in *Infanterie greift an*, and Rommel appreciated a head of state who saw everything through the eyes of a soldier, this special relationship had its roots in class. Neither man belonged to the Prussian military aristocracy that still dominated the High Command and whose members surrounded and advised Hitler. In Rommel he had found 'one of his own'.

While Rommel was making friends with Hitler he made a dangerous enemy in Martin Bormann. At Gdynia, Hitler decided to drive to the Baltic down a steep, narrow road that led to the sea's edge. Rommel allowed Hitler's car and an escort vehicle to set off down, then blocked the road. Martin Bormann in the third car demanded that he be allowed to pass. Rommel replied: 'I am headquarters commandant. This is not a kindergarten outing and you will do as I say.' According to Walter Warlimont, deputy chief of staff of the *Wehrmacht*, Bormann directed 'furious screams' at Rommel and 'swore in an outrageous manner'. Bormann was soon to become chief of the Nazi Party Chancellory and would carry with him a personal dislike of the upstart general.

Danzig was retaken, and on 19 September Rommel accompanied Hitler on the drive from the station at Goddentow-Lanz into the city, where the Führer broadcast to the nation. 'I was able to talk with him about military matters for almost two hours. He is extremely friendly towards me.' Their special relationship could no longer escape the notice of the Army High Command. On 23 September he told Lucie: 'I eat at his table twice each day now, and yesterday I sat next to him . . . relations with [Colonel Rudolf] Schmundt are strained. Apparently my relationship with the Führer is becoming too strong.'

This three-week period played a crucial role in forming Rommel's military career. He was able to study at first hand how the theory of blitzkrieg could be successfully applied by a rapidly moving 'arrowhead' of panzer divisions. But just as significantly his high regard for Hitler was confirmed by seeing him continually exhorting the troops from the front and at some personal risk: a command style that matched his own. Hitler had won Poland but he had also won Erwin Rommel. Rommel began signing his letters, 'Heil Hitler! Yours, E. Rommel'. He might never have been a card-carrying Nazi but he was now Hitler's man.

It was during one of their lunchtime conversations that he told the Führer he would like to be considered for an operational command, and that although his only experience was with the infantry he would prefer a panzer division. Such a request from any other officer with no tank experience would have been ignored; indeed, no other officer would have dared to ask it. But by now Rommel had a special place in Hitler's affections. When the last Polish resistance had collapsed and the train was heading back to Berlin, the Führer gave him a signed copy of *Mein Kampf* inscribed, 'To General Rommel with pleasant memories'.

★

1 George S. Patton as a cadet at the Virginia Military Institute, 1904

2 Erwin Rommel, pictured with Lucie, after he graduated from the Imperial War School in Danzig, 1911

3 Bernard Montgomery, in his new captain's uniform, shortly before leaving for France in 1916

4 Rommel plays with a fox cub in his dugout in the Argonne, 1915

5 Patton poses in front of a Renault light tank of the US Tank Corps in France, 1918

6 *(Left)* Monty wearing the red collar tabs and armband of a staff officer, France 1917

7 *(Below)* Patton's tank corps advancing towards Argonne

8 Adolf Hitler, Rommel and Martin Boormann (with goggles around his neck) at Maslowie during the invasion of Poland, September 1939

9 Rommel's 7th Panzer Division included these powerful Czech-built tanks, pictured during the advance through France, May 1940

10 7th Panzer Division crossing La Bassée Canal via a pontoon bridge

12 Patton and a tank of 1 Armored Corps at his Desert Training Center in California

11 Monty, commanding 12 Corps, talks to Winston Churchill about the defence of the south coast. Pictured with them is King Peter of Yugoslavia

13 Rommel watches the first tanks of his Afrika Korps unloaded from a boat at Tripoli, March 1941

14 Rommel's 'quarters' at his desert HQ, where he was determined to live in the same conditions as his men

15 Some of the first Sherman tanks to arrive in North Africa advance to the front

16 Rommel's pride is evident as he journeys to brief Hitler on the North Africa campaign

17 The Führer congratulates Rommel on his desert victories

18 Monty wearing his Australian hat complete with regimental badges collected as he visited the various units of Eighth Army

19 Monty briefs the commander of the Grenadier Guards holding a position on the Mareth line

20 Tank crew of Eighth Army. In the desert, dress regulations were not rigorously observed

21 At a forward dressing-station near El Alamein, a piper plays to British and German wounded

22 Patton leaves the USS *Augusta* to land on the beach at Fedala, near Casablanca, 8 November 1942

23 Patton watching the battle near Maknassy in Tunisia

On 30 September, twenty-six days after Britain's declaration of war on Germany, General Montgomery's 3rd Division – part of 2 Corps under the command of Lieutenant General Alan Brooke – disembarked at Cherbourg. From there they were transported three hundred miles in cattle trucks to take up a defensive position to the south and east of Lille in preparation for a German attack.

The London newspapers informed the nation that the British Expeditionary Force to France was superior in both the quality of its men and its equipment to any previous army sent to fight abroad. To Monty it was 'totally unfit to fight a first-class war on the continent of Europe' and 'its weapons and equipment were quite inadequate'. Half of his men were reservists without training of any kind and he had insisted that before leaving England each man at least fire a rifle and throw three live grenades. As for the country that had invented the tank, 'there was somewhere one Army Tank Brigade. For myself, I never saw it.' Although the BEF was concentrated along the Maginot Line of pillboxes, trenches and anti-tank obstacles, it was believed that German forces were most likely to attack through neutral Belgium. The Belgians would not allow Allied troops on to their soil until that actually occurred, and it was planned that immediately German troops crossed the border the BEF would do so too, moving sixty miles east to form a defensive line along the river Dyle. Monty alone among his fellow generals exercised his division in the required move to the new position, using a similar sixty-mile stretch of French countryside. Like Rommel, he at first took the rapid destruction of Polish divisions to be the work of the *Luftwaffe*, and put his troops through repeated night manoeuvres until moving camp in darkness became second nature.

His style of command was different too, and he deliberately set about making himself known to all those under him. During the day he left administrative tasks to his staff so that he could visit every unit in his division. He had a coloured light fixed to the roof of his car for use during night exercises 'so that the soldiers will know I am there'.

What his peers deplored as showmanship – nothing of the kind had been seen before – undoubtedly brought him to the notice of his men. But the particular affection felt by those who served under Monty had its origins in an almost surreal affair. In the middle of November the incidence of venereal disease in his division worried him. Forty-four cases had been reported in the previous four weeks and Monty discovered that the men were persuading (or paying) local girls to spend time with them in the beetroot fields that adjoined the camp. On 15 November he issued Divisional Order 179/A:

The whole question of women, V.D., and so on must be handled by the regimental officer, and in particular by the C.O. My view is that if a man wants to have a woman, let him do so: but he must take precautions against infection – otherwise he becomes a casualty by his own neglect, and this is helping the enemy. He should be able to buy French Letters in the unit shop. If a man desires to buy his French Letter in a civil shop he should be instructed to go to a chemist and ask for a 'Capote Anglaise'. The cases of V.D. we are getting are from local 'pick-ups'. There are in Lille a number of brothels, which are properly inspected and where the risk of infection is practically nil. These are known to the military police, and any soldier in need of horizontal refreshment would be well advised to ask a policeman for a suitable address.

The senior chaplain at GHQ in Arras complained about the improper nature of the order and Brooke seriously considered sending Monty back to England, but this was not a good time to remove a divisional commander. German troops were massing on the Belgian border and his division was on two hours' notice to break camp and move to counter an invasion. Brooke went to see Monty on 23 November and, in the latter's own words, gave him 'a proper backhander' that left no doubt he had barely escaped losing his job. When Monty said he thought his order had been 'rather a good one', Brooke 'began again and I received a further blasting'.

If Monty's seniors thought the order obscene, the reaction of his men was perfectly caught in a ditty written by a young soldier and given the title 'Mars Amatoria':

The General was worried and was very ill at ease,
He was haunted by the subject of venereal disease;
For four and forty soldiers was the tale he had to tell
Had lain among the beets and loved not wisely but too well . . .
No kind of doubt existed in the Major-General's head
That the men who really knew the game of Love from A to Z
Were his Colonels and his Adjutants and those above the ruck,
For the higher up an officer the better he can fuck . . .

While the incidence of venereal disease in Monty's division dropped sharply, the true effect was greater still. The 'VD affair' had much to do with the affection in which his men held him throughout the war years: a general who concerned himself with sexual diseases, condoms and brothels could be trusted with their lives.

His approach to the war was just as uncompromising. When the Prime Minister, Neville Chamberlain, visited 3rd Division on 16 December he said to Monty: 'I don't think the Germans have any intention of attacking us, do you?' Monty told him that they would attack in the spring. He was realistic about the chances of stopping them. He had his division continually practise 'a disengagement from close contact and a withdrawal to a position in the rear'. He insisted that his engineers improve their demolition skills, to blow bridges and crater roads behind the division as it fell back. His newly appointed intelligence officer, Kit Dawnay, summed it up: 'We assumed that German tanks would be involved, and our anti-tank rifle was useless, although we could use our artillery in an anti-tank role. Monty foresaw that we would have to withdraw, and apart from practising the limited counter-attack, we did nothing in the way of training for offensive action.'

★

Colonel Patton's first response to the German invasion of Poland was to reread *Achtung! Panzer!*. He saw that Guderian's theories of mobile armoured warfare had been successfully transferred from the page to the open terrain and cities of Poland. This was tank warfare as Patton had imagined it. His second response was a letter-writing campaign aimed at bringing his experience in commanding US tanks in the Great War to the notice of the top brass, and he asked Pershing to write, too, in his support. If they wanted a 'Rommel' he offered himself for the part. His posturing was a little premature. Britain and France had declared war on Germany; the US had declared its neutrality.

General George Marshall had become the new army chief of staff on 1 September and as such he would occupy Quarters One at Fort Myer where Patton commanded 3rd Cavalry. While Marshall's quarters were being redecorated, Patton asked him to stay with him. He told Beatrice, 'Gen. George C. Marshall is going to live at our house! I think that once I can get my natural charm working I won't need any letter from Pershing or anyone else.' The two men were soon on first-name terms. Patton ensured that their discussions on military topics included the merits of the tank in any future European war and the merits of George S. Patton as a tank commander in the Great War.

American newspapers had reported how German tanks cut effortlessly through the Polish cavalry and now reporters asked Marshall if the US Army had an armoured force of that kind. He arranged for them to visit Fort Benning to see an armoured manoeuvre, then ordered every tank in the army to be transferred there at once. Although there was a tank battalion, it was an *infantry* tank battalion. The difference may have been lost on the press but it made nonsense of the affair to Patton, who came to observe. The tanks were kept to the roads and used only as an infantry support; these were not panzers.

Marshall had to handle two influential lobbies, one for and one against the tank. General Chaffee insisted that the horse cavalry had been rendered obsolete by the panzer-led invasion of Poland

and called for the immediate creation of four armoured divisions for the US Army. General Lynch argued that the infantry saw no need for independent panzer-type divisions and that the army should concentrate instead on modernizing and extending its anti-tank defences. Marshall, caught between them, planned further manoeuvres to test and compare the tactical value of horse cavalry and tanks.

In manoeuvres in Louisiana, the infantry and cavalry were soundly beaten by tank units combined to form a provisional armoured division. After the exercise Patton and Chaffee were among a group of officers who met secretly in a school basement in Alexandria, Louisiana. They agreed that the US must have an armoured force independent of both the infantry and the cavalry, modelled on the German panzer divisions. The so-called basement conspirators sent their 'recommendations' to Marshall, uncertain whether the manoeuvres now offered him sufficient evidence to resist the anti-tank lobby or whether they had just blighted their own army careers.

Patton despaired at the slow pace of events. He needed the US to enter the European war and he needed to command an active unit. Neither seemed likely – until events in Europe forced Marshall's hand.

Hitler kept his word to Rommel, and on 6 February 1940 he was given command of 7th Panzer Division at Bad Godesberg. He had never commanded tanks in battle and Colonel Hans von Luck wrote: 'My infantry instructor from Dresden became our divisional commander. Much as we admired this man, we wondered if an infantryman could be a commander of tanks.'

Rommel, like Monty, trained his men hard, included night-time exercises and made a point of visiting every unit to make himself known to the men. He told them he was proud to lead a panzer division. He met the author of *Achtung! Panzer!* when Guderian visited the division and addressed the officers. 'You are the cavalry,' Guderian said. 'Your job is to break through and keep going.'

The division was comprised of 25th Panzer Regiment, two Schützen (motorized infantry regiments), an artillery regiment with thirty-six guns, an anti-tank battalion with seventy-five guns and an engineer battalion. The Panzer Regiment, commanded by Colonel Karl Rothenburg, had 218 tanks, and although these included the Mark III (with a 5-cm gun) and IV (7.5-cm gun), half of them were light, Czech-built tanks. The success of the German divisions has often been put down to the superiority of their panzers but at this point they had no technological advantage over the British and French. The benefit of the heavier shells fired by Rommel's best tank, the Panzer IV, was cancelled out by the thicker armour of the British Matilda Mark II, while the German tank's relatively thin armour left it vulnerable to both the Matilda and the French Somua S35.

The invasion of France would not be determined by the relative technologies of the tank but by the way in which it was used. The British and French viewed it as an infantry support vehicle and therefore spread their tanks thinly along the whole defensive line. The German Army, convinced by the concept of blitzkrieg, used panzer divisions as the spearhead of an attack co-ordinated with the *Luftwaffe* (to bomb ahead of the advance) and the infantry (to surge through the gap in the enemy line thus created). Blitzkrieg was not just about *speed* but also *mass*: achieving a rapid, deep penetration by massed armour.

For the invasion of France and Belgium, codenamed Fall Gelb (Operation Yellow), Rommel's division would advance as part of General Gerd von Rundstedt's Army Group A through the heavily forested valleys of the Ardennes, considered impenetrable by tanks. On their right General Fedor von Bock's Army Group B would enter northern Belgium along the 'traditional' route – the lowlands – and tempt the British and French forward to confront them. Then Army Group A would appear unexpectedly from the Ardennes, cross the Meuse and make a *Sichelschnitt* (scythe's sweep), striking first westward, then turning north across the rear of the British and French to trap the Allies in Belgium. Caught between these two army groups the enemy could then be destroyed. It was intended that Guderian would act as the forward edge of the scythe, with Rommel securing his right flank.

While many in the High Command had grave doubts about the plan, Rommel praised Hitler for understanding and implementing Guderian's concept of the tank attack. He told Lucie in April that 'If we did not have the Führer, I doubt there would be any other German capable of so brilliant a mastery of both military and political leadership.' Rommel's approval for Hitler was not given, as would later be said, in ignorance of the darker side of National Socialism. Two 'Party men' were attached to his divisional head-quarters: Lieutenant Karl Hanke, personal assistant to Goebbels, the propaganda minister, was now appointed ADC to Rommel; Karl Holz was editor of *Der Stürmer*, the country's most anti-Semitic paper, and acted as a link between Rommel and the attached propaganda unit. There is no evidence that Rommel found their presence objectionable; on the contrary, he and Hanke became close friends.

At the beginning of May the division moved to a training area near the Eifel mountains in north-west Germany for practice with live ammunition. Rommel stressed to his commanders that 'It won't be a walk-over as in Poland; the French and the British are quite different opponents.'

Fall Gelb was launched on 10 May. At 0430 the advance guard of Rommel's 7th Panzer Division crossed the Belgian frontier. In two days he broke through the outnumbered Chasseurs Ardennais and crossed the Ourthe river. In his first armoured encounter of the war at Marche, when his way was blocked by a handful of Renault and Hotchkiss tanks, 'prompt opening fire on our part led to a hasty French retreat'. He kept up the pace of his advance westward by ignoring his flanks, depending on his speed and firepower to crush French opposition. He commanded from the front and in person, writing to tell Lucie on the evening of 11 May that his voice was hoarse from continually shouting orders.

He reached the eastern bank of the Meuse near Dinant on

the afternoon of 12 May to find that the bridge had been blown and French infantrymen installed in buildings on the far bank. The next morning an attempted crossing in rubber boats failed under heavy enemy small-arms fire. Rommel positioned tanks along the east bank to fire on French positions, had houses set alight to provide a smokescreen, and was among the first men to cross in a rubber boat: 'I now took over personal command of the 2 Battalion of 7 Rifle Regiment and for some time directed operations myself.' While his riflemen engaged the French infantry, Rommel's engineers built pontoons to get his tanks across. Army Group A was crossing the Meuse at two other points but Rommel's division was first across, beating even Guderian's panzers to the west bank.

He now went ahead of the division with Rothenburg's 25th Panzer Regiment, moving rapidly west from the Meuse before the French had time to organize a counter-attack. His own command vehicle was a specially adapted Panzer III but he often rode in Rothenburg's Panzer IV. Now that they were through the Ardennes the open countryside allowed much quicker progress and by the evening of 16 May he had reached the 'impregnable' Maginot Line. His artillery provided covering fire while his tanks advanced and engaged enemy positions with direct, close-range fire, then his infantry followed up to assault and take the fortifications. It was here that in the Great War the Kaiser's army had been held for more than four years. Rommel hardly paused as his panzer division swept through, the tanks firing to both flanks as they sped on to the west.

Rommel led his division, covering fifty miles during the night, so that by the morning of 17 May he was ahead of both panzer divisions advancing on his flanks. The German blitzkrieg had become Rommel's blitzkrieg.

Montgomery's 3rd Division had led the way as the British took the bait Hitler dangled before them. On 10 May, immediately von Bock's Army Group B moved west into northern Belgium that country abandoned its neutrality and joined the Allies, and the BEF moved east to take up its prearranged defensive positions. In a surreal counterpoint to Rommel's lightning advance, a Belgian frontier guard attempted to halt Monty because he did not possess the necessary permits.

The division travelled through the night and by first light had reached its allotted position along the river Dyle near the medieval city of Louvain; a crucial sector, for if Louvain fell Brussels was certain to be next. Yet Monty found 10th Belgian Division (which possessed only horse-drawn transport) already there and its commander refusing to make way for him. Kit Dawnay felt that 'there was practically war between Britain and Belgium when dawn broke that morning'. While the two allies argued about who had been allotted this critical spot, German tanks surged through Belgium towards them. The Belgians refused to move and Monty placed his own men in line behind them. He established his headquarters in a small château at Everberg, the residence of the Prince de Merode, whose wine cellar was quickly discovered and raided by Monty's young staff officers.

On 13 May as the German offensive closed on Louvain, the commander of the Belgian division lost his nerve and signalled that he was 'moving out of the line'. Monty's 3rd Division immediately moved up to replace him. The enemy reached them a few hours later. For the next two days strong German forces several times fought their way across the Dyle and entered the outskirts of Louvain, only to be pressed back by fierce counter-attacks. Monty held his position under heavy shelling and boasted that he had 'bloodied their noses'. But German tanks had broken through the French lines on his right flank and were threatening to do as much on his left. The British sector risked being cut off and surrounded, and Brooke ordered a withdrawal.

At 1400 on 16 May 3rd Division fell back sixty miles to the west of Brussels. It was a withdrawal Monty had predicted and had exercised his division in many times, and it worked to perfection.

★

Rommel met the British in war for the first time on 21 May, three miles south-west of Arras when 1st Army Tank Brigade staged a counter-attack. Ironically he was without his own armour, having sent Rothenburg's panzers ahead while he remained with the rest of the division. German anti-tank shells failed to stop the Matilda IIs. They broke through his gun line and were only halted by heavy artillery fire and Stuka dive-bombers, although in his report to Berlin Rommel emphasized his own part in personally giving each of his guns its target. Thirty-six British tanks – one quarter of the total Allied force – were destroyed, but it was the nearest Rommel had come to defeat, the enemy having penetrated to within a hundred yards of his position.

It was a mark of the man that he immediately organized a counter-attack. He ordered 25th Panzer Regiment to turn back and attack the British tanks in the rear. He claimed that this was the only occasion during the campaign that he ordered his armour to move against the direction of the advance, and that was in order to attack. In this, Rommel's first tank battle with the British, he lost six Panzer IIIs, three Panzer IVs and a number of Czech-built tanks, but knocked out a further seven Matildas and the British fell back.

He played up his victory in the traditional manner of army commanders by overstating the strength of the enemy: 'Between 1530 and 1900 hours heavy fighting took place against *hundreds* [author's italics] of enemy tanks. Our anti-tank gun is not effective against the heavy British tanks even at close range.' While Hitler had expected the blitzkrieg to succeed, even he had been surprised by how quickly and how far Rommel's division had advanced, and this news of a near-defeat by 'hundreds' of enemy tanks convinced him that the panzers had gone too far too soon. He halted the advance temporarily, unwittingly giving the Allies more time to fall back on Dunkirk.

For his part in the action Rommel was decorated with the Knight's Cross of the Iron Cross. He told Lucie: 'Lieutenant Hanke, acting on behalf of the Führer, decorated me with the

Knight's Cross and gave me the Führer's regards.' After a two-day halt Hitler allowed his panzers to take up the advance and 7th Panzer Division moved towards Lille.

By 23 May Monty's withdrawal had brought him to Roubaix near Lille and virtually back to his start position. German forces were squeezing the BEF from the east and west, and three days later an evacuation was ordered.

At midnight on 27 May, King Leopold of Belgium surrendered his country's armed forces. This opened a gap between Brooke's 2 Corps to the south and the French Division Lourde Motorisée to the north, through which the German panzer divisions might press and outflank the BEF massing around Dunkirk. Brooke ordered Monty to move his 3rd Division from Roubaix and fill the gap. The movement required was the most difficult and dangerous of all: a flank march past the enemy's front. He had to move his division twenty-five miles overnight and have it entrenched by dawn when the first German attempt to break through was expected. His men had cursed the interminable night manoeuvres he put them through in England, but now they understood. The next day Brooke wrote in his diary: 'Found Monty had accomplished the impossible.'

All that could be done then was to fall back on Dunkirk by stages, delaying the German advance while the Royal Navy and a fleet of civilian ships ferried the men of the BEF to safety. The War Office ordered the most senior officers to return to England immediately, among them Brooke. He had to select one of his three major-generals to leave in temporary command of the Corps and Monty was the most junior. 'I had no hesitation in selecting Monty.'

Brooke was most distressed at the order to leave France before his men. Brian Horrocks, commanding a battalion of the Middlesex Regiment, came across an astonishing scene on the beach near Dunkirk: 'I saw two figures standing on the sand-dunes. I recognized General Brooke and General Montgomery. The former was under considerable emotional strain. His shoulders were bowed and it looked as though he were weeping. Monty was patting him on the back.' According to Monty, 'When he broke down and wept on my shoulder, I knew it meant his friendship was all mine.'

The last men of 2 Corps left Dunkirk during the night of 31 May, abandoning their equipment. Monty left early the next morning on a destroyer, HMS *Codrington*. The British press, encouraged by the new Prime Minister, Winston Churchill, called the evacuation a miracle. Montgomery knew it to be a terrible defeat for the army, yet believed that neither he nor his division had been defeated. Horrocks was impressed: 'He was always confident, almost cocky you might say. He was convinced that he was the best divisional commander in the British Army and that we were the best division.' That had a knock-on effect. The men believed, as did Monty, that if the whole of the BEF had been trained like them then the outcome might have been different. Dawnay agreed: 'The division felt itself quite confident about seeing the Germans off as far as it was concerned; it was a question of whether other people had let us down, put us in an impossible situation.'

As the British fell back on Dunkirk, Rommel's division took Lille and he accepted the surrender of a large part of the French First Army. A triumphant Hitler came forward to Charleville, and Rommel attended a conference of senior officers there. 'The Führer's visit was wonderful,' he told Lucie. 'His whole face was radiant and I had to accompany him afterwards. I was the only divisional commander who did.'

It was feared that the remaining British troops in France would escape Dunkirk-style from other ports and Rommel now raced ahead of his division with 25th Panzer Regiment, reaching the coast south of Dieppe on 10 June. The next day he took St Valéry and walked into the town with his tanks to take the surrender of French and British troops. He continued his advance, and on 17 June reached Cherbourg just as the last ship left with British troops, abandoning their vehicles on the docks.

On 21 June Hitler met with a French delegation in the Compiègne woods near Paris. A German officer had broken into a local museum and taken the railway carriage in which Germany had capitulated in 1918. Now Hitler beamed as he received the French surrender in that same carriage.

The invasion plans had allotted Rommel's 7th Panzer Division the role of flank guard to Army Group B as it advanced through Belgium and France. But by his forceful command and ability to apply the concept of 'lightning war' to real soldiering, Rommel had made his division the spearhead of the blitzkrieg. He defined his principles of operation as:

Always advance.

Manoeuvre to avoid the enemy rather than halt the advance to engage him.

Move quickly while firing on both flanks, particularly when ahead of neighbouring divisions – blanket fire into any area where enemy forces might be *before* they show themselves usually forces them to retire.

Following the successful German occupation of France it was possible that an invasion of the south coast of England would soon follow, and 3rd Division was ordered to take up defence positions along the Sussex coast between Brighton and Bognor. Sandbagged machine-gun posts were set up on seaside promenades but even Monty's own orders to his commanders, recorded in his HQ War Diary, retain a surreal air: 'All beaches, esplanades and amusement parks on sea fronts were to be cleared of all civilians by 1700 hrs daily, and no civilians would be allowed on them until 0500 hrs next morning.'

On 2 July the Prime Minister visited the division to witness a staged counter-attack on an airfield that had been 'captured by German invasion forces'. Here he met Monty for the first time and they dined together that evening at the Royal Albion Hotel in Brighton. 'Churchill asked me what I would drink and I replied

– Water. This astonished him. I added that I neither drank nor smoked and was 100 per cent fit; he replied in a flash that he both drank and smoked and was 200 per cent fit.'

Monty used their time at the dinner table to fundamentally alter the army's policy on defending against a cross-Channel invasion. The 3rd Division was deployed thinly along the south coast west of Brighton to fight 'on the beaches' in the event of enemy landings. Monty told Churchill that it was spread out 'like you spread butter on bread, being weak everywhere and strong nowhere'. He advised the Prime Minister that it was preferable to maintain a minimum force on the coast to harass and delay the invasion forces, and to hold the maximum inland as a mobile force for use in a counter-attack once the landing sites were known. Part of the justification for the division's present static role was the lack of transport post-Dunkirk, because the bulk of the army's lorries had been left in France. The solution, Monty argued, was to give him the promenade buses they could see from their dinner table.

Churchill was persuaded and the following day he told the secretary of state for war to requisition civilian buses – 'a large number even now plying for pleasure traffic up and down the sea-front at Brighton' – for Monty's use. Senior army officers were furious that a divisional commander should intercede with the Prime Minister to have their defence policy changed, but in the face of Churchill's insistence they could only go along with it.

The meal had a yet greater effect, which would prove itself later. There had been no rapport between the two men of the kind evident between Hitler and Rommel. On the contrary, the Prime Minister took a dislike to Monty as a man, while being mightily impressed by him as a soldier. He felt that Monty was insufferable but *right*.

Later that month Monty was given command of 5 Corps. His immediate superior was General Claude Auchinleck who made the mistake of telling Monty that he could also be *wrong*. Auchinleck had authorized the transfer of a number of BEF veterans out of 5 Corps. Monty wanted to retain them, but instead of taking the matter to his superior officer he wrote directly to the War Office. Auchinleck told him that 'going over the head of my Headquarters' was unacceptable. A few weeks later when Monty wanted several officers transferred to him from other units he again applied directly to the War Office. Auchinleck reminded him that this irregular procedure caused 'extreme annoyance'.

In the BEF Monty had properly co-operated with his superior officer, General Brooke, but he liked Brooke; he did not like Auchinleck and seems to have brought about confrontations for no good reason. When Auchinleck ordered that, as invasion was an ever-present danger, soldiers were to keep their firearms with them at all times, Monty asked: 'What happens when a soldier is with his girlfriend in the dark in the back row of the cinema? What does he do with his rifle then?' He announced that the order would not be enforced in 5 Corps. Auchinleck was furious. Such an insubordinate attitude to superior officers he did not like should have crippled Monty's career, and would have done so in peacetime, but the War Office overlooked it because his methods of training and command proved effective in an army that otherwise had little idea of how to prepare men for a modern battle. Monty never forgave Auchinleck for having the temerity to reprimand him and would later take his revenge in spectacular fashion.

In the US, on 12 July, General Marshall made much the same decision about Colonel Patton as the War Office had about Monty. Two days earlier, in response to the phenomenal success of the German panzer divisions in France, he had directed the War Department to create an independent armoured force comprised of two divisions. As the American press boasted, 'PANZERS FOR US TOO', Patton was given command of one of the two brigades of 2nd Armored Division to be formed at Fort Benning near Columbus, Georgia. Marshall, fully aware of Patton's awkward side, told him that 'I felt no one could do that particular job better.'

Patton assumed command of 2nd Armored Brigade in August.

He now commanded 5500 men with 383 (mostly M-2) tanks and 202 armoured cars, but the figures are misleading. More than half of his tanks were armed just with a machine-gun; only the new model had a 37-mm gun turret and additional armour. The *New York Times* described the men as 'a rabble of khaki-wearing civilians'. Patton agreed: 'Now we've got to make them attack and kill. God help the United States.' Most were new recruits and many were southerners. He never hid his prejudices and praised the men with 'light hair and eyes, the old fighting breed' as better than 'the subway soldiers' from New York and Pennsylvania (among whom there were many Jewish and black Americans).

For the same reason that Monty had a light fitted to his car, Patton now had a steamboat horn fitted to the scout car he used as a command vehicle, and he claimed it could be heard up to eight miles away. He applied his Great War tank experience and style of personal command to 2nd Armored. In France he had commanded a small brigade by visiting every unit personally and giving orders on the spot. Now he was struggling to make it work.

Perhaps his thoughts were not on the job. Beatrice had accompanied him to Fort Benning, but at the end of August 1940, while Patton was putting his division through its first major exercise, she left him. They had been married for thirty years and, despite the macho image he cultivated, he was shattered and sought desperately to win her back in a series of letters that leave no doubt as to the cause of their separation. On 31 August he wrote:

I just heard that Kilner who was with me in Mexico gassed himself on account of a disgrace he got into with another woman. I am inclined to think that he had more guts than I have or perhaps not, it is pretty hard to go on living and wishing one was dead when I realize that I have made it so hard for you who are innocent . . . I hope some day you may forgive me.

It is generally assumed that Beatrice left because of his affair with Jean Gordon in 1937 but his letters to her refer to a more immediate hurt. It is likely that a more recent affair had occurred or come to light during their first weeks at Fort Benning and that she left because of that. It appeared to be a permanent break: 'I have shattered all your ideals. I am shipping your books and things to Boston tomorrow.' Nevertheless he kept writing and his words of 3 September are typical: 'I have only myself to blame and you are the one I have hurt. I suppose the most charitable thing to think is that I was crazy. You are the only person I ever loved.'

He had a narrow escape when a light aircraft he was piloting developed engine problems, and told Beatrice, 'You almost got your wish that I die soon. I have been expecting a letter all day from you about starting our lives anew. But it has not come so I suppose it won't.' The letter came – and so did Beatrice. Her return saved the marriage and perhaps his career, for suddenly he became his fiery self again and his brigade, feeling the heat, was quickly hammered into shape.

It was during this period that the traditional epithet used by his officers, 'the Old Man', developed into the nickname by which he was to become best known. Sergeant Joe Rosevich, his secretary, recalled a particularly stirring speech to his commanders in which Patton emphasized that the combat effectiveness of an armoured division was determined by its 'blood and brains'. When reports appeared in the press this had become 'blood and guts', and because that seemed an even more apposite term for Patton than his own, it stuck. Patton became aware that the newspapers were calling him 'Old Blood and Guts' and commented: 'That's all right. It makes good reading. But it takes more than blood and guts to win battles. It takes *brains* and guts.'

In September he was promoted to the rank of general. He wrote to tell his friend Terry Allen, who had been promoted to general on the same day, that 'I guess we must be in for some serious fighting and we are the ones who can lead the way to hell.' In a letter to Eisenhower on 1 October he said that he was hoping for 'a long and BLOODY war' [his capitalization].

On 16 November Henry Stimson, the secretary of war, visited

Patton at Benning and heard his address to five hundred officers gathered in the infantry school there: 'It was full of cuss words and sharp remarks and they seemed to understand that.' Patton arranged a full-scale divisional exercise to prove that he could 'Rommel' too: a simulated attack by dive-bombers, an artillery bombardment and a tank attack, all in quick order, breaking the enemy line and allowing the infantry to follow through.

That demonstration had been arranged for his masters in Washington; it was convincing enough. For the exercise he mounted on 12 December his audience was the American nation and a great deal of advance publicity ensured that press were in attendance. Patton led two columns – 231 tanks, 895 other vehicles, and 6471 officers and men, with dive-bombers flying overhead at regular intervals – 170 miles from Columbus to Panama City, Florida. Moving day and night, the return trip culminated in a simulated attack on Fort Benning. Patton claimed it was the longest march ever made by an armoured division. As the column of tanks rumbled through small towns the townsfolk lined the streets to watch in awe. Newspaper reporters were as fulsome in his praise as the war correspondents travelling with 7th Panzer Division had been of Rommel, one of them even telling how Patton started stalled tanks 'by the power of his curses'. Patton and his 'American panzers' were big news across the US and captured the imagination of a public desperate for a home-born Rommel.

While Patton worked at creating his own image – from the 'warrior face' he put on for his men to the cross-country armoured marches he staged for the press – Rommel's image was created for him by Josef Goebbels (although Rommel himself lent a willing hand to it). War correspondents, like the portrait-painter Wolfgang Willrich, travelled with the invasion force as part of a Propaganda Company. Rommel often briefed them personally and took a hand in shaping what they wrote about him. One described 7th Panzer Division as Gespensterdivision (the Ghost Division) because it was both elusive and unpredictable, and the term was soon current throughout Germany. (There is some evidence that it was first used by the French – La Division Fantôme – and was picked up from them by a German reporter.) Goebbels's own newspaper *Das Reich* promoted Rommel with unreserved rhetoric: 'His tanks carve long, blood-stained trails across the map of Europe like the scalpel of a surgeon.'

A new word was coined: *Rommeln* (to Rommel) meant 'to make an audacious advance deep into enemy-held territory'. German troops even had a poem about him, '*Auf der Rommelbahn nachts um halb drei*':

> On the Rommelbahn at half three in the night
> Mighty ghosts rush by in the moonlight;
> Rommel himself is leading the race
> Everyone else is just keeping pace
> On the Rommelbahn at half three in the night.

Goebbels had given Rommel a Leica camera to take with him, and although he took many general photos he often asked an ADC to take a shot that included himself. He understood what Goebbels wanted: images of victory had to be 'personalized' and he enthusiastically played the part. He wrote home to ask Lucie, 'Please could you cut out all the newspaper articles about me?'

Only weeks after victory in France the propaganda ministry began filming a feature-length re-enactment of the campaign to be titled *Victory in the West* and shot at the original locations of the main actions. Goebbels persuaded Rommel to act his part again for the cameras. The men and tanks of his division were ordered to take part and French prisoners were made to play the enemy. The Ghost Division crossed the Somme once more, its tanks firing live ammunition as fearful Frenchmen walked towards the cameras with their hands held high.

Victory in the West premièred at the Ufa-Palast in Berlin in February 1941. The cinema was draped with swastika banners and senior members of the *Wehrmacht* and the Nazi Party attended. When the film went on general release, queues formed nightly at cinemas in Berlin and throughout the country. The Rommel caught on canvas by Willrich was now on the big screen and all Germany applauded its latest star. They did not know it yet, but his next role would thrill them all the more.

Part One Appendix
Original Rommel:
7th Panzer Division Invasion Narrative

As Rommel's 7th Panzer Division moved through France he began writing a history of his part in the invasion. He told Colonel Kurt Hesse, 'I am going to write a sequel to *Infanterie greift an*.' His son Manfred confirmed it: 'My father intended to publish another book. From the moment he crossed the frontier in May 1940 he kept a personal account of his operations, which he dictated daily to one of his aides. Whenever a lull allowed, he prepared a more considered version.'

Although this second book was never completed Rommel sent his account of the advance through France to Rudolf Schmundt, Hitler's adjutant, who replied, 'I presented your very clear history of your division to the Führer while he was at the Berghof. You can imagine with what great joy he read it.' Hitler wrote to him personally on 20 December: 'You can be proud of what you have achieved.' Rommel told Lucie of his 'eternal pride that the Führer, at a time when the burden of work pressed heavily upon him, found time to read my history and to write to me'.

Basil Liddell Hart, who championed mobile warfare in Britain, was most impressed:

No commander in history has written an account of his campaigns to match Rommel's. No other commander has provided such a graphic picture of his operations and methods of command. No one else has so strikingly conveyed the dynamism of *Blitzkrieg* and the pace of panzer forces. The sense of fast movement and quick decision is electrifying.

Rommel intended it to be a litany of his victories over the French and British, and although the colonel of one panzer regiment complained to army chief of staff Franz Halder that *his* successful advances had been included as if they were Rommel's own, the narrative is generally supported by the reports of others. It is the account of a general who was always at or near the front; Rommel had no need to invent victories in a dash across France that spoke for itself.

10 May.

In the sector assigned to my division the enemy had barricaded the roads, but we bypassed the blocks by moving across country. On first meeting French mechanized forces, we opened fire promptly and they withdrew. I have found that the day goes to the side that is first to plaster its opponents with fire – opening fire immediately into the area where the enemy is believed to be, instead of waiting to locate them accurately.

13 May.

There was no hope of getting my command vehicle down the steep slope to the Meuse unobserved, so I clambered down on foot through the wood. The 6th Rifle Regiment was attempting to cross the river, but heavy artillery fire was destroying their rubber boats one after the other. We had no smoke unit, so I ordered houses in the valley to be set alight to give the smoke cover we needed. A damaged boat drifted down with a badly wounded man clinging to it, shouting and screaming for help. The engineers were building pontoons and I wanted to get the Panzer Regiment across quickly. As soon as the first pontoon was ready I took my eight-wheeled command vehicle across. The enemy launched a heavy attack and shells dropped all around the crossing point.

14 May.

By 0900 we had got thirty tanks across and I ordered 25th Panzer Regiment to move round both sides of the wood 1000 metres north of Onhaye. I followed close behind in a Panzer III. Suddenly we came under heavy artillery and anti-tank fire and my tank was hit. The driver accelerated away but the tank slid down a steep slope and tipped over on its side. I had been slightly wounded in the right cheek by a splinter and it bled a great deal. I tried to swing the turret round to bring our 37-mm gun to bear on the enemy, but with the heavy slant of the tank it was impossible. We clambered out and ran as shells crashed nearby. Flames poured from the rear of the tank in front of us. I gave orders for the tanks to drive into the wood and force their way through the trees, to continue our advance. This tight combat control was only possible because I kept on the move and could give orders direct to the regimental commanders in the front line.

15 May.

I pushed my division forward with the 25th Panzer Regiment in the lead. The infantry was to follow up the tank attack in lorries. The artillery fired continuously on both flanks. I went forward in a tank so that I could direct the attack from the front and call in the artillery and dive-bombers when required to bombard the area ahead of our advance. Five kilometres north-west of Philippeville there was an exchange of fire with French troops occupying the hills. Our tanks fought the action on the move, with turrets traversed left, and silenced the enemy. On the southern edge of Vocedée we engaged and silenced a large force of French tanks. When we met a body of fully armed French motorcyclists coming in the opposite direction, most of them were so shaken at suddenly finding themselves in a German column that they drove their machines into the ditch. Without stopping we drove on at high speed to the hills west of Cerfontaine.

16 May.

The Panzer Regiment advanced on the French line of fortifications round Sivry. I rode in the command tank immediately behind the leading panzer company. One hundred metres from the line we came under heavy anti-tank and machine-gun fire, and two of our tanks were knocked out. Our artillery opened a heavy fire on my orders and was laying down smoke. Assault troops stormed the nearest concrete pillbox, throwing charges in through the firing slits. I ordered an immediate thrust

95

into the fortified zone, and the enemy anti-tank guns were silenced by a few rounds from a Panzer IV as we rushed by.

The way to the west was now open and the tanks rolled in a long column through the fortifications and past burning buildings. Our artillery dropped shells on the roads and villages ahead as we advanced, our engines roaring and tank tracks clanking as our speed increased. Enemy vehicles stood abandoned by the road. French troops took shelter in the ditches and hedges, and we could see the terror on their faces. On we went. Every so often I took a quick glance at the map by a shaded light, and a quick look out of the hatch. The flat landscape ahead was lit by the cold moon. We were through the Maginot Line!

Suddenly there was a flash from a mound 300 metres to the right – an enemy pillbox! Then more flashes from other points. I ordered the regiment to increase speed and we burst through this second line with broadsides to right and left. We drove through the villages of Poteries and Beugnies with guns blazing. All around were French troops lying flat on the ground. In Avesnes, which had been shelled by our artillery shortly before, the whole population was on the move in front of our tank column. Although there were French forces in the town I did not have the column halted, but drove on.

17 May.
We drove westward as the sun rose. French troops were so surprised by our sudden appearance that they laid down their arms without resistance. A French colonel was ordered to come with us, but he refused to get into the tank. After ordering him three times to get in, there was nothing for it but to shoot him. From Marilles onwards the main roads were crowded with fleeing civilians and to keep up our pace we drove through the fields alongside them.

21 May.
One of our Panzer IIIs was shot up as we came under fire from British tanks attacking southwards out of Arras. At the same time enemy tanks were advancing down the road from Bac du Nord. We were in a tight spot and I brought every available anti-tank gun into action. With the

enemy tanks so close, only rapid fire could save us, and I ran from gun to gun, giving each its target. Soon we put the leading enemy tanks out of action and about 150 metres away a British captain climbed out of a heavy tank and walked unsteadily towards us with his hands up. Our fire brought the enemy to a halt but they were firing into our position. Lieutenant Most sank to the ground beside me, blood gushing from his mouth. He was beyond help and died before he could be carried into cover. I gave orders for the Panzer Regiment to thrust south-eastwards to take the British armour in the flank. In fierce fighting, tank against tank, they destroyed seven heavy tanks and six anti-tank guns, at the cost of three Panzer IVs and six Panzer IIIs.

27 May.

British snipers were firing on our engineers who were building a sixteen-ton bridge near the demolished bridge at Cuinchy, picking off the men one by one. Under my personal direction a Panzer IV demolished every house near the bridging point. When we moved across the canal I discovered that the British snipers had installed themselves in the lock-house and our shells had killed them all.

28 May.

Heavy fighting developed when enemy tanks supported by artillery counter-attacked west of Lille. A hail of shells suddenly fell round my command post. I had a feeling they were our shells, probably 150 mm. I was making a dash towards the signals lorry to order the cease fire, with Major Erdman running a few metres ahead, when a shell landed close by. When the smoke cleared, he lay face to the ground, dead. He was bleeding from the head and from an enormous wound in the back. His left hand was still grasping his leather gloves. We sent up green flares to get the fire stopped, but it was a long time before the last shell fell.

5 June.

My tanks rolled round both sides of Le Quesnoy and came out on the wide plain to its south. On they went through fields of high-grown corn. A large concentration of enemy troops in the Bois de Riencourt was destroyed by the fire of the Panzer Regiment's tanks as they drove past. Heavy enemy artillery fire from the south-west crashed into the division, but was unable to halt our attack. Over a broad front and in great depth, tanks, anti-aircraft guns, field guns, all with infantry mounted on them, raced across country east of the road. Vast clouds of dust rose high into the evening sky over the flat plain.

8 June.

While crossing the main road from Rouen to Pont St Pierre at night, the tail of the 25th Panzer Regiment's column was fired on at a range of 100 metres by an enemy anti-tank gun. The tank crews were unable to hear the bursting shells above the noise of their engines, and to make them aware of this threat, I sent orders to the commander of the nearest armoured car to fire on the enemy with tracer. This soon brought the tanks into action and silenced the enemy gun, and we went on our way. The noise of our passage as we drove through villages wakened people from their sleep and brought them rushing out into the street. When we reached Sotteville at midnight – the first German troops to reach the Seine – tank brakes ground and screeched on the winding road down to the river.

10 June.

The sight of the sea with the cliffs on either side thrilled and stirred every man of us; also the thought that we had reached the coast of France. We climbed out of our vehicles and walked down the shingle to the water's edge until the water lapped over our boots. Close behind us Colonel Rothenburg came up in his tank, crashed through the beach wall and drove down into the water.

17 June.

The advance went on at a speed of 35–40 k.p.h. as we raced along the coast. There were French troops encamped on both sides of the road and we waved to them as we drove by. They stared in wonderment when they saw a German column race past. We drove in perfect formation through one village after another. In Flers crowds of people, both troops and civilians, looked on curiously as we dashed through the town. Suddenly a civilian a few metres from the column raced towards my car with drawn revolver, intending to shoot, but French troops pulled him down.

PART TWO

5. Conquests of the Deutsches Afrika Korps

Late in the evening on 7 February 1941 Rommel wrote to Lucie from his room at the Hotel Kaiserhof in Berlin. The previous day he had suddenly left their bungalow at Wiener Neustadt after a phone call from the Führer's headquarters on only the second day of home leave. For security reasons he was unable to tell Lucie about the new command Hitler had given him but he found a coded way of explaining why 'our leave' was cut short: 'Went to sleep last night thinking about my new job. It means that I can begin my rheumatism treatment next week.' He had suffered from rheumatism for some time and a doctor had said that as sunshine was the best treatment he should take a holiday in North Africa. Rommel was to be *Kommandierender General* of the newly formed Deutsches Afrika Korps.

The Führer had agreed to assist his Axis partner Mussolini by sending a German armoured corps as a 'blocking force' to help defend the Italian colonies of Cyrenaica (north-east Libya) and Tripolitania (north-west Libya). Italian forces in Libya had launched an attack eastward into Egypt intending to seize the Suez Canal, but the small British force there was quickly reinforced and a counter-attack, commanded by General Wavell, turned the Italians back. Wavell had pursued them across the border into Libya, taking Tobruk at the end of January and Benghazi on 6 February. All meaningful resistance crumbled as the Italians fell back on Tripoli.

Hitler personally selected Rommel to command the Deutsches Afrika Korps and told Mussolini that he was 'the boldest Panzer-waffen general that we have in the German Army'. He landed at the Castel Benito airfield in Tripoli only six days after the British had taken Benghazi and immediately asked the *Luftwaffe* commander to bomb the town. Informed that the Italians insisted no such action be taken as many Italian officers owned property there, Rommel appealed directly to Hitler and the raid was authorized.

He flew over the front in a light aircraft and studied enemy dispositions. If Wavell continued his advance beyond Benghazi the British would be unstoppable, but if he delayed until the men and panzers of the Deutsches Afrika Korps arrived then the situation would be different. Although Hitler had spoken of a 'blocking force', Rommel intended to do far more than stop the British reaching Tripoli. On 5 March he wrote to Lucie: 'A gala performance of *Victory in the West* was given here today. In welcoming the guests I said I hoped the day would come when we'd be showing *Victory in Africa*.'

Unwittingly, Churchill granted him the time he needed. The Prime Minister ordered Wavell to halt at Benghazi so that many of his troops could be transferred out of Africa, leaving him only the small force he required to hold his present position. Churchill would later regret weakening British forces at the precise moment that an ambitious Rommel and two panzer divisions were ordered into the desert.

As his men and tanks arrived in Tripoli, Rommel moved them straight to the front. Going forward himself on 13 March he experienced his first *ghibli*, a sandstorm so dense it darkened the sky: 'Immense clouds of reddish dust obscured all visibility and forced the car's speed down to a crawl. Sand streamed down the windscreen like water. We gasped in our breath through handkerchiefs held over our faces and sweat poured off our bodies in the unbearable heat.'

The British were not advancing. As it appeared that he had stopped them without engagement, he now wondered what a desert blitzkrieg might achieve against such an irresolute enemy. He submitted plans to Berlin to move rapidly into Egypt, take the Suez Canal (1500 miles from Tripoli) and chase the British out of

North Africa. On 19 March he flew to Berlin to convince the Führer that it could be done if he was given another two panzer corps. General Franz Halder, chief of the General Staff, was aghast: 'I asked where he thought we could find them. He said that was my problem.' Halder advised the Führer that Rommel's plans were impracticable. He knew, as Rommel did not, that Hitler was soon to invade Russia and that all available panzer corps would be required in the east. Rommel was told to hold his position and not to attack.

Back in Libya he immediately ordered an advance. He occupied Agedabia on 2 April – already beyond the halt line ordered by Halder – and crossed the Egyptian border. At 2200 the next day he retook Benghazi as the British fled (detonating 4000 tons of ammunition they had no time to take with them). He wrote to tell Lucie: 'The top brass in Rome and Berlin will be shocked. I have dared to go on, against orders, because I saw what might be won. The British are on the run. I cannot sleep for joy.'

The top brass were certainly shocked, but those in the High Command who had noted Rommel's special relationship with Hitler, and would have delighted in reprimanding him, nevertheless recognized that the Führer would revel in this latest 'lightning war'. A signal from Field Marshal Wilhelm Keitel, chief of the High Command, retrospectively authorized his advance to Benghazi. He was told not to make any further attack.

The following day Rommel moved forward again and captured the British base at Mechili. It was there, from among the fleet of vehicles left behind by the enemy, that he took the Perspex dust goggles that were to become his mark, fixed from that day forward to the peak of his cap. Even a general, Rommel told his aide Lieutenant Schmidt as he tried out the goggles, was allowed a little booty. Schmidt did not record whether a general was allowed his share of the whisky and gin that the British had also bequeathed to the Deutsches Afrika Korps.

By now the German Enigma code had been broken and Wavell was reading transcripts of the signals passed between Rommel and

Berlin. Ironically, this supposed advantage at first worked against him. Wavell had read an order from Berlin instructing Rommel to halt at Benghazi, and had taken no action to strengthen the defences at Mechili; he could not have foreseen the general's disobedience.

If Berlin could not stop Rommel, the desert might. Sand hurled at his tanks by each *ghibli* had by now penetrated the turning gear of the turrets; some had jammed while many turned only with an ominous grinding. General Johannes Streich urged him to halt the Deutsches Afrika Korps for 'essential maintenance'. Rommel ordered that they continue. The British had withdrawn to Tobruk and intelligence reports revealed that a larger than normal number of British ships was *en route* to the port. Taking his lesson on the enemy from their retreat to the coast in Normandy, he concluded that the British intended a mass escape by sea, and pressed on at full speed to prevent a second Dunkirk.

Rommel was again the talk of all Germany. In three weeks his dash across North Africa had recaptured all the territory the British had taken from the Italians. It was during this advance that Wavell's troops began referring to him as the 'Desert Fox', a term the Deutsches Afrika Korps proudly appropriated as Wüstenfuchs. The fennec was a small fox with a habit of burrowing quickly in the sand to escape predators, affording human occupants of the desert only an occasional fleeting glance. Its speed and ability to fade quickly into the landscape appeared to be characteristic of Rommel.

Germany celebrated his latest blitzkrieg but Rommel understood what the nation did not: that the vast expanses of desert he had 'taken' meant little if he could not destroy the enemy in Tobruk. He could have bypassed the town and pressed on into Egypt in accord with his principle of always advancing and not halting to engage fixed positions. But he dared not. The only road from Tripoli to the Egyptian border was the Via Balbia. Although pocked by the occasional bomb crater (in which, according to Captain Hermann Aldinger, 'it would be possible to lose a cow') this was his only overland supply route and its security was compromised while the British held Tobruk, as forces from that garrison could cut the road.

The Deutsches Afrika Korps approached Tobruk on 11 April and Rommel ordered an attack without taking the time to reconnoitre the defences, certain that the British were preparing an evacuation by sea. He is often credited with an uncanny 'feel' for what the enemy intended, but here he got it wrong. The additional ships were bringing in reinforcements and fresh equipment. Churchill had said that Tobruk must be 'held to the death without thought of retirement'.

The first attack by the panzers of General Streich's 5th Light Division was turned back by heavy shelling. The next day Rommel ordered them in again under cover of a sandstorm and again they were repulsed by artillery fire. In two days Streich lost 121 tanks. Another attack early on 14 April was met and turned back by the enemy's Matilda tanks, which now outnumbered the panzers. He told Major Ehlers what he could not say to Rommel's face: 'If the British had been daring, they could have rushed out from their fortress and captured the Afrika Korps headquarters, and that would have been the end of Herr General's reputation.'

When Rommel ordered him to attack yet again, Streich refused. The Korps diary noted that 'another attack was scheduled but did not take place'. Rommel's report to Berlin explained that the failure to take Tobruk was no fault of his own: 'My clear and specific orders were not obeyed by my commanders, and some commanders broke down in the face of the enemy.' He asked that more troops be sent.

Those among the General Staff who had been unable to put Rommel in his place were content that the British had done so. Halder felt vindicated: 'At last he is made to admit that his forces are not strong enough – we have known that all along. Anyway, he is not equal to his task. He stages raids and fritters away his forces.'

This last was a fair criticism. Rommel's own men believed he had been carried along by the momentum of his success and thrown them into the attack without properly assessing enemy defences. Another man might have acknowledged that much, and it is to his discredit that Rommel blamed others. Keitel, while acknowledging the Deutsches Afrika Korps was too weak to take Tobruk, believed that 'there is another problem too: the General's personality and his way of expressing it'. Halder thought it essential 'to head off this soldier gone stark mad' and signalled Rommel to 'hold ground gained' and take no further action until authorized by Berlin.

Halder's decoded signal reached Churchill's desk on 4 May and he pressured Wavell, whose main army was now separated from the besieged British forces in Tobruk, to mount an offensive to finish off the Deutsches Afrika Korps and relieve the garrison town. A convoy of new tanks, which should have taken the safe but lengthy route around South Africa, was ordered to sail through the Mediterranean (where U-boats prowled) so that the attack could be made 'at the earliest possible date'.

Wavell launched Operation Battleaxe at 0400 on 15 June. He now had twice as many tanks as Rommel, many of them the heavily armoured Matilda Mark II. The plan was for 7th Armoured Brigade to break through the German line at Halfaya Pass, then engage and destroy 15th Panzer Division.

The point of breakthrough was predictable. Rommel was aware that his anti-tank guns could not penetrate the Matildas and deployed 88-mm anti-aircraft guns in the pass, dug in at such an angle that their barrels could be elevated horizontally. These guns could rip a tank apart at a range of six miles, and of the first twelve British tanks to enter what became known as 'Hellfire Pass' eleven were destroyed within minutes. Rommel absorbed the attack, refusing the enemy a tank-on-tank battle and holding his out-numbered panzers in reserve while he weakened the British with his anti-aircraft guns.

The following day five fresh attempts to break through were

fought off with heavy enemy losses. At dawn on 17 June Rommel launched a counter-attack. By then Wavell's 7th Armoured had only twenty-one operational tanks left; Rommel had a panzer division he had not used. His panzers quickly forced the British to fall back. If the major tank battle Wavell planned for Battleaxe had taken place, a British victory would have been inevitable. Rommel had previously won ground by blitzkrieg; now he had proven he could hold what he had taken, although it was the 88-mm anti-aircraft gun that saved him. That evening Wavell signalled Churchill: 'I regret to report failure of Battleaxe.'

Hitler was elated. He promoted Rommel to *General der Panzertruppen* (General of Panzer Troops), giving him command of all land forces in North Africa. Rommel wrote to tell Lucie that 'I owe my rapid promotion to the Führer. I understand he is to present me with two splendid cars. You can imagine how very pleased I am.' The command of an army would reduce the direct control he presently enjoyed of the Afrika Korps and he indulged in no false modesty, wondering 'whether my promotion will not jeopardize the efficiency of the Korps'.

Churchill was frustrated. He had expected a major victory. He sacked Wavell, appointed General Auchinleck in his place, and ordered the new commander to fly to Cairo immediately. The Prime Minister wanted a new approach but Auchinleck's assessment was much the same as Wavell's: he felt he must build up his army's strength before going on the offensive. When Churchill insisted that 'war cannot be waged on the basis of waiting until everything is ready', Auchinleck told him: 'If an opportunity offers itself before October I will seize it; if not, I must collect the necessary tank forces.'

Auchinleck's caution was fed by Rommel's proven tactical skills and by his Wüstenfuchs reputation. A briefing compiled by British Intelligence reported that 'his very name and legend are in process of becoming a psychological danger to the British Army'. In an attempt to correct that, Auchinleck penned the most extraordinary order ever issued by a British officer with reference to an individual enemy officer:

To all Senior Officers

There is a real danger that our friend Rommel is becoming a 'bogeyman' for our troops in view of the fact that he is so much discussed. However energetic and capable he may be he is no superhuman. Even were he a superman it would be undesirable that our troops should endow him with supernatural attributes. You must make every effort to destroy the concept that Rommel is anything more than an ordinary German General. We must stop speaking of Rommel when referring to the enemy. We must speak of the Germans or the Axis troops, but never of Rommel. This matter is of great psychological significance.

[Signed] C. J. Auchinleck

Throughout July both sides concentrated on regrouping and the repair of damaged tanks, but for Rommel it was not only his panzer force that had been weakened. His army doctor diagnosed jaundice and prescribed bed rest, which he ignored, and a bland diet, which he attempted to follow, eating only what he described to Lucie as 'mush'.

On 30 July he flew to the Wolfsschanze (Wolf's Lair), the Führer's headquarters at Rastenburg in East Prussia, to discuss his plans for the rest of 1941. He told Hitler and the High Command that he wanted to take Tobruk, then advance on Cairo and the Suez Canal. They had heard it all before. Halder said the plan was impossible. Hitler agreed to an attack on Tobruk when the Deutsches Afrika Korps was ready, but nothing more. The map table showed German movements in Russia. It was obvious to Rommel that Hitler and the General Staff had their minds on the eastern front.

On 27 April 1941 Montgomery took command of 12 Corps, responsible for defending the south-east coast in Kent and Sussex, considered the most likely invasion area. The army he discovered was 'at ease': 'I burst in Kent like a 15-inch shell. Wives are being evacuated by train loads; it is just a matter as to whether the railways will stand the traffic. Heads are being chopped off – the bag to date is three Brigadiers and six Commanding Officers.'

He introduced regular cross-country runs for the men, much to the dismay of Spike Milligan (who would later become one of Britain's best-loved comedians). Officers were required to participate too, and when two complained that, considering their age and weight, the exercise might prove fatal, Monty replied that they should therefore run and die so that he could replace them all the sooner. He employed only officers who were 'full of binge', although he never explained to them what exactly that was. Among his men was a junior officer named David Niven (later to excel as an actor and raconteur):

Montgomery was a dynamic little man who demanded a fearsome standard of mental alertness and physical fitness. Just inside his headquarters was a large notice board: 'ARE YOU 100% FIT? ARE YOU 100% EFFICIENT? DO YOU HAVE 100% BINGE?' We never discovered what he meant by 'binge' because nobody dared to ask him.

In Monty's orders, *12 Corps Plans to Defeat Invasion*, issued on 12 May, he stated that to defeat the enemy 'in a real rough-house lasting for weeks' every officer and man 'must be 100% enthusiastic for battle and must possess 100% binge'. His task was to transform civilians into soldiers, which required physical fitness, mental alertness and a *fighting spirit*; it was this latter quality that he termed 'binge'. He wrote to tell Brooke who was now Commander-in-Chief, Home Forces: 'The trouble with our British lads is that they are not killers by nature; they have got to be so inspired that they want to kill, and that is what I have tried to do with this army of mine.'

In June he held a four-day exercise, codenamed 'Binge'. He was virtually alone among his peers in preparing his army for a modern battle: his exercises included movement by night, the co-operation of tank and infantry units, and the use of concentrated artillery fire in advance of an armoured advance. During Exercise Morebinge in August he made even greater demands on both officers and men. At his final briefing held at the Granada Cinema, Maidstone, on 12 August, he emphasized the need to prepare to fight for real. He was impatient with those who said the war would be won by the Russians on the eastern front or by the arrival of American troops: 'Anyone who thinks that the Russians or the Americans are going to win this war for us is making a very grave error. We have got to win this war ourselves, and we shall not do so without a very great deal of hard fighting.'

At Fort Benning, General Patton was impatient too. His fear was that the Russians or the British might finish the war before American troops reached the fighting. There was still no prospect of his 2nd Armored Division swapping the 'enemy' they engaged in exercises for the *Wehrmacht*. He took out his frustration on his driver, Sergeant John Mimms, who drove his staff car at 25 m.p.h., the regulation speed inside Fort Benning, and was told to 'hurry up'. Mimms increased his speed to 35 m.p.h. and Patton shouted, 'If you're scared to drive this God-damn thing, get out and let me drive.'

In US Army manoeuvres held in Tennessee in May Patton was in just as much of a hurry. One part of the exercise, scheduled to take a full day, was completed in three hours because 2nd Armored attacked so aggressively. He told his men they must 'think this is war – the next time, maybe, the bullets will be real'.

On 7 July 1941 he appeared on the cover of *Life* magazine standing in front of a tank and modelling a new uniform he had designed. The feature article, 'Patton of the Armored Force', praised the warlike training he put his men through, but it was the dark-green uniform that drew most attention. The gabardine jacket would conceal grease stains and the heavily padded trousers would cushion the repetitive bumps of a tank on the move. The helmet, which for this prototype model had been borrowed from the Washington Redskins football team, offered more protection than any existing headgear. The overall visual effect, however, was comical. Some called him the 'Green Hornet' (from a popular radio programme), others 'Flash Gordon'. The US Army decided not to introduce Patton's design for use by its tank crewmen.

An exercise in Louisiana in September, involving twenty-seven divisions and 400,000 men, was the largest ever held in the US. There were rules of engagement but Patton ignored them, and when his tanks ran out of petrol he ordered that they refuel at highway filling stations and paid with his own cash. In a similar exercise in the Carolinas in mid-November General Marshall observed that while other commanders used the tank as a battleaxe, Patton used it like a rapier: it was not force but speed that made his attacks successful. His men named themselves 'blitz troopers', paying simultaneous tribute to Rommel and to what Patton had made of *them*.

By the beginning of November, less than half of the troop reinforcements Rommel required for an assault on Tobruk had arrived. The five Italian ships carrying the fuel and ammunition he needed had been sunk by the enemy. He flew to Rome to finalize plans for his attack and to ask that five more ships be sent. He was reminded that intelligence reports indicated the British would attack first, but dismissed the idea of an imminent enemy offensive as ridiculous, perhaps because he had other plans for the next few days.

Lucie had travelled by train to Rome, bringing his 'brown civilian suit', and they booked into the Hotel Eden. They visited St Peter's and travelled on to spend Rommel's fiftieth birthday (15 November) in Naples. He left Italy to return to North Africa the next day but his plane developed engine trouble and had to make an unscheduled stop in Athens, where he was delayed for more than twenty-four hours.

The British offensive, codenamed Operation Crusader, began

late on 17 November. A simultaneous attempt was made to assassinate Rommel and leave the enemy without its most resourceful leader. British secret agent Colonel John Haselden had been working undercover in Beda Littoria, dressed in Arab clothes and keeping watch on the former Prefecture building. Staff cars came and went on a daily basis, among them a car identified as Rommel's. In London this was taken as proof that the building served as his headquarters and a commando operation was authorized to kidnap or kill the general. A British submarine landed twenty-nine commandos unobserved but as they approached the Prefecture building a sentry raised the alarm. In an exchange of gunfire and hand grenades, the attempt was beaten off. The building had briefly served as Rommel's headquarters until he had moved closer to Tobruk and was currently the office of General Schleusener, head of Supply. When he learned of the assassination attempt Rommel was indignant, not that the British had tried to kill him but that they had supposed his headquarters to be so far from the front.

He finally reached his army early the next morning, to be told that British tanks were advancing on Sidi Rezegh fifteen miles south of Tobruk. Still convinced that an enemy offensive was impossible he took this to be a reconnaissance in force and took no defensive measures. He realized his mistake when the BBC announced helpfully that General Auchinleck had 'started a general offensive in the Western Desert with the aim of destroying the German-Italian forces in Africa'.

By now Auchinleck had built up a formidable Eighth Army commanded by General Alan Cunningham, who had 770 tanks at his disposal but no experience of armoured warfare. Cunningham failed to capitalize on Rommel's slow reaction – a concentrated attack by British armour might have routed the unprepared Axis forces – and instead divided his army, sending 4th, 7th and 22nd Armoured Brigades to attack at separate points. Major von Mellenthin (Rommel's chief of Intelligence) observed that 'Cunningham has been obliging enough to scatter his armour.' In an opposite move 15th and 21st Panzer Divisions were united and told to 'destroy the enemy battle-groups'. The British helped by making elementary mistakes: the newly arrived 22nd Armoured made a head-on assault on a position protected by anti-tank guns and lost many of its brand-new tanks. Meanwhile the panzers attacked towards Sidi Rezegh and in the ensuing battle 7th Armoured suffered losses of more than fifty per cent. Rommel told a captured British officer: 'What does it matter if you have two tanks to my one, when you spread them out and let me smash them unit by unit?'

On 27 November as the battle raged around him, he found time to write to Lucie: 'It's our 25th wedding anniversary today. I want to thank you for all the love through the years which have passed so quickly. I think, with gratitude to you, of our son, who is a source of great pride to me.'

Observing the heavy British losses, Rommel led a large part of his force out from Tobruk in an effort to cut off Eighth Army's route of withdrawal and complete its destruction. Cunningham ordered a retreat. Auchinleck countermanded that, decided the man had 'lost his bottle' and replaced him with General Neil Ritchie. As the attack continued, 2nd New Zealand Brigade broke through the weakened force surrounding Tobruk and reached the garrison, ending the 242-day siege.

The Deutsches Afrika Korps was reduced to seventy service-able tanks, which engaged the British force at Bir el Gubi on 5 December. After a two-day battle, with further losses and his tanks short of both fuel and ammunition, Rommel came to the most painful decision of his military career so far: he ordered a retreat to El Agheila.

That same day Japanese planes bombed the US Pacific Fleet in Pearl Harbor and on 8 December the United States and Britain declared war on Japan. It was everything Patton had wished for. At the end of the month he addressed the officers of 2nd Armored Division. They had performed well in manoeuvres but still lacked battle experience. It was likely that in the year about to begin they would join the war in Europe and his task was to change 'war gamers' into 'killers':

Battle is not a terrifying ordeal to be endured. It is a magnificent experience wherein all the elements that have made man superior to the beasts are present: courage, self-sacrifice, help to others, devotion to duty. Those enemies, whom we shall have the honour to destroy, are good soldiers. To beat such men, you must not despise their ability, but you must be confident in your own superiority.

He wrote to tell Eisenhower, who had just taken up a War Department post in Washington, 'I have the utmost confidence that we will beat the hell out of those bastards – You name them: I'll shoot them.'

On Christmas Day Rommel wrote to Lucie: 'I opened my Christmas parcel in my caravan yesterday evening and was very pleased with the letters from you and Manfred, and the bottle of champagne. The night passed quietly.' Quietly, except for the carols sung by his troops, although their favourite song that Christmastide held more memories of their hometowns in Germany than of Bethlehem. Earlier that year the Reichsrundfunk (German Forces Radio) had begun shortwave broadcasts to the Deutsches Afrika Korps and now 'Lili Marlene' was on every man's lips.

Although Rommel had ended 1941 by losing ground to the Allies, Hitler made it clear in a personal Christmas message that he had not lost faith in his favourite general and that 1942 would bring new victories: 'I know that I can depend on my Panzer Division in the New Year.'

New victories were precisely what Rommel had in mind. A convoy reached Tripoli on 5 January with fifty-five new tanks. Another convoy was due in February. It would take time to rebuild Rommel's army and Hitler agreed to his plan to attack Tobruk in May. Meanwhile the German press and newsreels were featuring his progress in North Africa more regularly and he confessed to Lucie that he received many 'amorous letters' from female admirers. He wrote to congratulate Manfred on joining the local Hitler Youth troop, and when Lucie mentioned a poor school report he commented: 'The school ought to be proud to have a son of mine among its pupils.'

As Britain and America planned an Allied war effort, General Marshall was worried by the possibility that American troops would be thrown into the war 'at the deep end' against the battle-hardened Deutsches Afrika Korps. The US Army had no experience of desert armoured warfare. At the beginning of March 1942 he ordered Patton to establish a desert training centre to shape what would be in effect an American Afrika Korps capable of challenging and beating Rommel.

Patton piloted his own Piper Cub over huge tracts of California, Nevada and Arizona, looking for an uninhabited expanse of desert with conditions similar to those in North Africa. He chose an area 180 miles long by 90 miles wide near Indio, California, 200 miles east of Los Angeles. The United States Desert Training Center was immediately nicknamed 'Little Libya'.

Patton was confident he could produce an American tank corps able to engage and beat Rommel in the desert. He took command in Indio on 10 April. There were no barracks: the men and their general lived in tents, without electricity or running water. On manoeuvres they slept beside their tanks and survived on one canteen of water a day. There were regular large-scale exercises that included live fire. Patton left his command post to move continuously among the units by jeep, tank or half-track, emulating Rommel's style of command, not because it had worked for the German in North Africa but because it had worked for Pershing in Mexico and, in any case, it was as much *his* way.

Monty, like Patton, was now training men for war. On 19 May he began Operation Tiger, the largest training exercise ever held in England, on Salisbury Plain and involving 100,000 British and Canadian troops. While all previous exercises had simulated defensive procedures in the event of a German invasion of southern

England, Tiger was organized to test large-scale offensive manoeuvres against German-held territory.

Eisenhower was present as an American observer, and at one of Monty's briefings he lit a cigarette. 'Who's smoking?' Monty demanded. 'I am,' Eisenhower said, and was told, 'I don't permit smoking.' In his official report Ike described Monty as 'a decisive type who appears to be extremely energetic and able'. He had used a different term as he was driven away from the briefing. Kay Summersby, his driver, heard him talking in the back of the car to General Mark Clark: 'His voice turned harsh. I heard something about "that son of a bitch". He meant Monty. He was furious – really steaming mad. I sneaked a look in the rear-view mirror. His face was flaming red and the veins in his forehead looked like worms.'

General Ritchie's Eighth Army stood on an established defensive line at Gazala blocking the approach to Tobruk. He expected an attack to come through the minefields at the centre of the line, the direct route into the garrison. On 27 May Rommel made a strong feint at precisely that spot, but as enemy forces arrived there he led his panzers round the southern flank through Bir Hacheim to attack from the east. British positions fell back and the panzers moved forward into a gap between the minefields and Eighth Army. They were held there for several days and so thickly shelled that they named the area 'Hexenkessel' (the Cauldron). A British counter-attack on 5 June was repulsed with heavy losses on both sides, and after another five days Ritchie was forced to pull back. The panzers broke out of the cauldron as Eighth Army evacuated the Gazala Line and by 18 June Rommel's army was once again outside Tobruk.

He attacked the garrison two days later. Stuka dive-bombers cleared a path through the minefields and the panzers surged through. By 1600 most of the defences had been overrun. The British began exploding fuel dumps to prevent them falling into enemy hands and a thick pall of black smoke rose over the town. At 0800 on 21 June the garrison commander, the South African

General Klopper, formally surrendered. Eighth Army withdrew to Mersah Matruh with little of its armour left intact. Rommel issued a congratulatory Order of the Day: 'Soldiers of the Panzerarmee Afrika! A great battle has been won. The enemy has lost his powerful armour. Now we will shatter the last remnants of the British Eighth Army.'

He was interviewed by a war correspondent of Grossdeutschen Rundfunk (Greater German Radio) and his words were transmitted to Berlin to be broadcast to the nation that same evening. Hitler made him a field marshal but Rommel told Lucie that 'I would rather he had given me one more division.' He celebrated by opening a bottle of whisky procured by his staff from the abandoned Tobruk Naafi stores.

Reminded by the High Command that he must not attempt any further advance, Rommel argued that the British fuel and equipment captured in Tobruk enabled him to continue and made his first use of a field marshal's right of direct appeal to the Führer, who agreed. He was bursting with confidence and urged his army on towards Mersah Matruh. He boasted that he would be 'in Cairo and Alexandria by the end of June'.

The British thought that likely too. Senior officers evacuated Cairo, and in Alexandria the Royal Navy put to sea. On 25 June Auchinleck flew to the front and took over direct control of Eighth Army from Ritchie. He pulled the army out of Mersah Matruh just before Rommel's forces surrounded it, and ordered a retreat to the small coastal village of El Alamein. Here, there was a narrow gap of forty miles between the sea and the impassable Qattara Depression. At El Alamein, if time and the enemy allowed, he could organize a 'last stand'.

Rommel rushed his forces forward on an attempt to overtake the British before they could establish a defensive line across the gap, which he, too, recognized as the final point at which his advance on Cairo and Alexandria could be stopped. He sensed a victory greater than any he had won before.

★

Churchill had been on an official visit to Washington to discuss the war with Roosevelt when he heard that Rommel had taken Tobruk. The following morning the American press spelled out the consequences for him: 'ANGER IN ENGLAND. TOBRUK FALL MAY BRING CHANGE OF GOVERNMENT.'

Taking Tobruk opened the way for German forces to advance on Cairo and the Suez Canal, and to expel the British from North Africa. Churchill returned to London to face tough questions in the House of Commons. It was suggested that the problem was not an inadequate army but the predominance of upper-class officers in senior posts, and that 'If Rommel had been in the British Army he would still have been a sergeant.' Churchill, fighting for his survival, deflected attention from the failings of British generals by stressing the extraordinary qualities of their opponent: 'We have a very daring and skilful opponent against us, and, may I say across the havoc of war, a great general.'

The German press picked up Churchill's comments. The *Berliner Börsenzeitung* led with the headline, 'CHURCHILL SAYS: I BLAME ROMMEL.' Hitler was amused: 'People ask how Rommel has achieved such a worldwide reputation. It is because the British Prime Minister, speaking in the House of Commons, tells the entire world that Rommel is a military genius.'

For some time Hitler and Goebbels had been promoting the Rommel legend inside Germany and for good reason. The propaganda ministry had previously concentrated on the Russian campaign: Hitler expected a great victory and Goebbels, anticipating the propaganda benefits, kept the Russian front in the headlines, relegating North Africa to the inside pages. Then the Russian winter and enemy counter-attacks convinced both men that the campaign was in trouble. They had focused the attention of the German people on the Russian front; now they needed a diversion and found it in Rommel's advance across North Africa. Goebbels ensured that he received extensive press and newsreel coverage, and a nation that had been thrilled by his blitzkrieg across France easily transferred its attention to this latest 'lightning war'. Hitler considered the North Africa campaign secondary to the Russian campaign but Goebbels saw to it that the media covered the two as if the reverse was true.

Churchill needed a diversion too. He had to explain to the House of Commons and to the British people why their army in North Africa, which had chased Axis forces almost back to Tripoli, had then been pressed back the way it had come and was now making a 'last stand' at El Alamein. He despaired of his generals and blamed *them*, but that was not for public consumption. In suggesting that the problem was not inferior British commanders but a superior enemy commander he spread the Rommel legend (already established by Hitler and Goebbels inside Germany) around the world.

This strategy only worked for Hitler and Churchill because Rommel's achievements were both real and startling. Major von Mellenthin attributed the success of the DAK to two primary factors: in defence the superiority of the German 88-mm anti-tank gun, and in attack the magnificence of Rommel's tactics.

An aide, Heinz Schmidt, described Rommel in iconic pose standing in his command car and pointing ahead 'like a U-boat commander on his bridge'. Rommel himself compared desert warfare to doing battle at sea: if the panzers were battleships and the desert was their sea, then their purpose was not to take and hold huge tracts of it, but to attack and destroy enemy ships. Victory was not then determined by the area occupied but by a war of attrition against enemy forces: destroying enemy tanks while preserving one's own. Where equal forces met, Rommel's genius on the battlefield gave him the advantage. The tide would turn if enemy forces outnumbered his by a considerable degree. With the loss of his natural tactical advantage any war of attrition could go against him. That was about to happen at El Alamein.

Rommel's advance was seriously hampered by RAF bombing. By the time he reached El Alamein the British had a defensive line in place and he had only fifty-five panzers and thirty Italian tanks left.

His first attack at 0300 on 2 July was shrouded by a sandstorm that also hid Auchinleck's anti-tank positions until it was too late. Rommel lost another eighteen tanks. He resumed the attack later that day, and again on 3 July, but both attacks were repulsed and his losses had become unbearable: the Deutsches Afrika Korps had only twenty-six operational tanks left while Eighth Army had four times as many. He told Lucie: 'The enemy is too strong and we are too weak. I'm feeling tired and worn out.' The next day he pulled the panzer divisions out of the line to rest the men and salvage the repairable tanks. He was low on petrol, ammunition and, it must be said, spirit. He established a defensive line of his own protected by anti-tank and anti-aircraft guns.

For much of July, while Rommel was unable to mount any offensive action, Auchinleck launched attacks of his own. He concentrated his armour in attempts to break through the thinly spread line, particularly where it was held by weaker, Italian units, attempting to exhaust the enemy by continually switching the point of his probes. But on 19 July Rommel ordered a counter-attack by the remnants of the 5th Panzer Regiment and the British lost eighty-seven out of a hundred tanks.

Rommel and Auchinleck came to the same conclusion: neither of them currently had sufficient strength to win an outright victory and the confrontation was effectively a stalemate. Both men realized what Rommel told Lucie on 21 July: 'The build-up on the British side is faster than ours.' Because of that, Auchinleck resisted pressure from Churchill to mount a major attack and devoted himself to building up his forces, while Rommel saw an early attack as his best (and perhaps only) chance before the balance of power – calculated by the number of tanks available to each side – swung heavily against him.

Churchill wanted a decisive victory over Rommel. He had sacked Wavell because that had not been delivered and he decided to confront Auchinleck with the same choice: attack now and win, or be replaced.

★

After the fall of Tobruk Roosevelt offered Churchill whatever military assistance he required in North Africa and the Prime Minister asked for new Sherman tanks to strengthen Eighth Army. Marshall offered to send a fully equipped and manned armoured division and called Patton to Washington to command it. On the day he arrived Patton wrote a note to Marshall saying he needed *two* divisions. Marshall ordered him back to Indio. Patton telephoned to say that 'After a lot of thinking, maybe I *could* do the job with the forces your stupid staff is willing to give me.' Marshall told his deputy: 'That's the way to handle Patton.' When it became apparent that even one division could not be deployed in time to hold off Rommel, Churchill accepted Roosevelt's offer of 300 Sherman tanks to be crewed by the men of Eighth Army.

Britain and America calculated that their armies would not be ready to mount a cross-Channel invasion of occupied France until mid-1943. Roosevelt wanted American troops fighting somewhere in 1942 and agreed to Churchill's suggestion of an Anglo-American invasion of North Africa. The Vichy French forces in Morocco were not expected to resist too vigorously and should prove an easier blooding of untested American troops than an immediate introduction to Rommel's desert veterans.

Codenamed Operation Torch, an Anglo-American army would land in Morocco and Algeria, then advance eastward into Tunisia. Meanwhile the British Eighth Army would push through its defensive line at El Alamein and press the Panzerarmee back into Tunisia. Rommel would be trapped between Torch and Eighth Army. In return for Roosevelt's support, Churchill agreed that an American officer would command the Allied force: General Dwight D. Eisenhower. Three separate task forces would land simultaneously to take Casablanca, Oran and Algiers. Eisenhower had been Marshall's choice; so too was Patton. At the end of July he was recalled to Washington to command the Western Task Force, which would take Casablanca.

*

On 31 July Auchinleck sent a signal to London: 'Renewal of our effort to break enemy front not feasible owing to lack of resources and consolidations of enemy position. Opportunity for offensive operations unlikely to arise before mid-September.'

Churchill was furious. Sir Ian Jacob wrote in his diary: 'The Prime Minister is entirely fixed on the defeat of Rommel . . . He strode up and down declaiming on this point. "Rommel, Rommel, Rommel, Rommel," he cried, "what else matters but beating him?"'

Churchill flew via Cairo to Alexandria with Alan Brooke to confront Auchinleck. On 5 August they were driven to the head-quarters of Eighth Army at El Alamein. In the command vehicle Auchinleck's chief of staff, Freddie de Guingand, observed that 'the Prime Minister was eager for another offensive and wanted it soon. Auchinleck was a bit abrupt in the way he refused.' Churchill prodded the wall map and muttered, 'Attack, attack,' but Auchin-leck insisted it was impossible until he had built up his army, which would be mid-September. Churchill left abruptly and flew back to Cairo.

The next morning at breakfast he told Brooke that he no longer believed Auchinleck could 'do the job'. He would put General Alexander in overall command based in Cairo, and General Gott in command of Eighth Army at El Alamein. As Gott flew to Cairo for briefing, his plane was shot down by enemy fire and he was killed. Brooke immediately recommended Montgomery for Eighth Army. Churchill agreed and telegraphed the War Office in London: 'This post must be filled at once. Pray send him by special plane at earliest moment.'

On 5 August Patton flew to England to consult with Eisenhower, who had established an Allied Force headquarters in London. He and Ike met with senior officers of the Royal Navy and the US Navy to plan Operation Torch. Later in his room at Claridge's Hotel, Patton assessed his future allies. He liked neither London nor its people and wrote to tell Beatrice: 'If there ever were any pretty women in England they must have died – they are hideous with fat ankles.' *That* was to reassure her that his interests while abroad were purely military. But he had attended a US Navy conference on the invasion and wrote in his diary: 'They are certainly not on their toes. It is very noticeable that most of the American officers here are pro-British, even Ike. I am not, repeat not, pro-British.'

A telephone call from the War Office on 8 August ordered Montgomery to proceed immediately to Egypt to take command of Eighth Army at El Alamein. A Liberator was standing by to fly him and one ADC out that evening, first to Gibraltar, then on to Cairo. Monty arranged for his son to stay with Major Reynolds and his wife. He warned them that 'David's grandmother will want him to go and stay for his holidays. On no account is he to go. She is a menace with the young.'

Rommel calculated that he had the resources for one 'last-chance' attack on Auchinleck's line at El Alamein. It had to succeed because after that the 1400 tons of fuel captured in Tobruk would be gone and there was insufficient coming down his extended supply line to sustain any further action. If he could break through the Eighth Army he could take Cairo and the port of Alexandria, his supply problems would be eased and the Suez Canal would be within reach. He told Lucie that 'I must make best use of the next few weeks to prepare for it.' He would attack on 30 August. 'If this blow succeeds it could in part decide the whole course of the war.'

Bad weather delayed Monty's flight out and his Liberator finally took off after dusk on 10 August. He arrived in Gibraltar early the next morning. Enemy aircraft were active over the Mediterranean and transport flights were forbidden during daylight hours. After nightfall he flew out across the Mediterranean.

As Patton planned the Torch landings in Morocco, Rommel prepared for a decisive, final attack on the British line at El Alamein, and Monty flew to Cairo. These three men were on a collision course that would fundamentally affect their military careers and change the direction of the war.

6. Hounding Rommel – Monty and Patton Close In

'Well, Freddie, you chaps seem to have been making a bit of a mess of things,' was Montgomery's opening line on Thursday, 13 August 1942, when he took command of Eighth Army. He had stayed overnight at the British Embassy in Cairo and was then driven by Freddie de Guingand (now *his* chief of staff) to his headquarters in the desert. 'I have been doing a lot of thinking and worked out how I want this Army organized. You'll never win a campaign as it is at the moment.'

Eighth Army headquarters were established below the Ruweisat Ridge at the desert equivalent of Piccadilly Circus, where numerous camel tracks crossed. But it was the flies Monty noticed first. 'What's this?' he asked de Guingand. It was the officers' mess, a massive framed structure of mosquito netting that, built to keep out the flies, actually trapped them inside. 'Take it down at once.' The matter was urgent because that was where lunch would shortly be served. He dispensed with the flies but the constant buzz of RAF fighters continued overhead. Auchinleck had attempted to emulate Rommel by establishing his headquarters near the front; as a result it had been bombed by the *Luftwaffe* and required air cover throughout daylight hours.

During lunch Monty discovered that his predecessor had lived like the men, sleeping on the ground outside his command caravan and showing little regard for his own physical comfort. Monty was having none of that and sent his first signal to Cairo: 'Army Commander requires best available soldier servant in Middle East to be sent at once.' His second signal was: 'Montgomery assumes command of Eighth Army at 1400 hrs today.' Auchinleck was officially to hand over command in two days' time but in Monty's own words he 'seized command'. He admitted to 'an insubordinate smile: I was issuing orders to an Army which someone else reckoned he commanded.' Two years earlier in England Auchinleck had reprimanded Monty for doing much the same and had never been forgiven.

After a tour of the front he addressed his officers as the light dropped and the RAF fighters left. Not all of them liked what they saw. Auchinleck had stood tall, with the dark-brown desert tan that made him one of them. Montgomery was short and scrawny and his pale face told of the last two years spent in England. He had no experience of desert warfare. He had bought 'clothes suitable for the desert' in Cairo the previous day and was wearing the shirt and slacks for the first time. He mounted the steps of his caravan and had his fifty officers sit on the sand in a semicircle below him. His squeaky voice surprised them, but what he lacked in tone, stature and experience he made up for in 'attitude'. He said he did not like the 'atmosphere' he found there, which had caused 'a loss of confidence in our ability to defeat Rommel', and that he would create 'a new atmosphere':

Here we will stand and fight; there will be no further withdrawal. I have ordered that all plans dealing with withdrawal are to be burnt. If we can't stay here alive, then let us stay here dead . . . Now I understand that Rommel is expected to attack at any moment. Excellent. Let him attack. I would sooner it didn't come for a week, to give me time to sort things out. If we have two weeks to prepare we will be sitting pretty. Rommel can attack as soon as he likes after that . . . meanwhile, we ourselves will start a great offensive; it will be the beginning of a campaign which will hit Rommel for six right out of Africa. The great point to remember is that we are going to finish with this chap Rommel once and for all.

De Guingand described the effect of their new commander's talk on the officers as 'electric'. They were hardly to know that the

300 Sherman tanks Monty claimed were 'being unloaded at Suez *now*' were not due until September.

Monty's comments on the supposed plans for withdrawal that he had 'ordered to be burnt' caused a stir. He insisted that Auchinleck intended to withdraw Eighth Army to Cairo if Rommel attacked again, and that *he* cancelled all such arrangements and introduced a new policy: the army would hold its present position on the El Alamein line. The controversy over whether his battle plans were genuinely new or inherited from Auchinleck was to last longer than the war.

De Guingand confirmed that he brought to Monty's notice the several plans that had been drawn up and was told to 'burn the lot'. Monty implied that these had been plans for a withdrawal, although he did not inspect them. In fact, they were concerned with the further strengthening of the El Alamein line against Rommel's next attack and included advance planning for an offensive against him. One significant paper that survived, dated 27 July 1942, prepared for Auchinleck and entitled 'Appreciation of the Situation in the Western Desert', included:

Eighth Army may have to meet an enemy's sortie developing into manoeuvre by the southern flank from his firm front . . . Eventually we will have to renew the offensive and this will probably mean a break through the enemy positions about El Alamein. The newly arrived infantry divisions and the armoured divisions must be trained for this and for pursuit.

This paper forecast a defensive battle to be received at Alam al-Halfa to be followed by an offensive battle pressed out from El Alamein. That was precisely what was to happen under Montgomery's command. The suspicion must arise that his need for any future victory to be seen as resulting from plans that were his and his alone led him to misrepresent those that already existed. That previous plans had been drawn up under Auchinleck's direction no doubt added to his delight in rubbishing them.

Monty moved his headquarters to Burgh-el-Arab on the coast and de Guingand noted that 'we could walk out of our caravans into the sea. This made a great difference to everyone's morale, health and capacity for work.' Auchinleck had instructed his officers not to personalize the enemy by speaking of Rommel when they meant the Germans, but in his first address Monty did exactly that, and now he had a full-size copy of the Willrich portrait hung in his caravan to the right of his writing desk so that he could study the face of his enemy. He was aware that he had to compete with this world-famous opponent as a personality before he could compete with him on the battlefield. At the moment the men of Eighth Army knew a great deal about Rommel and nothing at all about their new commander.

While visiting the front on 14 August Monty quipped to an officer of 9th Australian Division that his own peaked cap was hardly suitable for the desert, and accepted the offer of an Australian slouch hat, which provided greater protection from the sun. In a moment of pure inspiration he asked for an Australian badge to fix on it. Later that day when he visited 5th Indian and 2nd New Zealand Divisions he asked for their badges, too, and fixed them to his hat. He was deliberately seeking 'the Rommel effect'. His army needed 'not only a master but a mascot': 'I deliberately set about fulfilling this second requirement. It helped for them to recognize as a person – as an individual – the man who was putting them into battle. But I readily admit that the occasion to become the necessary focus of attention was also personally enjoyable. I started by wearing an Australian hat.'

Monty the showman had arrived and word quickly spread among the men that they had a commander like no British general they had known before. While once the talk had been all of Rommel, now it was of Monty. His 'new atmosphere' was taking hold of Eighth Army.

Rommel began moving his panzers to their concentration area on 20 August in preparation for an attack on the El Alamein line at

Alam al-Halfa. He had a narrow window of opportunity as General Gause, his chief of staff, spelled out: 'The last chance was at the end of August, at full moon. Marshal Cavallero had promised that several tankers would arrive before then, and Field Marshal Kesselring had promised 500 tons of fuel a day by airlift.' It was known the British were due to receive additional tanks in the first week of September, so the end of August was the optimum time for an assault that Rommel believed might carry him through the enemy and on to Cairo.

He had two concerns. The first was that half of the five thousand tons of fuel he had expected by sea had been sunk by the RAF while the remainder had yet to leave Italy. He had more tanks than the enemy but insufficient fuel to exploit that advantage to the full, yet dare not delay the attack and allow British reinforcements to swing the disparity of armour in their favour. The second concern was his health. On 21 August his medical adviser, Professor Dr Horster, gave him a full physical examination and concluded in an official report to Berlin: 'Field Marshal Rommel is suffering from chronic stomach and intestinal catarrh, nasal diphtheria and circulation trouble.' Gause added a conclusion of his own to the signal: 'He is not in a fit condition to command the forthcoming offensive.' Rommel wrote to Lucie the following day: 'Much too low blood pressure, state of exhaustion, six to eight weeks' rest cure recommended. I have requested the High Command to send a substitute.'

His suggestion that he be replaced by General Guderian, the only man, in his opinion, who had the experience and understanding to command the panzerarmee, was rejected. Rommel would give way for no other substitute and had Gause signal that 'The Field Marshal's condition has improved to the extent that he can command the battle under constant medical supervision.'

Rommel knew that his chances of mounting a successful assault at El Alamein were slim. But he understood more than most field commanders the 'Grand Design' of which his operations were part. On the morning of the attack General Bayerlein noted that

Rommel left his truck with 'a troubled face' and said to Dr Horster: 'Professor, the decision to attack today is the hardest I have ever taken. Either the army in Russia succeeds in getting through to Grozny and we in Africa manage to reach the Suez Canal, or . . .' At that, according to Bayerlein, Rommel made a gesture of defeat.

German forces in Russia were moving south-east towards the Black Sea. This was the northern half of a great pincer movement – of which Rommel's African army moving east towards Cairo and the Suez Canal was the southern half – intended to take the oilfields of the Middle East, which were the lifeblood of the Allied war effort, and win the war for Germany. The British were aware of the threat and no one more so than Churchill, who persistently demanded that his generals in North Africa stop Rommel. The British Oil Control Board had estimated that the loss of the Middle East oilfields would require 13.5 million tons of oil to be transported from more distant ports, necessitating the construction of 270 extra tankers, and had concluded that, as this was impossible in the time-frame allowed, the British war effort would grind to a halt. One sentence in the board's recommendations to the Middle East Defence Committee read: 'You should strain every effort towards defeating Rommel.'

It is, then, ironic that while the battle at Alam al-Halfa was fought in good part for possession of the Middle East oilfields, its outcome was determined by the amount of fuel available to the Deutsches Afrika Korps. There was perhaps enough for one last attack, although much depended on the quality of the sand on Rommel's route of advance. This required the panzers to move thirty miles by night over minefields and 'unknown ground'. The latter might prove more of a problem than the former: tanks moving over 'soft' sand used up three times as much fuel as on 'firm' (stony) sand and the difference could prove crucial.

Patton left Eisenhower in London to draw up the overall plans for Operation Torch and returned to Washington on Friday, 21 August. In dingy offices in the loft of the Munitions Building on Constitution Avenue (the home of the War Department before it moved to the Pentagon) he shaped his part in the invasion set for 8 November. Intelligence on Casablanca and the landing sites was so lacking that he sent his staff out to purchase tourist guide-books on the area.

Torch was to be made up of three armies. Patton's Western Task Force would take Casablanca while the Center Task Force took Oran and the Eastern Task Force captured Algiers. Patton's force – 24,000 men and 250 tanks – was to land at Fedula fifteen miles north of Casablanca, with secondary landings at Port Lyautey to take Morocco's only airfield with a concrete runway and at Safi to block any move by the French garrison at Marrakesh. The early signs were not auspicious. At his first meeting with Rear Admiral Henry Hewitt, who would command the American naval force involved, the admiral was unable to tell Patton how many ships would be available: 'He did not strike me as a ball of fire.' Hewitt, having experienced the general's anger and his 'language', com-plained to his seniors that he could not work with Patton but was told that 'The General is indispensable to Torch.' Marshall admonished Patton: 'Don't scare the navy.'

It was perhaps a pique of nerves on both their parts. These two men were planning the largest foreign campaign ever conducted by the US, involving a sea journey of three thousand miles and an amphibious landing on enemy-occupied shores. Sixty-eight warships would conduct thirty-six transports loaded with troops, tanks and equipment from Norfolk, Virginia, to North Africa, across an ocean patrolled constantly by U-boats. There was no certainty that they would even reach the Moroccan coast, but if they did, the heavy swell off Casablanca, which produced waves rising to fifteen feet, might overturn landing craft and inflict heavy losses. Patton kept a straight face when informed that he should appoint a reliable second-in-command because there was a fifty per cent chance that he, Patton, would be drowned during the landing.

Intelligence reports indicated that, once ashore, American troops might be outnumbered by the enemy. However, these were French not German troops and it was unknown whether they would fight or surrender. Patton thought it 'better than even money we could land', and amused himself by composing an ultimatum to the French holding Casablanca, allowing them ten minutes to surrender, after which his forces would 'shoot the hell out of 'em'.

While planning Torch he had also to address and inspire his troops: men who had never left the shores of the US, never been under enemy fire and never killed. Such morale-boosting was a performance art at which he excelled. On each occasion he arrived in an army sedan with a wailing siren, wearing cavalry boots and riding breeches, and carrying a riding crop. General James Doolittle was present in Indio, California, when Patton spoke in 'a high-pitched almost feminine-sounding voice':

I can't tell you where we're going, but it will be where we can fight those damn Germans. And when we do, by God, we're going to go right in and kill the dirty bastards. We won't just shoot the sons-of-bitches. We're going to cut out their living guts and use them to grease the treads of our tanks. We're going to murder those lousy Hun bastards by the bushel.

According to Lindsey Nelson, who heard Patton speak at Fort Bragg, North Carolina, he went too far, saying, 'We'll rape their women and pillage their towns and run the pusillanimous sons of bitches into the sea,' at which the nurses from the camp dental clinic turned their backs on him and walked away. But there was no doubting his effect on the men. General Edwin Randle also heard him at Fort Bragg: 'At the end every man cheered, a genuine, spontaneous cheer. And there were cries of "More! More!" Never before had they heard a general talk like that. He made a deep impression.'

On the evening of 30 August Rommel wrote home to Lucie: 'I have worried so much about this day, but I am taking the risk, because I will not have another chance, in terms of the moonlight and the ratio of forces. So much is at stake. If I succeed, this may have a decisive effect on the course of the war.'

The attack began at 2300. He had 'his Italians' make a frontal attack on Monty's El Alamein line to occupy the British while the Deutsches Afrika Korps moved south and through the thirteen-mile gap (protected only by minefields) between its southern-most point and the Qattara Depression. The panzers would then swing northwards past the Alam al-Halfa ridge to the sea, trapping Eighth Army between the Italians to their front and the Germans in their rear.

It was predictable and he knew it. Auchinleck had planned his defences against it. Monty had pretended that no such plans existed and then prepared his defences against an identical offensive. But Rommel intended to execute the attack faster than the British could cope with: a rapidly changing tank battle that would bedazzle the ponderous enemy and leave them wanting.

Things went wrong from the start. The panzers had to advance thirty miles in seven hours and in darkness across minefields found to be deeper than earlier reconnaissance had suggested. Teams of troops clearing a way for the tanks made slow progress: there were not enough mine detectors available and some men had to probe with their bayonets. Tanks and trucks massed together behind them were lit up by parachute flares and became easy targets for the British artillery. It was 0800 before all of the tanks were through. Bayerlein, commanding the Deutsches Afrika Korps, later revealed that 'Rommel wanted to break off the battle the first morning as soon as it was obvious we had not achieved a sur-prise. I persuaded him to let me continue.' A healthy Rommel might have trusted his instinctive appreciation that the battle was already lost.

The panzers came under heavy bombing from the RAF and were shelled by artillery and the tanks of 7th Armoured Brigade dug in on the Alam al-Halfa ridge. Bayerlein admitted that 'the strength of the defences' overwhelmed them. Montgomery's biographers tell of how earlier he had pointed at Alam al-Halfa on a map and declared that *that* was where Rommel would attack. He strengthened the defences already prepared there by Auchinleck, who had made the same prediction.

After two days' constant bombing and shelling – at one point Rommel saved himself only by diving into a slit trench as the vehicle next to him received a direct hit – the Korps had no alternative but to withdraw. It had lost fifty tanks and as many artillery guns compared with Eighth Army's loss of seventy tanks and twenty guns: losses the British could bear but the Germans could not. The battle might still have swung Rommel's way if the tanks of 7th Armoured Brigade had pursued the panzers back through the minefield, but Monty knew better than to be drawn into open engagement with the master of mobile warfare.

Rommel's view was that he had been halted – he did not consider himself to have been defeated – not by Eighth Army but by the RAF and his own lack of fuel. On 4 September he explained to Lucie: 'My offensive had to be halted because of insufficient supplies and enemy air superiority, although victory would otherwise have been ours. I'm at the command post today for the first time since the attack began. I even managed to get my boots off and take a bath.'

The accumulated effect of Rommel's sickness, the shortage of fuel, RAF command of the air and well-prepared British defences have led some historians to conclude that Alam al-Halfa was for Monty a battle easily won and for which he can claim no great credit. But it was also a battle easily lost. Previous British commanders with the factors similarly weighted in their favour had been drawn into fighting Rommel's way and beaten. The German knew his strength and intended to outwit this new commander in the same way. Monty appreciated that *his* weakness was precisely that kind of quickly changing battle and that the terms of engagement had to be his own. There would be no tank-on-tank battle and no armoured cavalry charge as proposed to him by General Renton, commanding 7th Armoured Brigade, who asked when the tanks would be 'let loose'. While artillery shells and RAF bombs reduced the enemy, Eighth Army would remain still and allow Rommel to 'beat up against 400 tanks in position, dug in, and deployed behind a screen of 6-pounder anti-tank guns . . . and to suffer heavy casualties'. That Montgomery's tactics were correct was confirmed by Rommel himself:

British ground forces hardly put in an appearance. Montgomery attempted no large-scale attack to retake the southern part of his line, and would probably have failed if he had. He relied instead on the effectiveness of his enormously powerful artillery and air force, and harassing attacks by the 7th Armoured Brigade. There is no doubt that the British commander's handling of this action was absolutely right and suited to the situation, because it enabled him to inflict heavy damage on us in relation to his own losses, and preserve the striking power of his own force.

Monty was cock-a-hoop and wrote to his son David on 8 September to tell him that 'The battle I have been fighting with Rommel is over. I have defeated him and I expect you will see a good deal about it in the papers. I have enjoyed it all enormously.'

His success was, as he had imagined, splashed across the British papers, but for the Prime Minister it was hardly enough. Monty had beaten Rommel in a *defensive* battle. Churchill needed him to destroy Rommel in an *offensive* battle and he pressed for an attempt to be made soon. Monty replied that the additional equipment he needed would not be in place until October and later wrote: 'The Prime Minister signalled that the attack must be in September . . . I refused to attack until October; if a September attack was ordered, they would have to get someone else to do it. My stock was rather high after Alam-el-Halfa! We heard no more about a September attack.'

By now Rommel could name the new British general who had stopped him. General Gause briefed the junior officers: 'We know from our sources that Churchill was in Cairo at the beginning of

August and that on the twelfth Montgomery took command of the Eighth Army. A new wind seems to be blowing among our opponents. They will, undoubtedly, be working on an offensive, which this time could be decisive.'

As Rommel prepared his defences along the El Alamein line he was aware that the conflict to come would be one of *Material-schlacht*: a battle of attrition in which the side with the greater resources (and the greater ability to replenish them) must prove the victor. His supreme talent was for *Fingerspitzengefühl*, the instinctive response to rapidly changing battlefield situations. Here, both sides had taken up fixed positions and any form of mobile warfare in the open desert was unlikely to develop.

He set about strengthening his front-line defences, which ran forty miles from the sea to the Qattara Depression. The enemy would have to penetrate two parallel belts of mines separated by a gap scattered with bombs and shells (some of them wired for remote detonation) that became known as 'the Devil's Garden'. His chief engineer reported that 434,000 anti-tank mines and 14,000 anti-personnel mines were laid. Tank regiments were spread evenly along the line, stronger German regiments alternating with the weaker Italians. Rommel expected the minefields to delay the enemy long enough for his armour to be concentrated at the point of an attempted breakthrough. Losses were expected to be high on both sides.

At the same time he began a campaign to persuade the High Command in Berlin and the Comando Supremo in Rome that he must be sent more tanks, guns, ammunition and fuel. He communicated personally with Hitler and Mussolini to stress that they, not their field marshal, would determine the outcome at El Alamein by the extent to which they matched Churchill in providing the necessary resources. He received many promises.

At the beginning of September he visited the Siwa oasis, 200 miles to the south, where a small German detachment was kept to guard against any wide, outflanking move south of the El Alamein line. There was no pressing military reason for the trip and it is likely that he went to 'take the cure'. The German troops at Siwa called it 'Paradise': palm trees, water courses and the remains of Cleopatra's bath built in 30 BC and still fed by warm, carbonated springs. But that could not cure all his ills, and on 9 September he told Lucie, 'The doctor is pressing me hard to have a break in Germany and doesn't want me to postpone it any longer.'

A week later General Georg Stumme, who had most recently commanded a panzer corps in Russia but had no desert experience, arrived to relieve Rommel. Montgomery was expected to attack in October and Rommel told Stumme that he would 'cut short my cure and return if the British open a major offensive'.

He flew first to Hitler's headquarters at Rastenburg, where he was presented with his field marshal's baton. He briefed Hitler on his attack at Alam al-Halfa and the cause of its failure. When he stressed the superiority of the British in the air, Göring questioned its significance:

ROMMEL: The enemy win their battles from the air! They knock out my panzers with American armour-piercing shells.
GÖRING: That's impossible! The Americans only know how to make razor blades.
ROMMEL: We could do with some of those razor blades, Herr Reichsmarschall.

He said that the *Luftwaffe* in Africa must have more planes, his army must have more tanks – the British now had a two-to-one superiority – and he must have adequate supplies of fuel 'otherwise German troops will not be able to maintain our hold on Africa'. To his delight Hitler promised to send everything he needed, including forty of the new Tiger heavy tanks and 500 *Nebelwerfer* (multiple rocket launchers).

Goebbels organized an international press conference on 3 October at the propaganda ministry. The BBC had reported that the field marshal was sick and ostensibly his appearance was to deny it. But the German war effort was not going well and

Goebbels needed a morale-boosting story. He stage-managed a photo-opportunity for the newsreel cameras. Rommel took firm hold of a door handle and turned to declare: 'Today we stand one hundred kilometres from Alexandria and Cairo and have the gates of Egypt in hand.' He wrote later: 'With things as they were I could not give a true picture of the situation. In any case, by giving an optimistic account I hoped to bring about some postponement of the British offensive.' For Goebbels it did more than that, convincing the German people that reports of Rommel's exhaustion were British propaganda and that total victory in North Africa was within his grasp.

He returned to Lucie at Wiener Neustadt and together they left for his 'six-week cure' in the nearby mountain resort of Semmering. Reached by a mountain railway and set 3200 feet up in the Austrian alps, Semmering was surrounded by pine-forested hills and overlooked by the twin mountains of Rax and Schneeberg. It was a popular health resort and ideal for Rommel's recovery, but 'with my army in such a plight I was incapable of attaining real peace of mind'.

While he was there he received a letter from Stumme, attempting to put his mind at rest. The defensive minefields had been strengthened with several thousand more anti-tank mines laid. However, none of the supplies promised by Hitler had yet arrived. Rommel confided his fears in his wife: 'I wonder if the Führer told me all that to keep me quiet?'

Montgomery's plan was outlined in a five-page memo issued to his commanders on 6 October. The sea and the Qattara Depression ruled out any flanking movement and the attack would have to penetrate minefields that varied from three to five miles deep. These were laid with mostly anti-tank mines, triggered by a much greater pressure than anti-personnel mines. During the initial phase of the attack engineers equipped with mine detectors would advance to clear a path wide enough for the armour that was to follow: thus the offensive was codenamed Operation Lightfoot.

Monty and Churchill would later 'write up' Rommel's defensive line as particularly formidable, but it was not. Minefields five miles deep compared badly with those on the Russian front where a depth of ten to twenty miles was common. This lack of depth also meant that Rommel's troops and guns located behind the minefields could be reached by British artillery guns.

Two lanes would be cut through the minefields during the first night, allowing the tanks of 10 and 30 Corps to pass through by dawn the next day. Instead of continuing their advance the tanks would then dig-in and destroy Rommel's panzers as they attempted to dislodge them, gradually reducing the enemy until his losses were unbearable and he was forced to withdraw from El Alamein. Montgomery told his commanders that the battle would be 'a real rough-house' and 'the turning point of the war'.

Torch was scheduled for 8 November. Monty estimated that Lightfoot could defeat Rommel in less than two weeks and scheduled the operation to begin on 23 October, so that he would have defeated Rommel before the American-led landings in Morocco.

He knew that the attack would come as no surprise to the enemy and an attempt was made to deceive German Intelligence about where and when it would take place. The army consumed vast amounts of water and this was pumped to El Alamein through an underground pipeline. Soldiers began to dig a trench for an extension 'pipe' assembled from empty four-gallon petrol cans placed end-on-end beside it. Each night the cans were moved forward and the trench filled in, and the next day a further section was dug. This fake water pipeline, just visible from German positions on the Himeimat Heights, headed for an area thirty miles south of the actual point where Eighth Army would attempt to break through the enemy line. Just as important as fooling the Germans about the point of penetration was assisting them to miscalculate the date of the attack: the daily rate of construction indicated that the 'pipeline' would not be complete until 30 October.

*

At 0430 on 6 October American troops in landing craft stormed Solomon's Island in Chesapeake Bay. Patton watched this practice landing with mixed feelings. It went well despite some troops 'taking' the wrong beach, but there was no simulating the high waves to be encountered off the Moroccan coast and the artillery fire to be expected from shore batteries. During the next two weeks his assault troops, regiment by regiment, were put through the same exercise. Meanwhile 2nd Armored Division trained for a tank assault on Casablanca.

Patton went to see 'Black Jack' Pershing in Washington's Walter Reed Army Hospital on the morning of 21 October to say farewell and to thank him for giving him his first experience of action in Mexico. The eighty-one-year-old told him: 'I can always pick a fighting man and there are damn few of them. I'm happy they're sending you to the front. I like generals so bold that they're dangerous.' Patton knelt down and asked Pershing for his blessing. 'I kissed his hand, then put on my cap and gave him the salute.'

That afternoon he went to the White House, wearing his ivory-handled pistols, to meet the President. Roosevelt called him 'Cavalryman': 'I asked the General whether he had his old cavalry saddle to mount on the turret of a tank and if he went into action with his sabre drawn.' Patton was never short of a suitable riposte: 'Sir, all I want to tell you is this. I will leave the beaches either a conqueror or a corpse.' Roosevelt's parting comment – 'Well, everything's grand' – did not seem to Patton the most profound assessment of the operation that American armed forces were about to attempt.

The next day he left Washington. Beatrice accompanied him on the flight to Norfolk: 'We flew down the Potomac and right over the convoy. When G landed, he was so excited that he jumped right out of the plane and I had to call him back to say goodbye.'

Once alone, he wrote to Frederick Ayer, his brother-in-law, enclosing a letter to be given to Beatrice 'when and if I am definitely reported dead'. He told his army commanders that he wanted 'fearless leadership' and 'if you don't succeed, I don't want to see you alive'.

On the morning of 23 October America's foremost advocate of tank warfare and of audacity in battle set sail for Morocco to land in the rear of the German who had already proven himself a master of both in the North African desert. Patton had no time for humility: 'When I think of the greatness of my job, I am amazed, but on reflection, who is as good as I am? I know of no one.'

He wrote a message to the troops to be delivered after the fleet set sail – they were not yet aware of where the Torch landings would take place:

We are now on our way to force a landing on the coast of North West Africa. Our mission is first to capture a beach-head, second to capture the city of Casablanca, third to move against the German and destroy him. When the great day of battle comes, you must succeed. Americans do not surrender, for to retreat is as cowardly as it is fatal. The eyes of the world are watching us. The heart of America beats with us. God is with us. We will surely win.

That same day in North Africa, with Lightfoot to be launched after dark, a personal message from Monty was read out to the men of Eighth Army:

When I assumed command of the Eighth Army I said that the mandate was to destroy Rommel and his Army, and that it would be done as soon as we were ready. We are ready now. The battle which is about to begin will be one of the decisive battles of history. It will be the turning point of the war. Each one of us should enter this battle with the determination to see it through – to fight and to kill – and finally to win. And let no man surrender so long as he is unwounded and he can fight. Let us pray that 'the Lord mighty in battle' will give us the victory.

The Lord was expected to overlook Monty's earlier comment to his officers – a limp attempt at humour to put them at their ease before the coming conflict – that everyone must be imbued with the desire to 'kill Germans, even the padres, one per weekday and two on Sundays'.

Lightfoot and Torch were ready to go.

The opening bombardment of Montgomery's Operation Lightfoot was scheduled to begin at 2140 on the evening of 23 October. Patton's Western Task Force left Newport, Virginia, on the following morning to participate in Operation Torch but would take at least two weeks to reach North Africa. This begs the question: why were the two operations not timed to impact on Axis forces simultaneously, Monty's attack along Rommel's front coinciding with the landing of Allied troops in his rear?

The official reason given was that Lightfoot, by preceding Torch and delivering a victory over Rommel, would discourage Vichy French forces from opposing the landings that came later and make an immediate French surrender more likely. That was undeniably the case. What could not be openly stated was that any such surrender reduced the 'value' of Torch to the Americans, who saw it as an opportunity to 'blood' inexperienced American troops against French forces before they had to engage the Germans. Certainly Patton did not want the French to surrender too soon: 'There is a chance that we shall have, at least initially, a pushover. Personally I would rather have to fight – it would be good practice.'

The true reason why Lightfoot preceded Torch had more to do with what the British needed. On the face of it that was to defeat Rommel, however and *whenever* it could be done. Yet if that was so, then the simultaneous impact of Lightfoot and Torch offered the most economic (measured by the expenditure of men and matériel) means of doing so. An immediate offensive against Rommel's well-prepared defences was bound to be costly. Once the Torch landings had taken place, Rommel would have to contend with a large Allied army in his rear, which was nearer his base at Tripoli than he himself was at El Alamein. Caught between two armies, each larger than his own, it could be expected that he would leave his carefully prepared defences and fall back towards his supply base, fearing Torch more than Lightfoot because the former could cut him off from Tripoli. This forced move would make Monty's task much easier. Rommel had only 200 tanks to cover the walking withdrawal of 90,000 infantry, while Monty had 1100 tanks and 220,000 fully motorized infantry to follow and defeat him. In addition the RAF had command of the air. Monty had only to sit and wait until Rommel left his prepared defences, then pursue him into the 'pincer' to destroy him between Torch and Eighth Army at a much lower cost in lives.

So why the immediate offensive at El Alamein? So far the war had provided only a series of British defeats. The Empire was crumbling. The British people were losing faith in their generals and their army to win anything at all. Churchill shared their doubts and with greater reason: he was losing support in the House of Commons and his government might fall. The need was not just for a victory but for a *British* victory. Churchill had been pressing for Rommel to be defeated in September, and when Monty said he was not yet ready, in October – in any case, before the Torch landings in November.

Torch was predominantly an American affair. Churchill and the War Office saw that, after Torch, America would quickly become the senior partner in the alliance. The defeat of Rommel at any time *after* Torch would be seen as the US Army saving the British and that would not do. The British needed to defeat Rommel *before* Torch. The offensive at El Alamein was scheduled for that victory, but in the knowledge that if Monty failed to deliver, the Torch landings two weeks later would force Rommel to fall back anyway and Cairo and the Suez Canal would not be lost. Monty could boast about having no plans for retreat but he did not spell out why they were unnecessary.

Lightfoot preceded Torch because the British had to be seen to have defeated Rommel on their own account and before the first American troops set foot in North Africa. Churchill and

Montgomery were prepared to pay for it in tanks (many of which were, in any case, American and would be replaced by more American tanks) and in the lives of British soldiers.

There was an American corollary to this. Churchill may have needed a *British* victory to save his political career. But President Roosevelt knew that the American public had lost faith in the ability of Great Britain to fight and win, and feared American troops would be left to do it all, which translated into fewer votes for him. He needed a British victory too.

After dinner on the evening of 23 October Monty wrote to David: 'I shall be too busy to write to you for some time. I begin tonight a big battle which will help to end the war. I enclose for you a copy of a Personal Message I have issued to the army and it is being read to every man today. You should keep it carefully as one day it may be of value.'

It was still early afternoon in Norfolk, Virginia, when Patton rang Beatrice to tell her he was leaving for North Africa the following day. He felt good about the chances for success, he said, and that his whole life had prepared him for this operation. At 1445 he went aboard the cruiser USS *Augusta*, Admiral Hewitt's flagship. In his cabin he was still thinking of Beatrice and wrote a note to her: 'I will be thinking of you and loving you.' The fleet was to leave at 0800 the next morning. He noted in his diary: 'This is my last night in America. It may be for years or it may be forever. God grant that I do my full duty to my men and myself.'

In their villa at Semmering in the Austrian Alps, Rommel's health was improving after three weeks of mountain air, fresh food, rest and, not least, Lucie's presence. He referred to this time as 'our wonderful weeks'. As was their habit at Semmering they retired early.

Monty read a book and went to bed early too: 'At 2140 the Eighth Army which included some 1,200 tanks went into the attack. At that moment I was asleep in my caravan; there was nothing I could do and I knew I would be needed later.'

7. Operations Lightfoot and Torch

At 2140 on the evening of 23 October 1942 Montgomery's 744 artillery guns opened fire on the German line at El Alamein, supported by Allied air-force bombers. The barrage of exploding shells was the most intense of the war so far, measured by volume of high explosives and rate of delivery. Men on the ground at a distance who were deafened or suffered bleeding ear drums were the lucky ones. Its first target, the 62nd Italian Infantry Regiment, was annihilated.

Operation Lightfoot was to be a battle of attrition. Monty had no time for such a discreet term and told war correspondents it would be a 'killing match'. He had a military term for his commanders: 'crumbling'. The attack would not be pressed once Eighth Army armour reached a position close enough to trade shells with the Panzerarmee and methodically 'crumble' the enemy until he was too weak to prevent a breakthrough.

The statistics mattered. Monty had 195,500 men compared to the enemy's 104,000. He had 300 Sherman tanks (superior to the best of the enemy armour, the Panzer IV), 246 Grants and 421 Crusaders, a total of 952. Rommel had 230 German and 320 Italian tanks (the latter totally inadequate in armour and gun-power). Monty was confident Eighth Army must win because even if its losses were greater than those of the Panzerarmee (and that was likely, offence being more costly than defence) it could afford them and the Germans could not. Commentators who accuse Montgomery of fighting El Alamein as if it were a Great War battle miss the point: the tactical situation allowed no other way and in any case it was a battle he knew he could win.

The opening barrage lasted twenty minutes, and in the eerie

silence that followed, infantrymen and engineers climbed from their forward trenches and walked into the northern-sector minefields where Monty planned his main penetration. At first they made good progress. The use of 500 mine detectors speeded up an operation that traditionally involved prodding by hand or bayonet (and many still had to do so). Much of the enemy artillery had been put out of action. But it was not so much the tactics as the battlefield conditions that were now reminiscent of the Great War. The bombardment had thrown up great clouds of dust that hung thickly over the minefields, so that men and even whole platoons lost direction. Between the first and second mine belt they were delayed by barbed wire, and as they cut a way through, German machine-guns opened fire. Operation Lightfoot had been well named: engineers missed anti-tank mines and walked safely on while some of the tanks following close behind went up in flames. The channels that were cut through the minefields were too narrow, so that waiting tanks massed at the entry point and presented an easy target as flares burst above them and the surviving German artillery guns were brought to bear.

At dawn on 24 October General Stumme left his command post with Colonel Andreas Buechting to drive to the front as Rommel would have done to see for himself what was happening. As his car drove parallel to the line it came under anti-tank and machine-gun fire. Buechting received a fatal wound to the head. Stumme leaped out and hung on to the side of the vehicle, but as his driver, Corporal Wolf, swung it around, the general lost his grip and fell to the ground without this being noticed. The hapless driver accelerated away and only later discovered that he had 'lost' the army commander.

Stumme was missing. Montgomery was despondent: the engineers had failed to clear the required lanes through the minefields and his tanks were held up too far from the enemy to begin their 'crumbling'. He ordered the armour of General Lumsden's 10 Corps to 'fight its own way out'. When Lumsden argued that this might result in a massacre Monty applied what he called 'ginger' – presumably a variant of 'binge':

Lumsden was not displaying that drive and determination that is so necessary when things begin to go wrong; there was a general lack of offensive eagerness in 10 Corps. I therefore spoke to him in no uncertain voice, and told him he must 'drive' his Divisional Commanders, and that if they hung back any more I would at once remove them from command and replace them by better men.

An attempt was made in the afternoon but 'eagerness' was not enough. Sufficient German artillery and anti-tank guns had survived to keep the British armour pinned down. It became clear that Monty's tanks would not break through on the first day of Lightfoot as he had planned, granting the enemy time to redeploy forces to the point of attack. Rommel's plan for the defence had worked in his absence. Monty ordered that the armour make another attempt during the night.

Keitel telephoned Rommel at Semmering at 1500: 'There is bad news from Africa. The British have launched a big offensive. Stumme is missing, captured or killed. Are you well enough to go back?' Well enough or not, there was only one possible answer. Keitel said he would ask the Führer whether Rommel should return immediately or wait on further developments.

It was Hitler himself who rang next. He asked Rommel to go to the airport and wait there for a decision, but he did not want to interrupt the field marshal's treatment unless the situation was critical. Rommel rushed to Wiener Neustadt and a Heinkel-111 bomber was put on standby to fly him out. Keitel rang again at 2030 to say that Hitler was still assessing the situation and would speak to him in the morning; meanwhile General Ludwig von Thoma had assumed command of the Panzerarmee.

At 0200 on the morning of 25 October Monty was woken by de Guingand to face a potential mutiny. The 10th Armoured Division had advanced further but had come under such concentrated enemy fire that its commander, General Gatehouse, now wished to withdraw it back behind the minefield, and Lumsden agreed with him. Monty spoke with Gatehouse by field telephone and was told that the division 'faced annihilation by the enemy anti-tank guns'. He dismissed the general's protest, told him to continue the advance, and went back to bed. That order has been praised as 'steadfastness' and condemned as 'callousness', but Monty could do no other. He had chosen (or had had forced upon him) a battle of attrition, so Eighth Army losses had to be accepted and even sought out in order to inflict the necessary 'crumbling' losses on the enemy. The leading regiment of 10th Armoured Division was virtually wiped out before dawn but the division cleared the minefield. Monty recorded that 'by 0800 all my armour was out in the open'.

At that same time Hitler rang Rommel to say the latest situation report proved this to be Montgomery's big offensive. The field marshal should return to Africa and take command. His Heinkel took off immediately and landed at Heraklion in Crete at 1445 where he transferred to a faster Dornier 217 for the most dangerous part of his journey. The local *Luftwaffe* commander handed him a briefing paper from El Alamein. There had been a strong British attack on the northern and southern sectors of the line. General Stumme's body had been discovered; he had died of a heart-attack.

At an afternoon conference Monty's commanders again argued that the operation was proving too costly to continue and that the armour should be withdrawn. He compromised, shaken by the possibility that they might not obey an order to go on. The armour would be withdrawn from its present position and make a new attack on 28 October further north where enemy resistance would be less. Both he and his commanders knew that the Germans would redeploy their guns further northward too, but it defused a potential crisis of command. Colonel Richardson, a senior staff officer, felt that Monty came out best: 'The armour was stopped in the dog-fight area. Terrific credit goes to Monty that he was able to impose his will on his commanders and say, we will do this, we will not sit down and call it a day.' But it is an indication of the British losses involved that his commanders wanted to do so.

Rommel reached his headquarters at 2200. After two days of fighting, the mathematics of attrition were already telling. Although 15th Panzer Division had brought the British thrust to a standstill, it had lost 88 out of 119 tanks in doing so. (These statistics mask the horror of armour-piercing shells, which penetrated a tank and ricocheted around inside, killing the whole crew.) The army was running low on both fuel and ammunition, but an Italian tanker and a supply ship were *en route* to Tobruk.

Rommel was unknowingly at a further disadvantage. Montgomery was receiving decoded transcripts of Enigma messages and was in effect listening in on all communications between the field marshal and the Führer's headquarters. Rommel knew what a poor state his army was in: so did Monty. The RAF knew that the fuel Rommel needed was at sea and a bombing raid was ordered.

At 1125 he signalled all units: 'I have taken command of the army. Rommel.' The battle was not yet lost. The British had driven a wedge through the minefield at great cost and its front was too narrow to allow the decisive breakout that Monty needed. Rommel wrote a note to Lucie: 'Situation tense. Not easy after the beautiful weeks with you.'

On the morning of 26 October Rommel was in his operations truck by 0500 and at dawn he drove to the front to examine the state of his defences and enemy dispositions through his field glasses. In the Great War he had fought on mobile fronts and had missed the static battle of attrition of trench warfare, and was surprised now by its equivalent in the present war: 'Rivers of blood were poured out over miserable strips of land which, in normal times, not even the poorest Arab would have bothered his head about.'

British forces had made their most concentrated attack in the area of Hill 28 in the northern sector and had been met by 15th Panzer Division. Enemy probes along the southern sector had not been pressed vigorously and Rommel guessed they were meant to hold 90th Light and 21st Panzer Divisions there, and discourage him from transferring them to reinforce the northern sector. He ordered the move. During the day the British tanks again failed to pass through the narrow gap in the minefield and it appeared to Rommel that he had stopped them.

In fact, Monty had stopped them. He felt that 'Divisions were somewhat disorganized,' and ordered all major operations to cease so that they could be 'tidied up and things sorted out' before the offensive scheduled for 28 October. He had estimated 10,000 casualties over ten days of fighting. After less than three days the figure had reached 7000. But he would not cut it short, writing in his diary that 'If we fire 150 rounds a gun per day, we can continue the battle for three weeks.'

The next morning Rommel received devastating news: the tanker *Proserpina*, carrying the 2500 tons of fuel he desperately needed, and the transport ship *Tergestea*, bringing ammunition, had both been sunk as they approached Tobruk. It seemed to him that the battle of El Alamein had just been lost at sea. A quick calculation showed that he had enough fuel for one further manoeuvre. He could play to his strength and have Deutsches Afrika Korps fall back to draw the enemy armour into a mobile tank battle, but he knew Montgomery would refuse him that. He chose to mount a counter-attack on Hill 28; it had to succeed.

His panzers and infantry moved forward at 1500. British artillery guns opened up and soon RAF fighter-bombers were overhead. Tanks burst into flames and in the open country the troops suffered terribly. They could make no headway. Rommel returned to his truck and wrote to Lucie: 'Very heavy fighting. No one can imagine the burden that lies on me. Everything is at stake and we're fighting under the greatest possible handicaps.' However, all was not lost. A signal from Rome told him that another tanker had already left Italy.

Monty felt it had been a good day: 'The one thing we want is that the enemy should attack us. 1st Armd Div have today destroyed fifty enemy tanks (all burning) without loss to themselves.' The RAF was satisfied too: another target ship had been identified by the latest Enigma transcript.

Early on 28 October British forces began concentrating in preparation for Monty's new attack further north. He planned to reach the coast and get behind the enemy, trapping German forces between Eighth Army and the minefield.

Rommel visited the front and could see enemy forces massing. He ordered all surviving German units on the southern sector to move immediately to the north. The British attack would have to run on to a bank of anti-tank guns and could be delayed, but he was not certain it could be stopped. His army was exhausted while the enemy's losses were replaced from what appeared to be never-ending reserves. Once enemy armour got into the rear it would all be over. He wrote a despondent 'farewell' letter home:

> We are doing our utmost, but the enemy's superiority is over-whelming. If we fail, whether or not I survive the battle will be in the hands of God. Life is hard to bear for a vanquished man. Should I remain on the battlefield, I want to thank you and our son for all the love and joy you have given me. In our weeks in the mountains I realized how much you mean to me. My last thoughts will be of you.

The British artillery opened up at 2100 and then 9th Australian Division moved into the minefield. The battle raged through the night with heavy losses on both sides, and by dawn on 29 October the Australians had still not broken through. Monty was sorely disappointed and now feared a stalemate of the worst kind: that the German minefields, like the no man's land between the trenches in the Great War, might prove impassable by either side. British tank losses were mounting and the number of dead and wounded already exceeded 10,000.

In London Churchill called Brooke to Downing Street and demanded to know 'why Montgomery is fighting a half-hearted battle? Have we not got a single general who can win a single battle?' He sent a signal to Monty, reminding him that the Torch landings were imminent and that everything possible should be done to achieve a victory at El Alamein. He then called a meeting of the chiefs of staff, at which Brooke announced that Monty was about to launch a fresh, even bigger attack; he admitted later that *that* was pure fantasy designed to placate Churchill and, for all he knew, Monty had already been beaten.

Monty was not beaten and he believed the apparent stalemate would soon be broken for two reasons. First, he had planned Lightfoot from the outset to be a battle of attrition. The course of such a battle could not be calculated by individual engagements won or lost. A series of 'attempts' to penetrate the enemy line, all of which failed, would – if enemy forces incurred unbearable losses while the attacking force could replenish theirs – amount to an eventual victory. Second, Monty knew that this was happening because Enigma transcripts told him so. Rommel sent a daily summary report to Berlin to keep the High Command informed of his army's position, movements, resources and plans. Decoded transcripts reached Monty within an average twelve hours. Because he often rose early and Hitler did not always do so, he sometimes read Rommel's reports before the Führer.

The value of the Enigma transcripts to Monty is often asserted but their full detail and significance is revealed in the example below. This excerpt is from the summary report for 28 October transmitted from Rommel's headquarters to the High Command in Berlin at or soon after midnight that evening, taken from an original copy in the German Military Document Section of the US Defense Department. It was decoded at Bletchley Park and a transcript reached Monty before noon on 29 October.

At 2200 hours the enemy attacked in the northern sector. The main weight came in against the left flank of 155 GrenRegt and the right of 125 GrenRegt. Afrikakorps ordered to counter-attack with all available tanks if enemy breaks through.

The number of available tanks has dropped further, with 21 PzDiv suffering heavy losses. Available tanks: 15 PzDiv – 21 Mk III and IV; 21 PZDIV – 45 Mk III and IV; Littorio and Ariete Armd Div – 162 M

The petrol flown over on 27 October (391 tons) did not allow the Army to carry out extensive counter-attacks with mobile forces. The supply is now 1.3 consumptive units [one unit was the amount required for movements totalling 100 kilometres] and will soon reach a minimum, when the Army will be immobilized.

The 400 tons of ammunition which is to be sent over by 4 November is not sufficient – with the utmost economy about 500 tons are fired daily.

Ground reconnaissance confirms strong enemy forces massed in area west of a line from Alem el Qata to Burg el Arab. Therefore these orders were given:

(a) mobile reserve of 21 PzDiv to move from southern sector to north immediately
(b) 20 Corps to assume control of defence of southern sector
(c) 33 Recce Unit to move to area 5km SW of Sidi Abd el Rahman

All formations ordered to prepare to meet an attack from enemy armoured forces in minefield J. Artillery commander instructed to concentrate army and anti-aircraft artillery fire on enemy advancing from minefield J.

While Rommel's reports were of most interest to Monty, signals from the High Command in Berlin were more helpful to the RAF because they detailed when and where Italian oil tankers and supply ships would arrive in North Africa. The second tanker in which Rommel placed so much hope, the *Luisiana* carrying 1500 tons of fuel, was sunk as it approached Tobruk.

Monty knew that his Eighth Army was winning the numbers game with the Panzerarmee. In a battle of attrition time favoured the stronger side. He could afford to wait. But Churchill could not and Monty was forced to act: 'Torch is on 8 November. It is becoming essential to break through somewhere, and to bring the enemy armour to battle. We must make a great effort to break up his Army ... I have therefore decided to modify my plan.' He spent 29 October planning what he called Operation Supercharge. It was 'the Master Plan and only the master could write it'.

That was not exactly so. Earlier that day, soon after it became clear that the overnight offensive had failed, de Guingand spoke with Generals Richard McCreery and Charles Richardson. They agreed that as Enigma transcripts showed Rommel had concentrated his German divisions at the point of their present efforts, they should abandon this and attack forcefully further south where the line was now defended by his weaker, Italian divisions. A breakthrough there would enable them to swing north *behind* the panzers. De Guingand went to see Monty and returned to say, 'No, he won't have it.' McCreery wanted to challenge Monty about it and Richardson recorded what was said next: 'Freddie told Dick McCreery, "Look I will go and talk to Monty about it again – don't you, for goodness' sake, because if you do he won't do it. But if one can persuade him it's his own idea, so to speak, then I'm sure it's the right thing to do." He went back and had another go, and Monty did accept it!'

Monty wrote in his diary that 'At 1100 hours I changed my plan and decided to attack further to the south.' Operation Supercharge was scheduled for 31 October. It would be essentially a repeat of the day-one thrust through the minefields but against a much-reduced opposition.

When Rommel heard that the second Italian tanker had been sunk he told Lucie, 'The situation is serious again.' He had about ninety tanks left – the number changed by the hour as engineers raced to make damaged tanks serviceable. He knew it was not enough.

In the Atlantic the US fleet carrying the Torch task forces was accompanied by three battleships and forty destroyers and cruisers.

Air-force planes put up by the carriers circled continuously over-head. Patton's expectations of reaching North Africa lowered each time the *Augusta* began one of its zigzag manoeuvres to avoid possible U-boat torpedoes.

During the day he and his staff fired their carbines at an assort-ment of floating targets tossed into the ocean. He also used the time to address his officers and men: 'I have been giving everyone a simplified directive of war. Use steamroller strategy. Attack weak-ness. Hold them by the nose and kick them in the ass.' Monty would not have put it like that but it was what he had been persuaded to do in Supercharge.

Patton estimated that he was two-thirds of the way to Morocco. He started reading the Qur'ān each night better to understand Arab culture, but soon gave up and read a detective tale instead.

On 31 October Monty decided that he did not have everything in place for Supercharge to begin as scheduled that evening, and delayed it by twenty-four hours. Perhaps he had already sensed unease among the commanders of his armoured forces that they were being asked to undertake again the task they had 'protested' about on day one, because in that day's Army Commanders' Directive he wrote: 'This operation if successful will result in the complete disintegration of the enemy. Determined leadership will be vital; complete faith in the plan will be vital; risks must be accepted freely; there must be no "belly-aching".'

The necessity of that admonition was revealed in an exchange between General Freyberg, commanding the New Zealand Division, and General Currie, commanding 9th Armoured Brigade:

FREYBERG: We all realize that for armour to attack a wall of guns sounds like another Balaklava.

CURRIE: This might involve fifty per cent losses.

FREYBERG: It may well be more. Montgomery has said he is prepared to accept one hundred per cent.

The acceptance of one hundred per cent tank losses appears nonsensical but has its place in a battle of attrition: if each tank, prior to being hit, destroyed one enemy anti-tank gun, the attack against a numerically inferior enemy force could be counted a success.

Monty wrote three letters on the evening of 1 November. He told David that 'I am fighting a great battle with Rommel and I am enjoying it very much', and informed Brooke that the battle so far had been 'a terrific party'. To the Reynoldses, who were looking after David, he confided that 'It has become a real solid and bloody killing match.'

At 2200 the British artillery opened up on German positions and the RAF began an air bombardment. Engineers went ahead to clear mines, while behind them the tanks of 9th Armoured Brigade engaged enemy anti-tank guns across the line. By dawn the next day the brigade had lost seventy of its ninety-four tanks with only thirty-five anti-tank guns knocked out, but that was enough: the minefield had been breached. Rommel moved up 21st Panzer Division to counter-attack. Monty sent his 1st Armoured Division through the gap to hold them off. The two forces met at 1100 in a fiercely contested tank-on-tank engagement, which went on for two hours. Meanwhile 10th Armoured Brigade moved through the minefields and was met by the remains of 15th Panzer Division. Rommel ordered his 88-mm anti-aircraft guns to ignore the British planes overhead and concentrate on the armour. A further fifty-four British tanks were lost and almost as many panzers knocked out.

Eighth Army had breached the minefields but the Panzerarmee line held – just. Monty's total tank losses were huge but represented a bearable proportion of the number in reserve. Rommel's losses were slightly fewer but they represented virtually the whole of his armour. The battle of attrition had been concluded. He discussed the situation with von Thoma and they agreed there was now no chance of holding the British back: at Montgomery's next push their fragile line must shatter. He decided on a sixty-mile retreat to Fuka with his surviving thirty tanks forming an armoured rearguard to delay the British for as long as possible, and at 1950 he informed the High Command of his intention:

After ten days of hard fighting against overwhelming British superiority on the ground and in the air, the strength of the Army is exhausted. It will not be able to prevent a new attempt to break through, with strong enemy armoured formations, which is expected soon . . . There is only one road available and the Army, as it moves, will be attacked day and night by the enemy air force. In these circumstances we must expect the gradual destruction of the Army.

He ordered the retreat to begin at 2200 and wrote to tell Lucie that 'The dead are lucky, it's all over for them.'

Reports reached Monty at 0800 on 3 November that Rommel was withdrawing his infantry. Explosions across the front line were believed to be the enemy blowing up his ammunition dumps. Deutsches Afrika Korps still held the line but this rearguard could be expected to fall back next. Allied planes were sent to bomb the retreating columns as they marched. Monty set about regrouping his army in preparation for a final effort to get the bulk of his armour through the minefields and into the enemy's rear. In his Special Order of the Day he predicted the total destruction of the Panzerarmee:

The present battle has now lasted twelve days, during which the troops have fought so magnificently that the enemy is being worn down. He has reached the breaking point, and is trying to get his army away. The RAF is taking a heavy toll of his columns moving west on the main coast road. We have the chance of putting the whole Panzer Army in the bag, and we will do so.

A quite different signal from Hitler, decoded at Rommel's head-quarters at 1330, became known almost immediately and for good reason as the Führer's *Sieg oder Todt* (victory or death) order:

With me the entire German nation is watching your heroic battle in Egypt, with confidence in your leadership and in the courage of your troops. There can be no thought but to persevere, to yield not a metre of ground, and to throw every man and every gun into the battle . . . It would not be the first time in history that willpower has triumphed over the stronger battalions. To your troops you can offer only one road – that which leads to Victory or Death. [Signed] Adolf Hitler.

Rommel's initial response was fed by what remained of his special relationship with the Führer: he must obey a direct order. He dashed off a note to Lucie: 'I can believe no longer in a favourable outcome. I enclose 25,000 lire that I have saved from my pay. Our fate is now in the hands of the Almighty. PS Ask Appel [their bank manager] to exchange the lire.'

In London Churchill read Hitler's signal shortly after it had reached Rommel. He rang Brooke and suggested they order church bells to be rung across the country. Brooke convinced him that that might be a little premature.

The final breakthrough came at dawn the next day as the Highland Division advanced through a smokescreen and the weakened Deutsches Afrika Korps fell back with only twenty-two tanks left. During the morning the British widened the breach until it was enough for their armour to get through *en masse*.

Rommel had regarded Hitler's order as absolutely binding, but now he decided that a fighting withdrawal had to be preferred to a slaughter. For the first time he would disobey a direct command from his beloved Führer. 'With our front broken and the fully motorized enemy streaming into our rear, superior orders could no longer count. We had to save what there was to be saved.' He ordered an immediate retreat. It would be another twenty-four hours before Hitler rubber-stamped the move.

In a final rearguard action, von Thoma led the last tanks of his two Panzer Divisions against 1st Armoured Division, which had pressed through the line and might threaten the withdrawal. Some considered it a suicide mission. The German tanks were hit one by one until von Thoma's Mark II went up in flames. British troops found him standing beside it unhurt and he was taken at once to Montgomery's headquarters in a scout car.

As his army began again to fall back Rommel wrote one of his occasional letters to Gertrud, his illegitimate daughter. It had the tone of a last farewell:

> The enemy's strength is too strong and they are overwhelming us.
> It is in the hands of God whether my soldiers will survive. To you
> I send my heartfelt thoughts and wish you and your family all good
> things for the future.
> Your uncle Erwin

That evening Monty dined with von Thoma – the commander of Eighth Army and the former commander of Deutsches Afrika Korps – and enquired about Rommel's health. He told the German that 'I met Rommel once in August and I beat him; I have met him again now and I shall do the same thing.' He considered von Thoma to be 'a very nice chap' and was eager to know what the Germans thought of *him*, asking if they 'had a character sketch of the Eighth Army commander'. Von Thoma replied that they saw him as 'a hard, ruthless man who had introduced new tactics'. Monty was so pleased with the description that he later repeated it to news correspondents.

On the morning of 5 November, after breakfasting with von Thoma, Monty held a press conference. He wore a grey pullover that the correspondents had seen many times before, but with a new black tank beret, complete with his own rank badge and the badge of the Royal Tank regiment. The distant sound of exploding shells could be heard as he told the world he had beaten Rommel:

I drove two armoured wedges into the enemy and I passed three armoured Divisions through those places. They are now operating in the enemy's rear. Those portions of the enemy's armour that can get away are in full retreat. Those portions that are still facing our troops will be put 'in the bag' . . . After very hard fighting, the Eighth Army and the Allied air forces have gained a complete victory . . . But we must not think that the party is over. We must keep up the pressure. We intend to hit this chap for six out of North Africa.

Over the next few days two official photographs of Monty, the 'victor of El Alamein', were released to the press. The photograph taken on 4 November when von Thoma was first presented to him was intended to portray the victorious general gracefully receiving the beaten. In fact, von Thoma, who was markedly taller than Monty, retained a soldier-like demeanour that belied his defeat. It was Monty, in a long-sleeved pullover that clung to his scrawny frame and wearing no insignia of rank, who appeared out of place. When the photo appeared in the British papers he was criticized for dining with an enemy while his troops still fought and died. Churchill turned his wit against those who brought up the matter in the House of Commons: 'Poor von Thoma. I too have dined with Montgomery.'

Monty had written to tell David that 'I have now taken to wearing a black Tank Corps beret. The Australian hat was very good in the summer but I do not need it now the weather is cooler.' If he made the change of headgear seem a mere utilitarian choice, that was far from the truth. Adopting the beret had not even been Monty's own decision. It was a conscious and carefully stage-managed change, with the world's press as much as the men of Eighth Army in mind.

The two men at the heart of his public-relations effort were Captain Warwick Charlton, editor of *Eighth Army News* and *Crusader*, and Geoffrey Keating, head of the Army Film and Photographic Unit. Talking to John Poston, Monty's ADC, they said that the broad-brimmed Australian hat was most unsuited to the general's bird-like head and slender frame, and suggested replacing it with a beret. When Poston agreed, they looked for a suitable photo-opportunity to introduce it to the world's press. Monty posing in or beside his command caravan (where he actually worked) was rejected as pedestrian, and would compare unfavourably with photos of Rommel (complete with British sand goggles) in his command vehicle and directing the fight from the front line. It was decided to have Monty pose in the turret of a Grant tank, wearing his beret and binoculars as if commanding Eighth Army from there.

The 'new look' was introduced at the victory press conference. Charlton and Keating explained to Monty that if the beret first appeared at the time of his defeating Rommel, in the public mind it would henceforth signify victory. In modern parlance, they were creating the Monty 'brand' and associating it with the El Alamein victory. Monty understood but seemed concerned that the pressmen might miss the point, as one of those present, Denis Johnston, recorded: 'Monty received us in the morning sunshine wearing a tank beret with two cap badges. This piece of apparel seemed to be bulking in his mind, because his first words were: "Well, gentlemen, as you see I have got a new hat."' He spoke about his reasons for adopting this new headdress for several minutes before turning to the quite momentous matter at hand: his victory over Rommel. The picture of him in the turret of a Grant was used in newspapers around the world.

The route from El Alamein to Fuka was choked with Rommel's remaining tanks, artillery guns and trucks on the night of 5 November as he made his escape. The column stretched for miles and was bombed by the RAF as it moved. Speed was essential because Monty's armour was expected to be in hot pursuit:

It was a wild drive through the pitch-black night . . . Finally we halted in a small valley to wait for daylight. It was still a matter of doubt whether we would be able to get even the remnants of the Army away. Our fighting power was very low. Tanks and artillery had sustained such terrible losses that there was nothing left but remnants.

He was not exaggerating his plight: Deutsches Afrika Korps had only ten serviceable tanks left.

There are two supposed mysteries of Rommel's retreat. The first is how he found sufficient fuel to allow a long-distance withdrawal after declaring to Berlin and Rome that there was not enough to mount further counter-attacks along the El Alamein line. When he described his panzers as 'immobile on the battlefield', he was first deducting the amount required for a retreat to his first desert fuel dump, and in any case a retreat in a straight line along a hard-surfaced road used far less fuel than the back-and-forth movements over sand entailed in action up and down an extended front.

The second mystery is how, if Rommel's own figures are to be believed, he left El Alamein with only ten tanks intact, but at an early stage of the pursuit he had thirty. Tanks 'knocked out' on the battlefield ranged from those burned-out to those 'unserviceable' because, for example, of a damaged track. A number of lightly damaged tanks were recovered and taken with the Deutsches Afrika Korps as it fell back; these were repaired and became operational during the retreat.

The following morning the London *Daily Mirror* splashed across its front page 'ROMMEL ROUTED: HUNS FLEEING IN DISORDER'. Above a photograph of Monty, 'He Dished It Out', and above one of Rommel, 'He Couldn't Take It'. The paper said that 'Rommel's desert army is in full retreat with the Eighth Army in close pursuit of his disordered columns' and told 'the dramatic story of General Montgomery's smashing victory'.

For a moment it seemed as if the enemy column might be destroyed outright or forced to surrender, and Rommel killed or captured. In his plans for Lightfoot, Monty had described 10 Corps as a *corps de chasse*, suggesting that he would use it to pursue the fleeing army, but many of its units had been thrown against the line in the final days at El Alamein. Although de Guingand attempted to reassemble the scattered regiments and ready them for a pursuit,

Monty showed no interest in it or any sense of urgency and the 'moment of exploitation' passed. When General Briggs, commanding 1st Armoured Division, requested permission to prepare his tanks for a pursuit – which meant changing the weight ratio of fuel and ammunition taken aboard, long-distance mobile operations necessitating more of the former and therefore less of the latter – Monty refused.

Torrential rain began to fall on the afternoon of 6 November and continued through the night and into the next day. The desert became a quagmire and Monty claimed that this allowed Rommel's remnants of an army to get away: 'Only the rain on 6th and 7th November saved them.' He wrote to tell David that 'We had very heavy rain after I had driven the Germans back and that saved them from complete annihilation as many of my troops were bogged in the desert and could not move.'

Histories of the pursuit point out that the rain fell on Eighth Army and the Panzerarmee alike and caused them equal difficulties, and most condemn as absurd Monty's assertion that his forces were 'bogged down' while Rommel's forces moved further on. In fact, that is what happened. Rommel's main problem was the shortage of fuel. Therefore his tanks and trucks had to keep to the road in a single-file retreat; he had insufficient fuel to spread them across the desert. The road became impossibly jammed and every breakdown caused a further delay. The British, who had an abundant supply of fuel and could deploy their motorized forces across the sand, could make much faster progress and easily 'catch up'. The rain changed that. Monty had held an advantage that the weather took from him, forcing Eighth Army to keep to the road, too, and in this sense affecting its progress more than the Panzerarmee's.

Although Monty's much disparaged 'rain' alibi must therefore be allowed, the conclusion he draws from it – that this *alone* allowed Rommel to escape – cannot be the case. The crucial delay occurred before the rain came, and although it is conceivable that Eighth Army could have caught up (if deployed across the sand) that premise required not just the absence of rain but the presence of boldness in the army commander. Rommel certainly thought so:

The British commander showed himself to be over-cautious. He risked nothing in any way doubtful and bold solutions were completely foreign to him. So our motorized forces kept up an appearance of constant activity, to induce ever greater caution in the British and make them even slower. I was quite satisfied that Montgomery would never take the risk of following up boldly and overrunning us, as he could have done without any danger to himself.

On the morning of 8 November when General Gatehouse pleaded for permission to pursue the enemy he was (like General Briggs before him) refused. Monty did not quote the rain that by then had been falling for two days. He told Gatehouse there would be no 'mad rush' forward because such a pursuing force would be exposed to counter-attack by Rommel. It was General Montgomery, not the rain, that delayed the pursuit. He had predicted that El Alamein would be a battle of attrition and that he must win, and he had done so. He could not have known that Rommel had precisely ten tanks left, but the mathematics of attrition made it certain Deutsches Afrika Korps was incapable of seriously threatening a pursuing force.

Basil Liddell Hart had a score to settle with Montgomery, whose 1929 rewriting of the *Infantry Training Manual* had deliberately left out Liddell Hart's concept of the 'expanding torrent', but his assessment can hardly be doubted: 'The best opportunities had been forfeited before the rain by too much caution, by unwillingness to push on in the dark, and by concentrating too closely on the battle to keep in mind its decisive exploitation.'

In an entry in his diary written *before* the offensive, Monty admitted to himself, if to no other, that he had no intention of allowing any manoeuvre he could not plan in advance: 'The leadership is all right and the equipment is all right. The training is not all right and that is why we have got to be very careful.

Having made a successful break-in, we must not rush madly into the mobile battle with wide encircling movements and so on; the troops would all get lost.' He had decided from the outset that there would be no vigorous pursuit, although the possibility of the troops getting lost was perhaps secondary to the danger of Rommel beating him in the open desert. That he should still fear a Rommel who commanded only the remains of an army testifies to the genius of the German and the neurotic caution of his opponent.

Monty wrote to David: 'This battle is over and I have smashed Rommel and his army. It has been great fun and I have enjoyed it. I expect the papers in England are full of it and contain a great deal about me.'

They did, not only in England but around the world. The victory at El Alamein had been claimed by Montgomery, and a grateful British public was not going to argue with that. Churchill called him 'brilliant', and on 11 November Buckingham Palace announced that he had been awarded a knighthood.

It is probably more correct to say that Monty marshalled his resources brilliantly and managed the battle on the ground well enough not to lose it, while the RAF won it from the skies by inflicting such heavy damage on German forces and their supplies that in a battle of attrition the Panzerarmee had stood no chance. Certainly the primary lesson Rommel took from El Alamein was that victory would henceforth go to whichever side gained air supremacy and thus the ability to decimate enemy ground operations.

Monty was uncertain of his army – even after he had built up its matériel strength he felt that its training was insufficient for the task – and put down his caution to that. But in truth he was uncertain of himself and his ability to defeat Rommel in any kind of mobile engagement. He had no belief in his own generalship when applied to battle situations that could not be pre-planned and controlled from his command caravan. Fortunately at El Alamein there was no real scope for such ability and a desperate need for battle management. His caution would not allow the mobile engagements that the static front in any case forbade, while his supreme confidence in his own plans enabled him to press on when others faltered in a battle of attrition that involved sitting and absorbing losses in order to inflict similar losses on the enemy. The final verdict on Montgomery at El Alamein should be left to Rommel:

The British based their planning on the principle of exact calculation, which can only be followed where there is complete material superiority. They undertook no *operations* but relied solely on the effect of their artillery and air force. Their command showed its customary caution. They repeatedly allowed their armoured formations to attack separately, instead of throwing in the 900 or so tanks which they could have committed to gain a quick decision. Only half that number of tanks, acting under cover of their artillery and air force, would have sufficed to destroy my forces. These piecemeal tactics caused the British very high casualties.

In Berlin and Rome the defeat was blamed on the army and on Rommel's increasing 'defeatism'. It is hard to see how he could have done more with the resources at his disposal, or more than he did to communicate the significance of Allied superiority to the High Command and the Comando Supremo. He asserted both before and after the defeat that it was made inevitable by the failure of those in the twin capitals to deliver sufficient supplies to his army. He blamed Hitler and Mussolini, and to the officers of their respective High Commands who never visited the front to see for themselves he applied the German Army maxim, '*Weit vom Schuss gibt alte Krieger*' (Staying far from the battle makes for old soldiers). That was true enough. The Axis forces lost 32,500 men.

British casualties were 13,560. Almost before the sand had settled the Army Film and Photographic Unit began putting together a documentary to be called *Desert Victory*. It was Geoffrey Keating's idea and would be Monty's answer to Rommel's *Victory in the West*. The film used actual footage of the battle – four cameramen had been killed and seven wounded while filming – but some scenes had to be re-enacted and a cookhouse was set alight to provide the drifting smoke through which Australian troops could be seen advancing with fixed bayonets.

At 0200 on the morning of Sunday, 8 November, as the USS *Augusta* anchored off the Moroccan coast, General Patton went on deck to see the lights of Fedala and, fifteen miles to the south, Casablanca. The sea was unusually calm and he thanked the Almighty aloud for it. Fedala had to be taken first as a staging post for the advance on Casablanca.

Operation Torch began at 0530. Under cover of darkness troops filled a flotilla of landing craft and headed for the shore. There was no preliminary bombardment of enemy positions because it remained uncertain whether the French – nominally loyal to Henri Pétain, the Vichy French leader – would resist or welcome the Americans ashore. This matter was decided twenty minutes later when one of the French spotlights that continually swept the sea caught a landing craft in its beam and the shore batteries opened fire. The cruiser *Brooklyn* replied with all guns, putting the spotlights and the nearby batteries out of action, but now the French commander in Casablanca knew where the American fleet was and where the troops were landing. Action stations sounded on the French battleship *Jean Bart* and other warships and they left Casablanca at all speed.

By 0600 the first American troops were ashore and in the outskirts of Fedala. Two hours later the town was secure and Patton prepared to go ashore. At the same time the *Jean Bart* and two French destroyers opened fire on the transports and on the landing beach. Patton was about to disembark – a thirty-two-foot Higgins boat with his baggage aboard was already swinging precariously from the davits of the *Augusta*, waiting to be lowered into the sea – when the ship returned fire. The muzzle blast of its guns tore his boat from the davits, making matchwood of the Higgins and sending his baggage into the sea. More French ships appeared and opened fire on the *Augusta*. Patton was 'on the main deck just back of number two turret leaning on the rail when one hit so close it splashed water over me'. He was forced to remain aboard during a lengthy sea battle, watching from the bridge, and could not be put ashore in a landing craft until 1320.

He and his staff waded through the surf to find the beach cluttered with landing craft (which should have returned for more troops), supplies and equipment. His men were under fire from the French ships, casualties were mounting and many of the soldiers had dug foxholes in the sand. Despite Patton ordering them back to work and setting an example by refusing to take shelter as the shells came in, by nightfall only a fifth of the guns and transport had been landed. He stayed the night at the Hôtel Miramar in Fedala. It would not be possible to advance on Casablanca until everything had been brought ashore.

When he returned to the landing beach before dawn the next morning, the situation was just as bad: 'Boats were coming in and not being pushed off after unloading. There was shellfire, and French aviators were strafing the beach. Our men would take cover and delay unloading operations.' As he ran from unit to unit, giving orders, cursing and encouraging, he came across a GI crying on the beach: 'I kicked him in the ass and he jumped right up and went to work. I saw one Lieutenant let his men hesitate to jump into the water. I gave him hell. I hit another man who was too lazy to push a boat.'

By nightfall everything was ashore and on 10 November Patton's force reached Casablanca. The next day, 11 November, was Patton's birthday and he intended to make a gift of Casablanca to himself. The French troops defending the city outnumbered his own, and the Casablanca Division under General Desré was expected to put up a strong resistance, but Patton could call on naval gunfire and air support. For the young GIs there was the possibility of their first close combat and one of them told how 'Our officers wanted to give us a password we couldn't forget, so they told us to yell "George" at the first man we saw, and if he didn't yell "Patton" right back, to shoot hell out of him.'

The assault was scheduled to begin at 0730 but Patton was woken at 0200 by his intelligence officer to be told that Auguste Noguès, the French governor, and Admiral Michelier, in command of the Casablanca defences, were willing to negotiate. Patton sent word back that they had nothing to talk about: if he did not receive their unconditional surrender by 0730 a naval and air bombardment would be followed by a full ground assault. They surrendered at 0640 with only fifty minutes to go and when Patton had 'the bombers approaching their targets and the battleships in position to fire'.

His two ancillary landings had been successful and his task force had taken Casablanca with the loss of 337 American troops killed and 637 wounded. Although his bluff manner hardly suited a diplomat, in talks with the French and with the Sultan of Morocco he successfully 'delegated' command of the country to its previous French masters, thus achieving Allied control without disturbing the administrative status quo. That was undoubtedly the right thing to do in Morocco – when, after the war, he attempted much the same in a defeated Nazi Germany it was to have much graver consequences.

When Rommel's operations officer, Colonel Westphal, informed him that American troops had landed in Morocco and Algeria – a new enemy army in his rear – he knew that the Panzerarmee's retreat from El Alamein must be quickly followed by its withdrawal from North Africa. He suggested to the High Command a Dunkirk-style evacuation using aircraft, ships and even U-boats to save as many men as possible for the battles to come in Europe. Hitler replied that this was out of the question. Rommel told his interpreter, Lieutenant Armbruster, that 'I wish I was a newspaper seller in Berlin, then I could sleep at night.' Despite his motorized forces occasionally being brought to a standstill with empty fuel tanks until the next airlift arrived, he kept ahead of Eighth Army.

General Max von Pohl flew to North Africa on 15 November

to reinforce the message: 'We must hold Tunisia. Massive re-inforcements are already flowing into Tunis and Tripoli.' A new army under General Hans-Jürgen von Arnim was arriving but it would take time to build up its strength and Rommel should delay Montgomery by making a stand at El Agheila. Rommel saw an irony he decided not to share with von Pohl, that despite his earlier pleas, only now, at the moment when defeat had become inevitable, were extra forces pouring into North Africa. It was Rommel's birthday and von Pohl brought a gift from Lucie: a box of his favourite chocolate and almond macaroons.

Eighth Army was following the Panzerarmee along the coastal road. Monty took Benghazi on 20 November and approached El Agheila four days later to find the enemy holding there. He esti-mated that it would be at least two weeks before he could be ready to attack. He had to build up his supply of fuel and ammunition and because he was now 800 miles from his supply base at Alex-andria, his daily requirement of 1400 tons of fuel had not always been met.

Rommel had only thirty operational tanks and while making a show of holding the line at El Agheila he was already withdrawing his non-mechanized troops to Buerat. He counted on Monty's caution to allow him time to complete the move, and decided to fly to Berlin – without asking the permission he felt would be refused – and convince Hitler to allow an evacuation.

Patton regretted that his war in Morocco had ended so quickly and in any case it was not the French he wanted to fight. He lamented that during the taking of Casablanca 'personally I did not have an opportunity for engaging in close combat' and told war correspondents he wanted 'to meet Rommel in a tank and shoot it out with the son-of-a-bitch'. In the US, news of the war against Hitler had long been dominated by Montgomery and the British. Now the American press had its first American hero. Patton's scowling face was on the front pages. NBC called him 'the rootin', tootin', hip-shootin' commander of American Forces in Morocco' and said that he had 'enough dash and dynamite to make a Hollywood adventure hero look like a drugstore cowboy'.

But to Patton it felt as if the British still ran the war. On 17 November he flew to Gibraltar for a meeting with Eisenhower and accompanied him to lunch with the British governor, 'an old fart in shorts with skinny legs'. He was surprised to find that Eisenhower's chief of staff and all of his military advisers were British, and wrote to tell Beatrice that this American commander of the Allied force had even taken to using British words: 'He speaks of lunch as "tiffin", gas as "petrol" and anti-aircraft fire as "flak".'

Patton's Western Task Force had successfully completed its mission. General Lloyd Fredendall's Center Task Force (US II Corps) had taken Oran as planned. The Eastern Task Force – the British First Army commanded by General Kenneth Anderson and supported by American troops – had taken Algiers but made little progress in its intended move through Tunisia. Within a few days of the Torch landings the first German troops had arrived and by 10 December a second Panzerarmee was established in Tunisia, commanded by General von Arnim. Many of Patton's men were being transferred to Anderson and he became frustrated that he had no enemy to fight and that his army was being reduced to help the mainly British force in Tunisia: 'I'm afraid I will hold the bag while our troops are shipped to the British. I fear that the British have again pulled our leg. This was not the plan. I want to be Top Dog and only battle can give me that. The waiting is hard.'

Eisenhower asked him to visit the Tunisian front and find out why Allied tank losses were much higher than expected. He reached the First Army headquarters and from there drove out to units at the front. He arrived with 78th Division at 0830 to discover that the general in command was still in bed. Another British brigade commander was 'trembling all over. He told me this was due to fatigue. From the smell of his breath I could see that it was due to something else.' Patton gained the impression that one problem was poor British leadership. However, his report to

Eisenhower concluded that American tank losses were high because the Grant tank with light armour and a 37-mm gun was no match for the heavily armoured Panzer Mark IV with its 75-mm gun.

When Colonel von Luck spoke with Rommel on 20 November he noted that the field marshal's uniform was 'worn and dusty' and that the man himself appeared 'exhausted, as if the hard withdrawal actions, his deep disappointment and his illness had all left their mark on him'. He told von Luck that he was flying to Rastenburg to speak personally with the Führer because 'my word still counts for something'. He believed that Hitler was being misled by those around him and that his views could be changed by a clear presentation of the facts, particularly if they came from his (once) favourite general. There followed the first of a series of meetings that became a chief feature of Rommel's later military career, before each of which he expressed some hope that the Führer's mind could be changed and after each of which he was disappointed.

He flew to Rastenburg on 28 November, landing at 1520. It seemed to him that he was deliberately kept waiting, although Hitler might reasonably have been preoccupied with the Russian front where Sixth Army at Stalingrad had just been cut off by the enemy. At 1700 he was invited into the conference room to be met by Hitler's angry demand, 'How dare you leave your theatre of command without my permission?' Although the Führer quickly calmed down and listened intently to Rommel's report, the field marshal noted 'a distinct chill in the air'.

According to Captain Aldinger's account of the meeting Rommel 'laid the cards on the table'. He could hold on in North Africa if he was given all the resources he needed, but if they were not forthcoming he proposed a Dunkirk-style operation to save the men.

HITLER: How many men do you have?
ROMMEL: Sixty or seventy thousand.

HITLER: And how many did you have when the British offensive began?
ROMMEL: Eighty-two thousand.
HITLER: So you've hardly lost any.
ROMMEL: But we have lost nearly all our weapons, blown to pieces by the RAF. *Mein Führer*, my soldiers have hardly any ammunition or fuel . . . Africa cannot be held. We should transport as many men out as we can.
HITLER: I will not allow it. We must maintain a bridgehead in Africa at any cost. There is to be no talk of an evacuation.

When the field marshal asked whether it would be better to lose Tunisia or the whole of his Panzerarmee, Hitler lost his temper and yelled back that the army was of no concern to him. Rommel later told Lucie that he suddenly saw the Führer's contempt for the men who fought and died if they delivered anything less than total victory.

Perhaps Hitler regretted that their differences had come between them. After Rommel had left the room he went after him, put an arm around his shoulders and said, 'You must excuse me, I'm in a very nervous state. But everything is going to be all right. I will see that your army receives everything it needs.' Rommel no longer believed his promises.

While Rommel was at Rastenburg, Monty flew to Cairo to discuss his plans for the offensive at El Agheila with Alexander and spent what he called 'a very pleasant weekend' at the British Embassy. He revelled in his fame as the victor of El Alamein: 'My appearance at St George's Cathedral for the Sunday evening service, where I read the lesson, caused quite a stir. It is a strange experience to find oneself famous and it would be ridiculous to deny that it was rather fun.'

That weekend he also had his first sitting for war artist Neville Lewis. Keating photographed Lewis at work and prints were sent to the British and American press accompanied by a suggested caption: 'Desert Rat takes time off from chasing Desert Fox to sit for his portrait'. Sitting for a portrait was virtually *de rigueur* for victorious British generals; Monty was the first to have a photograph taken while he was in the act and cabled around the world. Always a showman, he had now become a master of PR.

Back in the desert, he felt that by 10 December Eighth Army was strong enough to attack the Panzerarmee. When the British artillery opened up that evening, always the precursor to a Montgomery offensive, Rommel withdrew the last of his armour and left the position to the enemy. Two days later he remarked in his diary that the British had occupied El Agheila 'and claimed to have taken 100 prisoners, and I haven't lost a single man!' Monty wrote to tell Brooke that 'We turned the enemy out of a very strong position by aggressive tactics . . . he got a very severe mauling.'

Rommel fell back towards Tripoli in an effort to join up with von Arnim. Monty was keeping pace with him but saw no need to force an unnecessary engagement with a Panzerarmee heading straight for the First Army. In Morocco Patton was frustrated at losing his troops to the fighting in Tunisia while he watched from the sidelines. But as Rommel began a final fight for survival in the desert, both Monty and Patton were about to discover that their North African war was far from over – and about to discover one another.

8. Fractures in the Anglo-American Alliance

On 14 January 1943 President Roosevelt and the British Prime Minister, Winston Churchill, flew into Casablanca for discussions, with their respective chiefs of staff, at the Anfa Hotel. Patton, as the area commander, was responsible for their security. Determined to impress, he greeted the President at the airfield and had several thousand troops line the roads for their drive to the hotel. He felt it something of an insult that Roosevelt's Secret Service bodyguards rode in the car with their weapons drawn 'to protect the Commander-in-Chief from being assassinated by his own troops'.

Patton was taking no chances either. He ringed the Anfa with a battalion of infantry and anti-aircraft guns, and prepared a private air-raid shelter for the two VIPs he codenamed A-1 (Roosevelt) and B-1 (Churchill). Alan Brooke travelled with the Prime Minister and met Patton for the first time: 'I had already heard of him, but must confess that his swashbuckling personality exceeded my expectation.'

The week-long conference was intended to plan future operations and decide how the war was to be won. The allies had different ideas of what should be done next. Marshall, for the US, believed that the Mediterranean was 'a blind alley' and wanted American forces in North Africa moved to southern England to prepare for a cross-Channel invasion. Brooke, for the British (and presenting Churchill's opinion), argued that once they had completed the job in North Africa, Anglo-American invasions of Sicily and Italy should attack Hitler's 'soft underbelly'. There was, of course, a third partner. Stalin, although he did not attend the conference, indicated that a second front must be opened as soon as possible by an Anglo-American invasion of France, to relieve the Red Army by forcing Hitler to transfer forces from the eastern front to the west.

The French, while not officially represented, had observers there and they, too, preferred an early invasion of occupied France. Only the British argued for operations in the Mediterranean area first. Patton noted in his diary that when Roosevelt was speaking privately with a French general and admiral, 'B-1 came in without being asked and hung around, started to leave, and then came back. The whole thing was so patent a fear on the part of the British to have the French and Americans alone together, that it was laughable.'

The conference concluded with what Churchill called a compromise but which felt to most Americans like a British victory. There would eventually be a cross-Channel invasion but it would be preceded by the invasion of Sicily and Italy. In Washington the secretary of war, Henry Stimson, suspected that this decision served the best interests of the British Empire, which had extensive interests in the Mediterranean, rather than the best strategic policy for the war: 'The British are getting away with their own theories and the President must be yielding to their views as against those of our own Chief of Staff.'

General Marshall, US Army chief of staff, had certainly been outmanoeuvred by Churchill, and perhaps the frustration he dared not voice was expressed when the Prime Minister ran out of cigars and the American restocked his supply with the best he could find at short notice: five-cent 'White Owls'. Churchill lit one, took two puffs and put it out.

Eighth Army had advanced 1200 miles from El Alamein in twelve weeks. As it approached Tripoli Rommel decided the city was not worth fighting for and pulled out. He received 'a grave reproach from Rome because we do not hold out longer'. Mussolini accused him of continually withdrawing under pressure from Montgomery, but Rommel believed his one chance was to join forces with von Arnim and *then* turn with the combined strength of their two Panzerarmees to fight both the British and the Americans.

Advance units of Eighth Army entered Tripoli at 0400 on 23 January and Monty joined them later that morning:

I drove into the city with Leese and we sat in the sun on the sea front and ate our sandwich lunch. Our ADCs and police escort sat not far away, also having lunch. I asked Leese what he thought they were talking about after three months of life in the desert; he reckoned they were speculating on whether there were any suitable ladies in the city. I decided to get the Army away from Tripoli as early as possible.

Before he could do that, Monty had to put on a show for Churchill, who flew in to take the salute at a victory parade of troops and tanks through the city's main square.

The Allies had imagined crushing Rommel in Tunisia between Torch and Eighth Army, but now von Arnim's Fifth Panzer Army held Tunis, and Rommel was moving to join up with him. Roosevelt wanted to give American troops an easy blooding by pitting them against, first, the Vichy French and then the survivors of an army fleeing from El Alamein. Hitler had decided unexpectedly to make a real fight of it, not because he wished to retain a hold in North Africa as such but because he feared that Tunisia under Allied control might be used as a base from which to invade southern Europe. That possibility was of particular concern to his Axis partner Mussolini.

Churchill and Roosevelt had agreed that Torch, as a predominantly American force, should have an American commander. Eisenhower was based in Algiers 400 miles from the battlefield and as his most pressing responsibilities were political, on 24 January he delegated all operations on the Tunisian front to General Anderson commanding the British First Army. At the same time he placed Fredendall's US II Corps of 32,000 men under Anderson's command.

General Everett Hughes, Eisenhower's deputy, thought this was a mistake and wrote in his diary that 'It will soon be a British war and Montgomery will be in command.' Less discreetly, Patton told everyone who would listen that he was 'shocked and distressed' at American troops being placed under British command, and word reached Ike. He called Patton to his headquarters on 3 February and told him: 'George, you are my oldest friend, but if you or anyone criticizes the British, by God I will send him home.'

While Patton was threatened with being sent home, Rommel had been *ordered* home. He received a signal from the Comando Supremo on 26 January informing him that, due to his poor health, he was relieved of his command effective at a time of his own choosing. He was undeniably ill. He admitted to Lucie that he was having recurrent severe headaches and was taking sleeping draughts. But he knew it was not his health so much as his continuing retreat that worried Mussolini. He ignored the order: he was relieved at a time of his choosing and for the moment he would retain command, and continued to do so when his replacement, General Giovanni Messe, arrived. That he should be dismissed was bad enough; to be replaced by an Italian was intolerable.

He reached the former French defensive line at Mareth – which stretched for thirty miles from the sea to the Djebel Matmata escarpment and was virtually a second El Alamein – and here he *could* make a stand. He was also near enough to Tunis to plan joint operations with the Fifth Panzer Army. He calculated that Monty would not have his whole army at Mareth until the beginning of March and, in any case, would not make an immediate offensive on such a strong position: 'Montgomery had an absolute mania for bringing up adequate supplies behind his back and risking as little as possible.' While Monty built up his strength Rommel was free to attack the inexperienced US II Corps in his rear. He would leave 15th Panzer Division to make a show of strength along the Mareth line while he led out 21st Panzer Division, rearmed with von Arnim's newly arrived tanks.

★

Monty made use of the delay too – he estimated that he would not have everything in place for an attack on the Mareth line until the middle of March – by holding a study week in Tripoli, beginning on 14 February. He sent invitations to all senior Allied officers. Patton was the only American general to accept and Monty interpreted that as a snub. Patton took no satisfaction from knowing that it was, in fact, because he was the only American general in North Africa with no fighting to do. He flew to Tripoli in a B-17 bomber to make his own assessment of 'the victor of El Alamein'.

Monty began his first address with the usual requirement that while he spoke there was to be no smoking and no coughing, then spoke for two hours on the art of modern warfare. It was a long time for the chain-smoking Patton to abstain and General Leese noted that he chewed gum and yawned throughout the lecture. Nevertheless Patton was impressed by what Monty had to say. In his diary he described him as 'small, very alert, wonderfully con-ceited, and the best soldier I have met in this war. General Briggs says he is the best soldier and the most disagreeable man he knows.'

In his second address the following day Monty described his victory at El Alamein and spared himself no praise. General Brian Horrocks found it most impressive:

When it was over I found myself walking back with Patton to his billet, so I seized the opportunity to ask what he thought of it. With a twinkle in his eyes, he answered in his Southern drawl, 'I may be old, I may be slow, I may be stupid, and I know I'm deaf, but it just don't mean a thing to me!' From that moment he developed an almost pathological dislike of Montgomery, who, I must admit, did little to heal the breach.

In a letter to Brooke dated that same day Monty had been so little enamoured of the American that he could not recall his name: 'Only one American General has come, the Commander of an Armoured Corps, an old man of about sixty.'

*

On 14 February while Monty lectured and Patton listened, Rommel attacked the US II Corps positions at Sidi Bou Zid and the Kasserine Pass. Fredendall's forces were badly deployed and easily overwhelmed. The panzers destroyed 112 out of 120 American tanks and inflicted 6000 casualties. The American 37-mm anti-tank gun hardly caused a dent in the panzers' armour and Fredendall commented wryly that 'The only way to hurt a Kraut with a 37-mm is to catch him and give him an enema with it.'

The Americans were shattered. Eisenhower's naval adviser wrote in his diary that 'Proud and cocky Americans today stand humiliated by one of the greatest defeats in our history. This is particularly embarrassing to us with the British.' An Enigma transcript revealed that the Germans, too, having engaged the Americans for the first time, disparaged the fighting skills of their new enemy.

It is unlikely that Rommel expected to make any permanent gains. His medical adviser was urging him to return to Germany and he had already been officially 'dismissed'. But for a short time von Arnim and his political masters appeared willing to let him use the newly arrived panzers, and it was in Hitler's interest that Rommel departed North Africa having won a final victory (however vicarious) and with his propaganda value intact. Having occupied a large area of the Kasserine valley, he disappeared back through the pass during the night of 22 February. He had made his point against the Americans and now he needed to regroup his forces and return to his defensive position along the Mareth line before the British arrived there, as he intended to leave them with a painful reminder of the Wüstenfuchs too.

Monty explained the German withdrawal from Kasserine quite differently, taking a grain of truth and making of it a bushel: 'In response to a very real cry for help on the 20th February to relieve the pressure on the Americans, I speeded up events and by 26th February it was clear that our pressure had caused Rommel to break off his attack against the Americans.'

Eisenhower sacked Fredendall and gave command of the

demoralized II Corps to Patton. It was everything Patton had hoped for – the chance to confront and beat Rommel – although II Corps was under British control. He declared openly that 'those mealy-mouthed Limeys' would not push him around as they had Fredendall, and wrote in his diary on the evening of 4 March: 'Well, it is taking over rather a mess, but I will make a go of it. I think I will have more trouble with the British than with the Boches.' Ike briefed him the following day and stressed again that criticism of the British must stop. Patton was unrepentant: 'He has sold his soul to the devil on Cooperation, which I think means we are pulling the chestnuts for our noble allies.'

Patton took command of II Corps on 6 March, with General Bradley as deputy corps commander, and was aghast at what he found: 'No salutes. Any sort of clothes and general hell.' Morale had hit bottom after the defeat at Kasserine. The Corps was to assist Montgomery in an offensive on the Mareth line to be launched on 19 March. That gave him insufficient time for retraining so Patton 'did a Monty', presenting himself as a larger-than-life figure to the men in order to stamp his personality on the Corps. He drove around in a command car complete with wailing siren and an armed escort. Any man who failed to salute an officer, and anyone who was found with his shirt sleeves rolled up, incurred a heavy fine. Nor did the officers escape his attention. When he visited the 1st Division headquarters and found it surrounded by slit trenches for staff officers to 'retire to' when under fire, he asked the most senior officer present to indicate which trench he would use, then walked over and urinated in it. His approach was crude; his message was unmistakable.

When he spoke to the men he used barracks language, although Bradley felt this was not naturally his own and was put on for effect: 'Whenever he addressed men he lapsed into violent, obscene language. I was shocked. He liked to be spectacular; he wanted men to talk about him. Yet when he was hosting at the dinner table, his conversation was erudite and he was well-read, intellectual and cultured. Patton was two persons: a Jekyll and Hyde.' That was the point: Patton's men cursed him but they knew in battle it would be 'Hyde', not 'Jekyll', who was leading them, and they wanted it no other way.

On the same day that Patton took command of II Corps, Rommel attacked Eighth Army at Medenine with his rearmed panzer divisions. Monty had planned an assault on the Mareth line similar to his attack at El Alamein, and he defended Rommel's pre-emptive attack at Medenine as he had at Alam al-Halfa. He fought a defensive battle 'the same way as I had at Halfa . . . I refused to move to counter any of his thrusts. I refused to follow up when Rommel withdrew.' This static defence worked perfectly and de Guingand described the effect of the concentrated use of artillery as 'devastating – the enemy completely failed even to penetrate our positions'. Rommel attacked with 140 tanks and left with fifty-two. Monty avoided any mobile engagement and did not lose a single tank.

Rommel had learned nothing from Alam al-Halfa and was surprised that Montgomery, the meticulous battle-manager, was ready for him at Medenine, but he realized the consequences of his defeat: 'The attack broke down in the break-in stage and there was no opportunity for fluid action. Montgomery had grouped his forces well and made all his preparations. The battle was lost . . . A great gloom settled over us. The Eighth Army's attack was now imminent. For the Panzerarmee to remain in Africa was plain suicide.'

When he took off from Sfax airfield at 0750 on 9 March, Rommel was never to return to North Africa, although at the time he intended to do so and there is no evidence to support Eisenhower's later assertion that 'he escaped, foreseeing the inevitable and desiring to save his own skin'. Leaving von Arnim in command, he flew first to Rome to talk to Mussolini, then on to confront Hitler. He hoped that a final, personal appeal to the two leaders would convince them the Allied armies could not be beaten in Tunisia and it was necessary to save German troops for use in the fight soon to come in Europe.

In Rome Mussolini accused him of painting too black a picture and left Rommel 'heartily sick of all this everlasting false optimism'. For his part the Italian was disgusted by Rommel's 'defeatist attitude' and at the last minute decided not to present him, as previously planned, with the Medaglia d'Oro al Valor Militare (Gold Medal for Military Valour). Rommel later told Lucie: 'It was a beautiful thing, made of porcelain and gold. A pity I could not have kept my mouth shut a little longer.'

Göring was in Rome too and Rommel 'hitched a lift' in his special train to the Wehrwolf, Hitler's forward HQ in the Ukraine, where he did pick up 'a beautiful thing': the Führer presented him with the *Brillanten* (diamonds) to his Knight's Cross, making him only the sixth to receive the award. But there was the same discrepancy of viewpoint when he pleaded with Hitler to be allowed to organize an immediate evacuation, leaving the equipment but saving the men:

I was invited to take tea with Hitler and so was able to talk to him in private. He appeared upset about the disaster in Stalingrad, but insisted our prospects there were not all dark. He was not receptive to my arguments about Tunisia and seemed to dismiss them by saying I had become a pessimist. He ordered me to take sick leave and get well so that I could command a counter-attack on Casablanca. He simply could not see what was happening in Tunisia.

Rommel returned to Lucie at Wiener Neustadt and they went to stay at the Semmering mountain resort. Hitler ordered his absence from North Africa to be kept secret so that 'the Rommel effect' might continue to work against the Allies.

The field marshal had received his *Brillanten*, and on 12 March Patton was promoted to lieutenant general – his two silver stars became three. Ambition will out and that same night he wrote in his diary: 'Now I want, and will get, four stars.' Better than that, he would earn the fourth against a German army still commanded (he believed) by Rommel.

Monty had discussed his plans for the Mareth offensive with Alexander, who had overall command, during his visit to Cairo, and accordingly on 17 March Alexander ordered Patton's II Corps to make a diversionary advance towards Gafsa to engage Italian forces that might otherwise be used to reinforce Rommel's defensive line. Patton did not like the idea of a supporting role on the flank of Monty's 'big push' but Alexander specified that he was 'not to pass large forces beyond the line Gafsa–Fondouk'. The restriction was not unreasonable: if he did so and Eighth Army broke through at Mareth, he would be crossing Monty's line of advance.

Patton told his commanders they must be 'victorious or dead' and advanced on the hills surrounding Gafsa. By 22 March the Italian troops had withdrawn after only a token defence and he saw that a rapid dash forward (although his orders forbade it) would take him behind the Afrika Korps and enable him to cut the German supply line. Von Arnim saw the danger too and sent 10th Panzer Division to block him. It was not Alexander's order that prevented the advance but von Arnim's panzers.

Patton deployed his artillery and tank-destroyers in the hills and as the panzers arrived they were caught on the open plain east of El Guettar. After a fierce, day-long battle the Germans pulled back with the loss of thirty-two tanks. The US Army had its first North African victory over German forces and Kasserine was avenged. Patton believed mistakenly that in this engagement he had beaten Rommel.

Meanwhile Monty had broken through the Mareth line where the Italian General Messe was in command. In his message to Eighth Army Monty had said they would destroy the Panzerarmee, then advance through Sfax to Tunis and drive all German forces out of North Africa. After an artillery barrage from 300 guns (firing 36,000 high-explosive shells), supported by 600 bombers, a frontal assault was made by tanks of 50th Royal Tank regiment. A counter-attack by 15th Panzer Division repulsed the British with heavy losses on both sides. As at El Alamein, Monty's original plan had failed and he next ordered an outflanking movement (which some of his commanders believed should have been done from the outset), supported by another 'blitz attack' from fighter-bombers. The 15th and 21st Panzer Divisions needed help but the 10th could not be recalled from El Guettar, it being essential to prevent the Americans breaking through and getting behind the panzers. On 24 March von Arnim ordered Messe to withdraw from the Mareth line to Sfax and, if the enemy pursued forcefully, on to Tunis.

Monty had broken through but, as at El Alamein, he refused to rush after the fleeing enemy. He took to his bed with tonsillitis and, anyway, as he told Alexander, 'It was essential to regroup the two armies, First and Eighth, so that the attack on Tunis could be made with the maximum strength.' The retreating panzer divisions were able to get away without further harassment. Both Monty and Patton believed it was Rommel who had once again slipped from their grasp. (Even after it became known that the field marshal had left North Africa, de Guingand continued to refer to the enemy as 'Rommel's troops'.)

The Americans were wrongly blamed for allowing 'Rommel' to escape total destruction. This severe criticism came not from the British – Monty acknowledged that they had contributed to his success by keeping 10th Panzer Division away from Mareth – but from the American press. Expectations had been raised by the victory at El Guettar and further puffed up by Patton's personaliz-ation of his feud with Rommel, and correspondents who eagerly anticipated the biggest story of the war so far – Patton's destruction of the Deutsches Afrika Korps – turned on him for not delivering it. Even Roosevelt told Eisenhower he was disappointed that Patton's II Corps had not broken through at El Guettar as Mont-gomery had at Mareth. Marshall in turn attacked the American press for 'stories that attributed almost exclusively to American units the blame for not securing a more decisive victory over Rommel and cutting him off during his retreat'. Captain Butcher, Eisenhower's ADC, wrote in his diary that 'The public was not aware that Patton had a limited objective probably dictated by Montgomery's desire to have a clear path for the Eighth Army.'

Patton felt that he had been deliberately stopped to allow Monty through: 'Having spent thousands of casualties making a breakthrough, we are not allowed to exploit it. The excuse is that we might interfere with the Eighth Army. One can only conclude that when the Eighth Army is going well, we are to halt so as not to take any glory. And Ike falls for it! Oh for a Pershing. *Sic Transit Gloria Mundi*.'

His frustration had some foundation in fact. Alexander rightly forbade his advance because when Monty broke through, Patton would be crossing Monty's line of advance as he pursued the enemy – but Monty had not immediately pursued the enemy. However, his protest was disingenuous. He had every intention of disobeying those orders and it is likely that, although he bloodied 10th Panzer Division at El Guettar, he could not break through them to continue his advance. The panzers later withdrew as part of the general withdrawal from the Mareth line, but not because of pressure from Patton.

Monty's success at Mareth hit the British headlines just as the documentary film of his victory at El Alamein, *Desert Victory*, was released. Audiences were enthralled by the battle footage that even included the dead and wounded, although those shown were mostly German. It was an immediate box-office success and its production cost of six thousand pounds was taken in the first three weeks. It was just as popular in the US (and its director David McDonald would eventually win an Oscar). Churchill sent Roosevelt a personal copy and the President replied that 'Everybody in town is talking about it.'

They were also talking about Patton. He counted a total of forty-nine correspondents and photographers following him around, and their stories fed headlines across the US. He was the only high-profile American hero they had and criticism of his failure to 'destroy Rommel' quickly faded. On 12 April he was on the front page of *Time* magazine and a three-page feature entitled 'Fight Against the Champ' told how Patton said he looked forward to meeting Rommel (the Champ) on the battlefield: 'It would be like a combat between knights in the old days. The two Armies could watch. I would shoot at him. He would shoot at me. If I killed him, I'd be the champ. America would win the war. If he killed me . . . well . . . but he wouldn't.'

The War Department produced large posters and displayed them in defence factories across the country:

OLD BLOOD AND GUTS ATTACKS ROMMEL!

'Go Forward'

'Always Go Forward'

'Go until the last shot is fired and the last drop of gas is gone and then go forward on foot'

Lieut. General George S. Patton, Jr.

DARE WE WORKING HERE DO LESS?

The British were not so effusive in praising Patton or any of their American allies. It was known that Rommel placed little value on the fighting skills of his Italian troops, and British officers began calling the Americans 'our Italians'. After the American defeat at Kasserine, senior British officers had exchanged knowing nods. Alexander wrote to tell Brooke: 'They simply do not know their jobs as soldiers and this is the case from the highest to the lowest, from general to the private soldier.' He felt that they were 'ill-trained and rather at a loss', and 'not professional soldiers as we understand the term'. Monty chipped in: 'It was lack of proper training allied to no experience of war, and linked with too high a standard of living.'

Alexander now established training schools for both First and Eighth Armies but run solely by British officers. He sent British liaison officers to 'advise' senior American officers. The former were veterans of the long British campaign in North Africa and the idea made sense, but it was insensitive in the extreme. Patton's first words to his liaison officer, Brigadier Holmes, were 'I think you ought to know that I don't like Brits.' For his part, Holmes was 'sneering and supercilious'. Patton would later cross swords with Monty in Sicily and in Normandy, and this has been presented as largely a personality clash, but its true significance is revealed when seen within the wider context of a growing Anglo-American mutual 'dislike' that had its roots in North Africa in early 1943.

Patton's first significant skirmish with the British was triggered by one part of his situation report of 1 April: 'Forward troops have been continuously bombed all morning. Total lack of air cover for our units has allowed the *Luftwaffe* to operate almost at will.' Air cover was a British responsibility in that it was controlled by Air Marshal Coningham. The Empire struck back: Australian-born Coningham suggested that Patton was using the air force 'as an alibi for lack of success on the ground' and continued, 'It can only be assumed that II Corps personnel concerned are not battleworthy.'

The British had said as much among themselves but now it had been said to an American, and Patton was a bad choice. He was furious and a meeting between the two men produced only an exchange of insults. Air Chief Marshal Tedder wrote that 'There was grave danger of very serious political repercussions . . . this was dynamite and could have led to a major crisis in Anglo-American relations.' Tedder visited Patton at Gafsa to find him 'wearing his fiercest scowl', and apologized that Coningham had 'gone beyond the facts'. At that precise moment three Focke-Wulf 190 fighter-bombers strafed Patton's headquarters and a bomb exploded just outside; part of the ceiling fell in. Tedder got the point – at this time and place the *Luftwaffe* was most certainly operating at will – and acknowledged as much with an ironic smile as he brushed the plaster from his shoulders. Patton quipped that 'If

I could find the sons-of-bitches who flew those planes I'd mail them each a medal.'

Under pressure from Tedder, Coningham cabled Patton to express his regret that his message might have been 'interpreted as a slight to American forces'. He claimed that a garbled transmission had changed his original text, which 'assumed that a *few* corps personnel concerned are not battleworthy' to make it read 'that II Corps personnel concerned are not battleworthy'. No one believed that, but the fists were lowered.

The second skirmish was triggered by Monty on 10 April when Eighth Army, following Messe's retreating troops, entered Sfax. Some time earlier, during a meeting at which General Bedell Smith represented Eisenhower, Monty had said that he would take Sfax by 15 April and asked if Bedell Smith wished to make a bet of it. The American, perhaps not realizing that Monty was in earnest, said that if he got there by that date Eisenhower would give him anything he wanted. Monty replied immediately that he wanted an American plane: 'A Flying Fortress complete with American crew to remain on the US payroll, my personal property until the war ends.' On 10 April having taken Sfax five days earlier than the bet specified, he immediately sent a signal: 'Most Immediate. Personal from Montgomery to Eisenhower. Have arrived Sfax. Recall our bet. Despatch Flying Fortress.'

Ike was not happy but he felt that the bet had to be honoured and a B-17 Flying Fortress with its American crew was handed over. Later Monty used it to fly to and from Cairo while planning the invasion of Sicily (codenamed Operation Husky), and to fly to and from London. The affair caused some ill-feeling on the American side – Brooke admonished Monty for his insensitivity – and was a particular irritation for Patton: Monty was flying around in an American plane while Patton was allowed no plane at all for his personal use.

Patton's third skirmish with the British came on 11 April. Alexander announced that the role of II Corps in the final Allied offensive on Tunis was to support the main effort with British forces by protecting the flank. He also said in passing that one of his corps commanders, General Crocker, had reported that US 34th Infantry Division, while under British command, had shown itself to be 'no good'. This secondary role had Patton protesting to Alexander that his American troops should have an equal role with the British. The comment about 34th Infantry riled him even more and that night he wrote in his diary: 'God damn all British and all so-called Americans who have their legs pulled by them. I bet that Ike does nothing about it. I would rather be commanded by an Arab. I think less than nothing of Arabs.' When on the following day there was no American response to the insult, he blamed Eisenhower for allowing a British general to openly belittle an American division: 'Ike is more British than the British and is putty in their hands. Oh, God, for John J. Pershing.' He confronted Eisenhower about it but felt that he got nowhere: 'That son-of-a-bitch Crocker publicly called our troops cowards. Ike says that since they were serving in his Corps that was OK. I told him that had I so spoken of the British under me, my head would have come off. He agreed, but does nothing. He has an argument, probably provided by the British, for everything.'

To be fair to Eisenhower, he understood his main task as maintaining peace between the Allies. He was good at it too, a fact recognized by Montgomery, though from him it was a backhanded compliment: 'Eisenhower knows practically nothing about how to make war, and definitely nothing about how to fight battles. He is quite good at the political stuff.' And Patton's assertion that Eisenhower 'does nothing' was mistaken. According to Butcher, he went to see Alexander and pointed out that the British war effort was heavily dependent on American tanks, shells and equipment, and that if the American public believed its troops were being insulted, then they might well demand that the American war effort be turned against Japan and leave Great Britain to deal with Hitler as it felt able. Alexander promised to have a word with Crocker. There was, however, no public apology from either man.

Ike may have had 'the political stuff' in mind when on 15 April he ordered Patton to hand over command of II Corps to Bradley and return to Casablanca. Patton was to command US Seventh Army in Operation Husky and it made sense that he should be removed from the present fighting to prepare for Sicily. But Monty, who was to command the British Eighth Army in Husky, was for the time being left in place so that he could lead the assault on Tunis.

That evening Patton criticized Eisenhower in public and in his diary, and surprisingly it was the latter that gained the wider circulation. The diary page for 15 November, in which he called Ike 'an ass' for telling Alexander that 'he did not consider himself an American but an ally', was stolen, sold to a war correspondent, and appeared in papers across the US.

Removing Patton eased Anglo-American tensions. It also allowed Bradley, for whom Ike had high hopes, to be tested in command of II Corps. But it took Patton out of the field just as the final victory over 'Rommel' was about to be won. The laurels would go to Montgomery. Patton's disappointment led him to see a plot where there was none, although his discontent with what appeared to be Eisenhower's pro-British bias was shared by several of his peers. On 16 April he wrote: 'Bradley, Everett Hughes, General Rooks and I, and many more, feel that America is being sold. I have been loyal to Ike and have taken things from the British that I would never have taken from an American. If this trickery to America emanates from Ike, it is terrible.'

Just as disappointing to Patton was the news that during his recent encounters with the enemy Rommel had not been in command. Bradley noticed that 'He was possessed of the idea that he, George Patton, was here to do battle with Rommel.'

Montgomery told in his *Memoirs* of how he planned the final capture of Tunis. Seeing that US First Army was better positioned to break in than Eighth Army he sent some of his own units to strengthen them, and 'the first troops to enter Tunis were those of our own 7th Armoured Division'. Alexander wrote simply in his *Memoirs* that: 'Of course, Montgomery had nothing to do with the attack on Tunis.' *He* had decided on the transfer of British troops to First Army. Montgomery was still suffering from tonsillitis and was in bed when Alexander visited to inform him of what was being done. First Army attacked at 0300 on 6 May.

In Semmering Rommel had kept up his campaign for an evacuation by writing to Hitler and the High Command. They took Kesselring's advice that Tunis could be held. Rommel knew better because von Arnim kept him informed: the army had only seventy tanks left and so little fuel that an attempt was being made to distil more from Tunisian wine. Then, unexpectedly, on 8 May Hitler called Rommel to his Rastenburg Wolfsschanze and had the grace to say, 'I should have listened to you. Africa is now lost.' He had that day ordered all German and Italian troops to be evacuated by sea.

It was too late. At 1200 on 12 May von Arnim sent his final signal: 'All ammunition gone. Arms destroyed. In accordance with orders the Afrika Korps has fought until it can fight no more. The DAK will rise again. Heia Safari!' The Allies took 100,000 German and 148,000 Italian prisoners. It was for Rommel 'terrible to know that all my men had found their way into Anglo-American prison camps'.

On 17 May Monty flew to London in the Flying Fortress accompanied by his two ADCs and Geoffrey Keating, his photographer. He was welcomed as the conquering hero. As his plane approached RAF Northolt it was met by two squadrons of Spitfires. He stepped out of the plane to cheers from the airmen assembled to greet him, and Keating took photographs. One observer described him as wearing 'his battle dress with a sweater sticking out a foot below his jacket, and his usual beret bearing the Tank Corps as well as the General's badge'. Monty was driven straight to his rooms at Claridge's, booked in the name of 'Colonel Lennox' to ensure that the Germans had no foreknowledge of his arrival.

The next day he was invited to tea at Buckingham Palace and wore his desert dress (having checked with palace officials that this would be acceptable to the King). That evening he went to the theatre to see *Arsenic and Old Lace*, and when at the end the audience stood and cheered they had their backs to the stage and their eyes on Montgomery in his box. The adulation continued on the street as he left the theatre; he wrote that he 'got completely mobbed by the crowd'. By now the true identity of 'Colonel Lennox' was well known and he was met by more cheers and applause outside Claridge's.

De Guingand had noticed that while in Africa Monty 'became more dictatorial and more uncompromising as time went on', and felt that this visit convinced him he could get away with it: 'He realized that he was a power in the land and that there were few who would not heed his advice. He realized he could afford to be really tough to get his own way.'

On 20 May Eisenhower and Alexander took the review at the Allied Victory Parade in Tunis and alongside them stood a number of senior British officers. Bradley and Patton had to watch from a stand they shared with French officers and civilians. Patton was outraged and called it 'a God-damned waste of time'. Bradley thought much the same: 'It seemed to give the British overwhelming credit for the victory in Tunisia. For Patton and me the affair merely served to reinforce our belief that Ike was now so pro-British that he was blind to the slight he had paid to us and, by extension, the American troops who had fought and died in Tunisia.' According to Everett Hughes, Patton said that 'He told Ike that some day soon a reporter would send home as fact the story of American capitulation to British terms.'

On 2 June Monty travelled to Algiers to prepare for the invasion of Sicily. His plane flew low to avoid enemy fighters and it seemed to Air Marshal Broadhurst that they skimmed the Atlantic waves 'at nought feet' and Monty 'sat there reading *The Times* newspaper as if nothing was happening. I was absolutely scared stiff.'

General Brooke had flown out, too, and took this opportunity to have a 'quiet word' with Monty, who was pressing hard to be appointed commander of both armies for the invasion of Sicily. He stressed that the operation was to be executed by Monty and Patton commanding the British and American armies respectively and as equals. Brooke was concerned that Monty's high-handed approach in dealing with the Americans in North Africa – now visibly represented wherever he went by the Flying Fortress – had earned him considerable ill-will. He told Monty what he called 'a few home truths . . . to bring home to you the importance of your relations with allies . . .' He wrote in his diary:

Montgomery requires a lot of educating to make him see the war as a whole outside the Eighth Army orbit. A brilliant commander and trainer of men, but liable to commit errors, due to lack of tact, lack of appreciation of other people's outlook. It is most distressing that the Americans do not like him, and it will be a difficult matter to have him fighting in close proximity to them. He wants guiding and watching continually and Alex is not sufficiently strong and rough with him.

Alexander was at least aware of the growing tension in the Anglo-American alliance and said as much to Patton:

Alex said that it was foolish to consider British and Americans as one people, as we are each foreigners to the other. I told him my boisterous method of command would not work with the British, while his cold method would never work with Americans. He agreed. I found out that he has an exceptionally small head. That may explain things.

Deep beneath the political and military foundations of the Anglo-American alliance, moving almost imperceptibly but with massive force, the tectonic plate of one (declining) empire rubbed against that of another (emerging) power. This friction, felt throughout the officer class, found its most volatile expression whenever Monty and Patton were thrown together, and in Sicily it would register for the first time on the operational equivalent of the Richter scale. The war between the Allies was about to become very personal indeed.

9. Patton's Race to Messina

The final plan for the Allied invasion of Sicily was agreed between Montgomery and General Bedell Smith, Eisenhower's chief of staff, in an Algerian lavatory on 2 May 1943. They then went to consult Eisenhower, who liked it, and Operation Husky was approved. Patton was not consulted.

The original plan had been drawn up by British chiefs of staff in London. The strategic objective was not merely to occupy Sicily but to destroy the German and Italian forces holding the island. It was anticipated that as the Allies advanced enemy troops would fall back on Messina in the north-east corner of the island and only three miles across the strait from mainland Italy. The aim was to take Messina before those troops could be ferried to safety. Monty's British Army would land on the south-east coast and take the port of Syracuse, while Patton's US Army landed on the north-west coast and took the port of Palermo. The British would then move north and the Americans east to take Messina.

Monty called the intended operation 'a dog's breakfast' (the term he applied to every plan prior to his own involvement) and advised that 'It has no hope of success and should be completely recast.' Eisenhower called a conference at his headquarters in Algiers to consider Monty's objections. Alexander's plane was delayed by low cloud; Patton's plane was grounded by heavy rain. Monty had no time to wait for either of them:

I went to look for Bedell Smith. He was not in his office and I ran him to ground in the lavatory. So we discussed the problem then and there . . . I said the American landings near Palermo should be cancelled and the whole American effort put on the south coast, with the object of securing the airfields . . . The Eighth Army and the Seventh US Army would land side by side. We then left the lavatory and went off to consult Eisenhower, who liked the plan.

Monty argued that as German forces could be expected energetically to oppose the Allied landings, these should be concentrated rather than divided. Once a bridgehead had been established in the south-east corner of the island the two armies could separate to advance on Messina by different routes. Monty would thrust immediately north to take the ports of Syracuse and Augusta and continue along the east-coast road, passing between the sea and Mount Etna to take Messina. Patton would land at Gela on his left flank and thrust inland to take three crucial airfields. Ike suggested that, as these airfields would be strongly defended, British forces should assist the Americans in their capture, but Monty insisted he needed the whole of Eighth Army for the push up the coastal road. Patton's Seventh Army was then to move north through the centre of the island to protect Monty's left flank as he advanced on Messina.

Eisenhower had the task of explaining the new plan to Patton, who took it well: he would do as he was ordered. Later he told his headquarters staff that 'This is what you get when your Commander in Chief ceases to be an American and becomes an Ally.' To his diary he confided the suspicion that Monty's plan was framed 'to make a sure thing attack for the Eighth Army and its "ever-victorious General"'.

Monty's most outspoken critics were his fellow British officers. Admiral Cunningham, who would command the naval operation, informed the First Sea Lord that 'Montgomery is a bit of a nuisance,' and even suggested to Patton that he protest his secondary role in the operation. The American replied: 'No, Goddamnit! I've been in this Army thirty years and when my superior gives me an order I say, "Yes, sir!" and do my Goddamn best to carry it out.' Air Chief Marshal Tedder, in charge of air operations, thought it unwise not to allow part of Eighth Army to assist in the capture of the airfields and told Patton: 'It is bad form for officers to criticize each other so I shall. Montgomery is a little fellow of average ability who has had such a build-up that he thinks of himself as Napoleon – he is not.'

Both Cunningham and Tedder may have been influenced by their personal dislike of the 'little fellow'. A more thoughtful indictment came from Monty's friend and supporter, Alexander. In his official report on Husky he wrote:

The risk was unevenly divided and almost the whole of it would fall on the Seventh Army. In other ways also it might well seem that the American troops were being given the tougher and less spectacular task; their beaches were more exposed than the Eighth Army's, and the Eighth Army would have the glory of capturing the more obviously attractive objectives of Syracuse, Catania and Messina, names which would bulk larger in press headlines than the obscure townships of central Sicily. I felt that this division of tasks might cause some feeling of resentment.

Despite this insight Alexander did nothing to redress the balance. Eisenhower, as Supreme Commander, had delegated to him command of Operation Husky, but he failed to command that part of the Allied force most in need of a restraining hand: Montgomery.

Having had 'his plan' adopted, Monty again attempted to oust Patton and have himself put in command of both armies. On 5 May he signalled Alexander: 'Only one commander can run the battle. It seems clear that Eighth Army HQ should command and control the whole operation.' In his diary he made clear precisely what he meant by 'Eighth Army HQ': 'I should run Husky.' Alexander appeared unable to refuse Monty anything and agreed. Eisenhower did not: on 7 May he ruled that the American force must operate as a separate army. Monty and Patton would invade Sicily as equals, although they were, of course, invading to Monty's plan.

Montgomery's involvement in the planning of Husky must be

judged in terms of its three separate aspects. First, he was undoubtedly correct in opposing the original 'separate-landings' plan and demanding an initial concentration of invasion forces to secure vital ports and airstrips. It was an example of Monty the battle manager doing what he did best.

Second, his attempt to be placed in command of the two armies was not fuelled by any national or personal rivalry. The problem was Alexander. Even Patton commented on his 'lack of force'. Alexander's weakness as a commander left a vacuum that Monty felt had to be filled if Husky was to be successful. He was perhaps a little too eager to fill that space himself and would have delighted in outranking Patton.

Third, in relegating Patton's army to the role of 'flank guard' to his own 'proven' army he failed to realize the progress that the Americans had made since the débâcle at Kasserine, failed to see how such a relegation would be interpreted, and failed to understand that his own behaviour in North Africa had helped shape that interpretation. Of course, these failings would be forgiven (by all except Patton) if his plan for the conquest of Sicily *worked*.

On 10 May Rommel went for tea with Goebbels in the garden of his Berlin home and the two men discussed the anticipated Allied invasion of Sicily. Goebbels found that 'Rommel has the lowest opinion of the Italian troops there and says that as soon as the British and Americans land the Italians will end all resistance.' Just as Monty belittled the battle capabilities of his American allies on the basis of his North African experience, so Rommel had only disdain for the Axis allies he had fought alongside in the desert.

He travelled on to the Rastenburg Wolfsschanze, where Hitler had a new task for him. The Führer was looking beyond Sicily and feared that if an Anglo-American army reached mainland Italy the Italians would be more interested in saving themselves than acting as a buffer between the enemy and the Reich. He ordered Rommel to prepare plans for the occupation of Italy in the event of that country defecting from the Axis and joining the Allies.

24 Rommel being interviewed by Lutz Koch

25 Patton giving one of his rousing pre-battle speeches

26 Monty visited every unit of Eighth Army and addressed the men

27 Sherman tanks, led by a Crusader command tank, pass through Mersa Matruh during the advance from Alamein

28 Monty salutes a tank unit during the victory parade in Tripoli

29 British troops land near Pachino on the south-east coast of Sicily, 10 July 1943

30 Montgomery and Patton discuss the progress of the Sicily campaign in the latter's new headquarters, the royal palace at Palermo

31 Monty discusses the campaign with Patton at Catania

32 Patton visits an observation post in a Sicilian village

33 Canadian infantry move cautiously through the narrow streets of Campochiaro

34 Patton talks to Colonel Bernard as his army approaches Messina

35 The longest Bailey bridge built during the advance in Italy stretched 1,200 feet across the river Sangro

36 Rommel inspects the Atlantic Wall defences

37 Rommel in his powerful Horch staff car during his inspection of the coastal defences

38 Panther tanks near Bayeux. Aware that these were superior to all Allied armour, Rommel wanted them moving closer to the coast

39 Rommel returned to Herrlingen to celebrate Lucie's birthday and is seen here on the day before the Allied landings in Normandy

40 Monty visits his son David at the home of his guardians, Major and Mrs Tom Reynolds

41 In the days following the Normandy landings, massive quantities of troops, tanks and supplies were brought ashore

42 General Brooke, Winston Churchill and Monty at the latter's HQ in Normandy

Rommel wrote to tell Lucie that he was 'delighted with the new job'. That was in good part because he blamed Mussolini and the Comando Supremo for his dismissal from North Africa and could imagine no sweeter revenge than commanding German troops as an occupation force in Italy. He gathered a small staff and began to plan Operation Achse (Axis).

While Rommel worked on that, Hitler was planning a major counter-attack in Russia. There were insufficient forces available to run both operations simultaneously and the two men 'waited in suspense to see what the next weeks would bring'. Rommel felt they had put North Africa behind them and that his relationship with Hitler had been rejuvenated: 'We talk about everything under the sun.'

Monty used the final few weeks before the invasion to motivate the men, making use of what he called 'personal command' and his peers dismissed as showmanship, though this latter response says as much about the traditional distance kept between a commander and his men in the British Army as it does about Montgomery: 'I visited every unit; saw and was seen by every man, and spoke to them. My custom is to have the men gather round the car; I then talk to them and ask questions about them. The troops like this informal and friendly way of making contact with them and it has excellent results.'

General Sidney Kirkman confirmed that these 'results' included a devotion to Monty among the men that reached Hollywood proportions, evidenced by the scene outside a cinema in Suez in which he held a conference for his officers:

When we came out there were a hundred soldiers waiting to see Monty! He was a film star. It's like, in the Peninsular War, a glimpse of Wellington. And there they were looking at a film star. Of course, Monty at once seized the opportunity, stood up in a jeep, and made a speech to them. That is, as far as Monty was concerned, the art of command.

Such 'personal command' generated confidence in his leadership and ability to beat the enemy, and helped to 'binge men up for the battle'. Monty was under no illusions about what that might involve: 'The present plan – my plan – ensures concentrated effort. But it will be a hard and bloody fight and we must expect heavy losses.'

Patton, too, saw his primary task as motivating his army, although his talks to the troops were a little more forthright. Addressing the men of 45th Division on 27 June he advised them to be wary when Germans or Italians raised their arms as if to surrender. According to General Albert Wedemeyer (a Washington observer attached to Patton's staff): 'Patton said sometimes the enemy would do this, throwing our men off guard. The enemy soldiers had on several occasions shot our unsuspecting men or had thrown grenades at them. He warned the 45th Division to watch out for this treachery and to "kill the sons-of-bitches unless they were certain of their intention to surrender".' A private described more accurately the sense the troops took from that – 'Fuck them: no prisoners!' – although there is no record of Patton using precisely those words.

He was as frank with Beatrice as he was with his men, writing on 5 July to tell her that 'when you get this you will either be a widow or a radio fan. I trust the latter. In any case I love you.' The next day he boarded Admiral Hewitt's flagship, the *Monrovia*, from which he would command the American landings. Monty was to command the British landings from Malta.

General Alfredo Guzzoni, commanding Axis forces in Sicily, knew the Allies were coming. His reconnaissance planes had sighted several large formations of enemy ships during the afternoon of 9 July and at 1830 he informed Rome that an invasion was imminent. He did not know where the Allies would land but the ships appeared to be heading for the west coast, and when parachutists were reported in that area he was certain.

It was a simple but effective deception. Allied ships sailing from

Malta headed for Sicily's west coast and only after dark changed course for their true location. Allied planes dropped dummy parachutists. The aim was merely to fool the enemy as to the site of the landings. One of the two German divisions on the island was stationed in the west, and if it could be kept there long enough for Patton's army to land and push inland, it would find the Americans blocking any attempt to interfere with Monty's army advancing on Messina along the east coast.

The landings were preceded by two airborne assaults. Of 137 gliders carrying British troops and pulled by American Dakotas, only twelve landed in the drop zone while thirty-six came down in the sea (drowning 252 men) and the rest were scattered over much of the island. Gliders carrying American troops were hit by friendly fire. Nevertheless British paratroops took and held a crucial canal bridge south of Syracuse, while the Americans captured a road junction east of Gela and delayed enemy forces who might otherwise have reached the beaches. The amphibious landings began two hours before dawn on Saturday, 10 July, with four British and four American divisions (a total of 180,000 men) landing abreast along the beaches of the Gulf of Gela and the Gulf of Noto, the largest amphibious assault ever attempted (and greater even than the initial force that would later storm the Normandy beaches).

By mid-morning Monty knew that Eighth Army's landings had been virtually unopposed. The small number of Italian troops in the area, awed by the size of the invading force, was (in Alexander's words, after speaking with Monty) 'driven like chaff before the wind'. By that evening a bridgehead had been consolidated and Monty's signals to his two corps commanders contradict his later reputation as a commander unable to 'grasp the moment'. He told General Leese, commanding 30 Corps, to advance 'with all speed' inland. To General Dempsey, commanding 13 Corps, who was already on the coastal road pushing north, he signalled: 'Air recce shows no enemy movement closing on you. Operate with great energy towards Syracuse.'

That evening he left Malta on the destroyer HMS *Antwerp* and went ashore in Sicily at 0700 the next morning armed with rations for five days and a bottle of whisky for himself, and a large stash of cigarettes to give out to the men. By then Syracuse had been occupied and the cranes and quay facilities found to be undamaged. Monty visited the port in the afternoon. His army had met little resistance and was surging ahead of schedule.

Reports from Patton's front merely confirmed his suspicions and he wrote in his diary: 'On my left the American Seventh Army is not making very great progress at present; but as my left Corps pushes forward that will tend to loosen resistance in front of the Americans.' This was crucial because, according to Monty's plan, he needed the Americans to press inland and secure Caltagirone, then advance along the Caltiagirone road inland to the north coast, cutting the island in two and preventing enemy forces in the west from moving across to take Eighth Army in the flank.

Monty's rapid progress slowed as he met heavy German resistance defending the approach to Catania. At 2200 on 12 July he sent a signal to Alexander that was to change both the agreed plan and the whole nature of the campaign:

Have captured Augusta. Intend now to operate on two axes. XIII Corps on Catania and northwards. XXX Corps on Caltagirone. The transport and road situation will not allow two Armies both carrying out offensive operations. Suggest my Army operates offensively northwards to cut the island in two and that the American Army holds defensively.

Monty intended to divide his army. One half (keeping to his original plan) would push northwards up the coastal road through Catania and (if it could break through the blocking forces) on towards Messina, passing to the east of Mount Etna. The second half (following his new plan) would cut across the rear of the enemy forces held on the coast by the Americans, then push northwards on the Caltagirone road, passing to the west of Mount Etna. Monty's Eighth Army would converge on Messina from both sides of Etna. The problem was that the original plan had allocated the Caltiagirone road to Patton. Monty now wanted it for himself.

He 'told' Alexander, ostensibly the overall commander of Husky, what he intended to do, and later explained why it was necessary: 'The battle required to be gripped firmly from above. I was fighting my own battle and the Seventh American Army was fighting its battle; there was no co-ordination and the enemy might well escape; given a real grip on the battle I felt we could inflict a disaster on the enemy and capture all his troops in Sicily.'

The change of plan required Alexander's permission but when he dithered – eager as always to go along with Montgomery but aware the Americans would react badly – Monty took the road anyway. His new plan made tactical sense and he expected it to cause a big problem for the Italians and Germans. He either did not see or did not care that it would cause a big problem for the Americans.

'PATTON LEAPED ASHORE TO HEAD TROOPS AT GELA,' ran the headline of the *New York Herald Tribune*. 'PATTON LED YANKS AGAINST NAZI TANKS IN SICILY,' proclaimed the *Los Angeles Herald-Express*, and its correspondent with the invasion force told how the general 'leaped into the surf from a landing boat and, personally taking command, turned the tide in the fiercest fighting of the invasion of Gela'. That was exactly what the American public wanted to read. Unfortunately the war correspondent responsible for both accounts was reporting from aboard ship, not from the beach, and was going by hearsay.

Patton had remained on the *Monrovia* throughout the first day of the invasion. The landing met little resistance, apart from an attempted counter-attack by Italian tanks, and his troops occupied Gela by noon. He went ashore at 0930 on the second day, wading through the surf in his polished cavalry boots, and packing his ivory-handled pistols. By then the panzers had arrived too. The Hermann Göring Panzer Division was based just north of Gela and was described by Colonel Oscar Koch (Patton's intelligence officer) as 'hot mustard'.

Patton watched from a rooftop command post as the panzers launched a counter-attack, surging across the open plain north of the town. They were repulsed by the tanks of 2nd Armored Division and the men of 1st Infantry Division supported by naval gunfire called in by Patton. His advice to the men as he left was not tactically profound but perfectly in tune with their own sentiments: 'Kill every one of the Goddamn bastards.' There is no reason to think that his presence at the time of the enemy counter-attack made any difference to its outcome. When he was later awarded the DSC for 'extraordinary heroism at Gela on 11 July', an award perhaps influenced by exaggerated press reports of his part and America's need for a hero of its own to rival Monty, he confided to Beatrice that 'I rather feel I did not deserve it but won't say so.'

Elsewhere along Patton's front his forces were pressing inland, meeting stronger resistance as they approached the three airfields at Ponte Olivio, Biscari and Comisco. While Italian troops quickly fell back from the landing beaches, they put up a ferocious resistance at these landing strips and it was at Biscari that two 'incidents' occurred on 13 July. The 45th Division had taken a number of casualties from Italians who made as if to surrender, then opened fire. Patton wrote to tell Beatrice: 'These Boches and the Italians are grand fighters, but have now pulled the white flag trick four times. We take few prisoners.' After forty Italians surrendered at Biscari, Captain John Compton had them shot by a firing squad of two dozen men. Colonel King later saw 'three mounds of bodies' and had no doubt that 'they had been prisoners of war slaughtered while being moved to the rear'. When word of this reached General Bradley he reported it to Patton, who replied: 'Tell the officer to certify that the dead men were snipers or had attempted to escape or something, as it would make a stink in the press and also would make the civilians mad – anyhow, they are dead so nothing can be done about it.'

★

News of the invasion reached Hitler at Rastenburg during the morning of 10 July. Rommel wrote in his diary: '1200 – war conference with the Führer. The British and Americans have invaded Sicily with paratroops and landing craft.' Four days earlier Hitler had launched Operation Citadel, throwing 2000 tanks against Stalin's 3000 in a battle that dwarfed the tank battle of North Africa.

From 2130 until after midnight Rommel was in private conference with him, telling him that the Italians could not be trusted and urging that, despite the scarcity of forces, the German occupation of mainland Italy be ordered immediately. However, Hitler believed from intelligence reports that Mussolini had his army and people under control and decided for the moment to take no action.

On the morning of 13 July, Bradley's II Corps was breaking across country to the Vizzini–Caltagirone road allocated to Seventh Army. Once there they could make good progress north into the centre of Sicily and on to the north coast, cutting off all Axis forces in the western half of the island and preventing any interference with Monty's flank.

Patton's lunch was interrupted when Alexander arrived to tell him that the plan had been changed. Montgomery wanted to move his 30 Corps up along the west side of Etna to take Messina using the Caltagirone road, which would now be transferred to Eighth Army. The Americans would have to move further west and continue to protect Monty's flank. For the moment Patton held his tongue, but Alexander had visited to break the news in the company of a number of British officers and it struck him as 'noteworthy that the Allied Commander of a British and an American Army had no Americans with him. What fools we are.'

He signalled Bradley, whose II Corps was by then nearing the road. Bradley was aghast: he had lost men to enemy fire as they cut through enemy-held territory to take it:

Just before we got there, we got an order giving the road to the British. Here we were within 1000 yards of the road. We had to back off, come clear round to the beach in trucks. I was very peeved. We should have been able to use that road . . . It confirmed my suspicion. Only Montgomery was to be turned against Messina. There was no glory in the capture of hills, docile peasants, and spiritless soldiers.

A staff officer at Patton's headquarters put it even less discreetly: 'We can sit on our prats while Monty finishes the goddam war.'

We should not judge the Caltagirone road affair by the reputations that Montgomery and Patton acquired on the basis of their war service as a whole. Monty as a commander is, for good reason, characterized as cautious and slow-moving, but he *could* act differently, as the first two days of the Sicily campaign proved. When his troops landed they were supplied with ammunition and fuel for an advance of ten miles. Yet in forty-eight hours Eighth Army advanced forty miles and advance units more than eighty miles. Monty wrote in his diary on 12 July: 'No transport was available for troop lifting and the infantry have to move entirely on foot, and also fight, in the hot Sicilian sun. This is exhausting. But they respond splendidly to my calls on them; I am keeping up the tempo of the operation as if I "let-up" the enemy will seize the opportunity to pull himself together.'

Patton is known for the rapid progress he made *later*, but Montgomery had to judge by previous evidence alone, which was the slow progress the Americans had made at first in North Africa. Now in Sicily he compared the rapid progress his own Eighth Army had made with the slow progress he expected from Patton's army and which, at that moment, seemed to be confirmed by reports reaching him from the American front, and easily concluded that 'their' road should be given to him.

Nevertheless General Leese felt that Monty was wrong to take the Caltagirone road:

It was an unfortunate decision not to hand it over to the Americans. Unknown to us, they were making much quicker progress than ourselves, largely owing to the fact that their vehicles all had four wheel drives. We remembered the slow American progress in the early stages in Tunisia, and did not realize the immense development in experience and technique which they had made. If they could have driven straight up this road, we might have ended this campaign sooner.

Montgomery justified his decision in terms of the number of lives saved: 'To persist in the advance to Catania would have meant very heavy casualties . . . the object could be achieved with less loss of life by operating on the Adrano axis.' But when the campaign was over, General David Belcham broke ranks and contradicted his master: 'If Monty and Patton had made a co-ordinated drive around Etna to Messina they would have got there more quickly with less loss of life.'

That part of Monty's army still following his original route up the east coast now met stiff German resistance and came to a halt at Catania. The other half of his army, Leese's Anglo-Canadian corps advancing northward along the Caltagirone road, also made slow progress as strong German forces moved in to block it.

Patton was furious that, having taken the road from the Americans, Monty could not now take it from the Germans. 'If we wait for them [the British] to take this island while we twiddle our thumbs, we'll wait forever.' He believed that Monty was 'trying to command both armies and getting away with it'. General Clarence Huebner, the only American on Alexander's staff, agreed. He visited Patton's headquarters and (according to Patton's chief of staff, Hobart Gay) reported that 'A determined effort was being made to place General Patton in a secondary role.'

Patton decided to confront Alexander personally and demand a greater role for Seventh Army. During the flight to Alexander's headquarters in Tunis on 17 July he wrote in his diary: 'Neither he nor any of his British staff has any conception of the power and mobility of the Seventh Army. Nor are they aware of the political implications. I shall explain to General Alexander that it would be inexpedient politically for the Seventh Army not to have equal glory in the final stage of the campaign.'

He complained to Alexander that US troops had been forced into a subsidiary role and stressed the likely reaction among the American people. He rubbed in the fact that 'Montgomery attacked Catania with a whole division yesterday and only made 400 yards', and that if the Americans had been allowed to use the Caltagirone road as planned 'we would be on the north coast by now'.

Alexander gave in to Patton just as he had earlier to Monty – convinced perhaps by the arguments, but also aware that Eighth Army's advance had stalled on both the coastal road and the Caltagirone road – and gave him permission to move one force north to reach the coast and cut the island in two and another force west towards Palermo. The precise words used were not recorded but British members of Alexander's staff claimed he agreed to 'a reconnaissance in force *towards* Palermo', and it is unlikely Patton said specifically that he intended to *take* Palermo. The Americans treated the meeting as a victory. General Geoffrey Keyes, Patton's deputy, wrote that 'God and the British willing we are still a Republic.'

The following day Patton gave Keyes command of 3rd Infantry Division, 82nd Airborne and 2nd Armored Divisions and told him to take Palermo. At the same time he bet British Air Vice Marshal Wigglesworth that his troops would be in the city by 23 July, Patton's bottle of whiskey against Wigglesworth's gin.

Keyes made rapid progress. His 3rd Infantry Division marched more than a hundred miles through mountainous terrain in seventy-five hours. This dash to Palermo puzzled the Germans, and Kesselring was happy to observe that 'Patton just marched and captured unimportant terrain.' They mistakenly believed the advance was about beating *them*. On 19 July Patton told Beatrice that 'Monty is trying to steal the show and may do so, but to date we have captured three times as many men as our cousins. If I succeed,

Attila will have to take a back seat.' General Hughes recorded that 'George thinks he is doing OK. Sicily is half taken. Says our cousins got a bloody nose!' Even some British officers enjoyed a joke at Monty's expense: Brigadier Sugden commented that Patton 'had got Eighth Army bloody well surrounded'.

Alexander realized from the direction and speed of Patton's advance that he intended to *take* Palermo, and sent a signal requiring a redeployment of American units, which (without mentioning the city) necessitated abandoning the present advance. The message was received by Gay, who replied that it had been 'garbled in transmission' and requested that it be re-sent. By the time it reached Patton his advance troops had reached Palermo.

The city was taken on 21 July, and the first Americans to enter were welcomed by cheering Sicilians. There is no evidence to support the claim often made that Patton ordered Seventh Army to halt outside so that he could lead the first column in. In fact, most of the American war correspondents and photographers entered with Keyes before Patton arrived. When *Life* magazine printed a feature on the capture of Palermo in August it included six photographs: Keyes appeared in every one and Patton in none.

He established his headquarters in the royal palace and told Beatrice that now he lived like a king, eating in the state drawing room and 'using a toilet previously made malodorous by constipated royalty'. The cardinal of Palermo visited to pay his respects and crowds outside the palace waved American flags. Patton thought that 'I could have been elected Pope . . . all of the Dagos cheered.' Knowing how fickle their enthusiasm for their new conqueror was, he added: 'I wonder when one of them will try to kill me.'

Montgomery's official reaction was magnanimous: 'Many congratulations to you and your gallant soldiers on securing Palermo and clearing up the western half of Sicily.' Privately he told his staff that he hoped the American could now turn from glorifying himself to defeating the enemy.

Patton was surprised and suspicious when he received a signal from Monty saying he would be 'honoured if you will come over to discuss the capture of Messina'. Flying to Syracuse airfield on 25 July for their lunchtime meeting he admitted to feeling 'like a little lamb' – not so much a reference to meekness, which could never describe Patton, as to the innocent being led to the slaughter. 'When we arrived, no one paid any attention to us until I finally spotted General Montgomery.'

It was a short meeting around Monty's staff car with a map of Sicily spread across the bonnet. Patton expected him to reaffirm that Messina was to be taken by the British and perhaps insist that the Americans restrict their activities to Palermo. Monty surprised him, suggesting that Eighth Army 'take a back seat' while Seventh Army struck for Messina. Patton was confused. Despite Monty's brusque attitude he had in effect offered to rein in his troops and leave Messina to the Americans. After the meeting he wrote: 'I felt something was wrong, but have not found it yet.' Monty had not offered him lunch and had given him 'a 5-cent lighter – someone must have sent him a box of them'. He thought it must be a trick. He had already planned an advance along the north coast towards Messina, and now he told Bradley: 'I want you to get into Messina just as fast as you can. I don't want you to waste time on these manoeuvres even if you've got to spend men to do it. I want you to beat Monty into Messina.'

'These manoeuvres' were Bradley's proposed outflanking move-ment that might have reduced the casualties involved in a frontal attack on the city but would have taken more time. It is a serious indictment of Patton that he was prepared to 'spend men' to take Messina quickly, not because greater speed might change the outcome – nothing could now prevent German troops escaping across the strait to mainland Italy – but because it might take him into the city before Montgomery.

Historians have generally taken Monty at his word and given two likely explanations. Either he now realized Eighth Army could not take Messina alone or, as it was clear that Axis troops would escape and taking Messina was no longer the great prize it might have been, he chose instead to rest his troops for the later invasion of Italy. However, Monty's subsequent behaviour does not support either of these theories.

Patton felt that he was being hoodwinked but could not see how. It is possible that his gut feeling was correct, and in two ways. First, Monty was placing himself in a win-win position. He had praised Patton's achievement against the Italians but knew that the experienced German troops blocking the road to Messina would prove altogether more difficult to break through. If Patton reached Messina first, it was because he (Monty) had planned it that way. But if the American was delayed by enemy action and Monty (as in all probability he expected) entered the city first, so much greater the humiliation inflicted on his ally. As will be seen, Monty did not 'stand down' his army after the meeting with Patton but actually attempted to speed up his advance on the city.

Second, on the same day that Monty 'gave' Messina to Patton he wrote to Earl Mountbatten, the chief of Combined Operations in London. He told him that planning for the Normandy invasion needed someone to 'take hold' of it and generously suggested that Alexander be moved to London for the job. That would leave Alexander's present post vacant: 'It is the intention to carry on against Italy and knock her out of the war. Someone will be required to command the field armies for Eisenhower; we shall probably have three Armies, two American and one British. So if Alex goes home, as I think he must, then I must stay here and take on his job – and knock Italy out.'

Patton expected to be involved in the invasion of Italy, too, and Monty had included Seventh Army as one of the two American armies likely to be used. But if he had been aware of the letter to London, he would have realized that Monty was playing the long game and expected to end up as Patton's senior for the Italian campaign. Monty was not ensuring that Patton took Messina: he was ensuring, with a future appointment in mind, that *he* was not seen to fail to take Messina.

★

Sicily had fallen and Hitler feared that the Allies would invade Greece next. On 23 July he ordered Rommel to fly to Salonika and take command of all Axis forces, which included the Italian Eleventh Army, supported by 1st Panzer Division and three infantry divisions.

He arrived in Salonika two days later. 'Terrific heat. Much work to be done before Greece can be regarded as a fortress.' He settled into his room at the Hotel Méditerranée but was woken at 2315 by a call from General Warlimont at Hitler's headquarters. Mussolini had been arrested; no one was sure what was happening in Italy. He was to report back to Hitler immediately. (In fact Mussolini had been deposed and replaced by Marshal Pietro Badoglio who immediately began secret peace negotiations with Eisenhower.)

Rommel arrived in Rastenburg at noon the following day just as the Führer's situation conference began:

All present, including Dönitz, von Ribbentrop, Himmler, Goebbels. The Americans have occupied the western half of Sicily and have broken through. We can expect Italy to get out of the war, or the British to undertake major landings in Italy. Lunched with the Führer. In Rome it has got to the point of actual violence against the Fascist Party. I hope to be sent there soon.

Hitler told Rommel to begin infiltrating German troops into Italy in anticipation of Operation Achse.

Rommel set up a secret headquarters in Munich to be known only as 'High Command Rehabilitation Unit', from where he could direct the 'quiet invasion'. He would send four German divisions into northern Italy, supposedly to bolster Axis defences against any Allied invasion of the mainland. Then, immediately Hitler ordered Achse to be executed, a further sixteen divisions would be placed under his command and all pretence would be dropped.

On 26 July Monty signalled Patton: 'Would like to visit you on 28 July. Would arrive airfield 1200 hours in my Fortress.' The mention of 'my Fortress' was unnecessary and even Johnny

Henderson, Monty's ADC, felt it was 'as if someone sent a telegram saying, Can I come to lunch in my Rolls-Royce?' Monty had an American plane complete with American crew at his personal disposal, and Patton did not. In his reply Patton failed to point out, and in fairness had no reason to know, that the landing strip at Palermo was too short to take a Flying Fortress.

The plane approached the city just before 1200 and not until its wheels were already on the landing strip did the pilot realize there was not enough of it. He braked hard on one side only, while revving the engine on the opposite side to swing the plane abruptly about, and it came to rest on its side. Monty and his staff got out, badly shaken but unhurt. Monty's pride, however, had been severely dented and he told Brooke in a letter that 'I very nearly got killed in my Fortress the other day trying to land at Palermo to see Patton. He said it was OK for a Fortress, but it was far too small.'

'He said it was OK', carrying just a hint that Patton might deliberately have misled him, expressed Monty's feelings about the American rather than any actual words between the two of them. Henderson, who also 'nearly got killed', wrote that 'they [Monty's American crew] hadn't cleared whether the runway was long enough'.

Patton had sent an aide to the airfield to meet Monty, preferring (in their competing one-upmanship) to greet him on the steps of 'his' palace in Palermo where an honour guard and a military band were intended to rub in the pomp and splendour. Their discussions were, as always, perfectly civil and they talked over the details of the American advance along the northern coast to take Messina. Monty urged the Americans on and Patton kept his doubts for his diary: 'I can't decide whether he's honest.' He still suspected some kind of trickery and immediately after the meeting he signalled General Troy Middleton, commanding 45th Division: 'This is a horse race in which the prestige of the US Army is at stake. We must take Messina before the British. Please use your best efforts to facilitate the success of our race.'

★

The advance on Messina was marked not merely by the ferocity of the fighting but the brutality of Patton's visits to two field hospitals. On the afternoon of 3 August he visited the 15th Evacuation Hospital near Nicosia. He spoke to many of the wounded men and, visibly moved by the heavy bandaging and lost limbs, commended them on their bravery. When he came to Private Charles Kuhl who was 'whole and unbandaged', Patton asked him why he was there and was told, 'I guess I can't take it, sir.' The general exploded with anger, called the man a coward and ordered him to leave the tent at once. When the soldier remained seated Patton slapped him across the face with his gloves. Colonel Leaver, commanding the hospital, wrote that the general then 'raised him to his feet by the collar of his shirt and pushed him out of the tent with a final kick in the rear'.

Two days later Patton wrote a memo to all Seventh Army commanders:

It has come to my attention that a very small number of soldiers are going to the hospital on the pretext that they are nervously incapable of combat. Such men are cowards and bring disgrace to their comrades, whom they heartlessly leave to endure the dangers of battle while they themselves use the hospital as a means of escape. You will see that such cases are not sent to the hospital.

Exactly one week later he visited the 93rd Evacuation Hospital. Among the badly wounded men waiting to be admitted he discovered Private Paul Bennett. Reports filed by Colonel Currier, the hospital commander, and one of the nurses present record what was said between the two men:

PATTON: What's your problem, soldier?
BENNETT: It's my nerves, sir. I can't take the shelling anymore.
PATTON: Your nerves? Hell, you are just a coward, you son-of-a-bitch. Shut up that Goddamn crying. I won't have these brave men here who have been shot seeing a yellow bastard sitting here crying. You're a disgrace to the army and you're going back to the front to fight.

You ought to be lined up against a wall and shot. In fact, I ought to shoot you myself right now.

Patton reached for his pistol. The soldier was still sobbing and Patton slapped him across the face with such force that Currier intervened: aware of the pistol still in the general's hand but wary of physically restraining a superior officer, he pressed himself between them. As Patton left he told Currier: 'Get that coward out of here. I won't have these yellow bastards in our hospitals. We'll probably have to shoot them some time anyway.'

The hospital commander sent a full report of the incident to Bradley to be forwarded to Eisenhower. Bradley was disgusted by Patton's behaviour but not prepared to be party to his downfall; he 'filed' the report in a locked cabinet. But the incident was too big to cover up. A nurse informed her boyfriend, a captain in the army's public-affairs department, and he passed the story to four American war correspondents attached to Seventh Army. Instead of immediately plumping for the headlines it guaranteed, they took the story to Eisenhower and offered to 'bury' it on one condition: Patton must be fired.

On 11 August Rommel visited Rastenburg for a conference with Hitler, Jodl, Himmler and Göring. 'The Führer agrees with my views on Italy. He intends to send me in very soon. He says the Italians are playing for time, and then they will defect.' He asked Rommel and Jodl to visit the Italian leaders and assess the situation. Despite the elevated company Rommel sat next to Hitler at lunch. It meant a lot to him and he reported it in his next letter to Lucie: 'He has complete confidence in me.'

Four days later Rommel and Jodl flew to Italy, landing at Bologna airfield at 1100. They had arranged for an entire battalion of SS troops to meet them and accompany them to the villa where discussions with the Italian Comando Supremo would take place. Both men carried loaded pistols in their holsters and the SS ringed the villa while they were inside.

As the Italians still felt towards Rommel the keen antagonism they had developed in North Africa, Jodl did most of the talking. General Mario Roatta replied on behalf of the Italians. When Jodl announced that 'All new German troops arriving in northern Italy are to be under the command of Field Marshal Rommel', Roatta demanded that they be moved to southern Italy, leaving only Italian troops in the north. According to Roatta, this was so that German units would be better placed to counter an Allied invasion, but both Jodl and Rommel believed it was because the move would cut off their troops – and Rommel – from escape back to Germany when Italy changed sides, trapping them between the Italians to the north and the Allies landing in the south.

Following this meeting General Vittorio Ambrosio wrote to Hitler, demanding that Rommel be withdrawn from Italy. 'The Italian High Command procured his recall from Tunisia and he is still affected by events there. It is most inappropriate for him to assume command in Italy.' The protest was ignored.

An advance patrol of Seventh Army reached Messina on the evening of 16 August and there is no doubt that in this case Patton did give orders that no formed units were to enter until he arrived to lead them. Bradley noted that he 'held our troops in the hills instead of pursuing the fleeing Germans to get as many as we could. The British nearly beat him into Messina because of that.' He led a column in at 1030 the next morning. In the jeep behind Patton's, General Porter was impressed by 'the wild applause of the people'. Bradley complained that his 'triumphal entry' was inappropriate and that he 'steamed about with great convoys of cars and great squads of cameramen'. The next day he was front-page news in the US, and a national hero.

Patton had taken Messina with 7500 American casualties; Monty had failed to reach Messina but had taken 12,000 British casualties. Although all German troops had fled, Seventh Army rushed reinforcements into the port. General William Eagles, second-in-command of 3rd Division, claimed this was 'to see that the British did not capture the city from us after we had taken it'.

Some British historians have suggested that the 'horse race' to Messina was a fixture of Patton's invention (or paranoia) and that it found no place whatsoever in Monty's calendar. None has felt it necessary to refer to Freddie de Guingand, Monty's chief of staff, who not only accepted the appropriateness of Patton's analogy but added to it:

The days of the Axis in Sicily were numbered and considerable traffic was taking place across the Straits. It now became a race as to which Army would reach Messina first. Patton was coming along the northern coast road at a great pace. He had his whip out, and brought off a very successful seaborne landing behind the enemy, which accelerated things. On August 15th we occupied Taormina, but the going was bad. The coast road clinging to the steep hills made demolitions an easy matter. Bridges had to be built every few hundred yards and progress was slow. *In order to speed up our advance*, Montgomery ordered a seaborne landing to be carried out at Ali on the night of August 15th/16th. [Author's italics]

In its context, and in view of Monty's knowledge (from Enigma transcripts) that the bulk of the German forces and equipment had already been ferried across from Messina to the mainland, 'to speed up our advance' suggests that Monty's seaborne landing was ordered with the primary aim of reaching Messina before Patton. It very nearly worked. De Guingand again: 'The Americans won the race and entered Messina on August 16th, and some of our troops who had landed at Ali joined up with them the next day.'

It is easy to catalogue Patton's feelings about Monty and the 'race' because he verbalized them regularly and colourfully. Monty was naturally less verbose and expressed his feelings in more subtle ways that were less likely to be noted by those around him. Patton told his officers exactly what he was thinking or feeling. Monty did not, although his body language and general demeanour made it clear to those who knew him well, and the animosity can sometimes be glimpsed in the detail. His ADC Johnny Henderson knew him better than most and recorded one significant encounter between Montgomery and Patton when the fighting was over:

After Monty and Patton had greeted each other in the usual friendly manner, the party sat down. Monty, however, remained standing. He was clearly in command. The first item on the agenda was 'fraternization' – the problem was how far soldiers could go when they were talking to Italian girls. George Patton quickly came to the point. 'I say fraternization ain't fornication – that is, if you keep your hat on and the weight on your elbows.' His comment caused much laughter. Monty joined in but he did not much like the attention shifting towards Patton. The rest of the meeting took place in an altogether cooler atmosphere.

On the same day that Patton entered Messina, Eisenhower wrote to tell him that reports had been received of the slapping incidents: 'I am well aware of the necessity for hardness and toughness on the battlefield. But this does not excuse brutality, abuse of the sick, or exhibition of uncontrollable temper in front of subordinates.' Ike ordered him to apologize individually to the men he slapped, and publicly to their divisions. Patton did so, but meant not a word of it: in his diary that evening he lamented that he was forced to 'soft-soap a skulker to placate the timidity of those above'.

That was unfair to Eisenhower. Any other commander would have dismissed him at once, but Ike considered him indispensable to the war effort and called him 'one of the guarantors of our victory'. He explained this to the reporters who demanded Patton's dismissal in return for their silence and they agreed (for the moment) to co-operate.

Writing to General Marshall on 24 August, Ike pointed out that Patton's success in Sicily had to be attributed 'to his energy, determination and unflagging aggressiveness'. He acknowledged that those same qualities had a darker side:

George continues to exhibit those unfortunate personal traits which during this campaign caused me some most unfortunate days. His personal abuse of individuals was noted in two specific cases. I have had to take the most drastic steps, and if he is not cured now, he never will be. Personally, I believe he is cured because he is so avid for recognition as a great military commander that he will ruthlessly suppress any habit of his own that will jeopardize it.

An even more serious matter had to be dealt with when the campaign was over. Captain John Compton (who had ordered a firing squad to shoot forty Italian prisoners) and Sergeant Horace West (accused of killing thirty-six prisoners) were tried by a military court. Both men quoted speeches by Patton as part of their defence, arguing that although he did not order the killings, his talks to the men created an atmosphere and an expectation that influenced their behaviour and 'inspired' such acts.

Patton was known throughout his military career for making vulgar and profane speeches to the troops. The foul language can be forgiven if it had the right effect on the men, but here was a claim that in some instances it had a tragically wrong effect. Colonel Martin and several who served under Patton, in their own written notes of his addresses, refer to 'pillaging their towns', 'raping their women', and 'killer divisions' that took no prisoners. The words certainly figured in Patton's vocabulary but it is likely these phrases record the sense the hearer took from him rather than the terms he actually used. More damning was the testimony of men who repeated *Patton's* words. Colonel Homer Jones of Seventh Army gave evidence of the speeches he heard: 'General Patton indicated that if the enemy waited until our troops were about to capture them, and then offered to surrender, the American soldiers could not afford to take the chance – that it was too late and that they should kill them.'

Captain Howard Cry testified that 'He just said not to fool around with prisoners. He said that there was only one good German and that was a dead one.' Lieutenant Duncan confirmed that 'He said we would not take any prisoners.'

Compton was acquitted and West found guilty. Patton told war correspondent Alexander Clifford that he 'felt partly responsible'.

Of the two objectives of the Sicily campaign – to occupy the island and to destroy or capture all enemy forces stationed there – only the first had been achieved. Messina had not been taken quickly enough to prevent 40,000 German and 60,000 Italian troops escaping across the strait. But given that those forces would necessarily fall back on Messina, so close to mainland Italy, their total destruction was never feasible. Considering the wider strategic benefits of taking the island – the fall of Mussolini and the defection of Italy to the Allies – the campaign must be counted a significant step in the war against Nazi Germany. The amphibious landings were a rehearsal for Normandy and a great deal was learned and applied, particularly by Montgomery.

More immediately, he had to lead the assault on mainland Italy. While Patton kept a low profile in his palace at Palermo, Montgomery had established himself in his new official residence: the palace at Taormina from which he could see the landing sites across the Strait of Messina. He wrote to tell the Reynoldses, his son's guardians, that 'the view is wonderful' and that he had begun to collect canaries 'of which I have always been fond'. Although his Flying Fortress had been 'written off' at Palermo, Eisenhower had replaced it with a Dakota and this new plane came equipped with its own jeep. Monty was delighted.

He invited Ike to a lavish dinner at 'his palace' and laid on fine wine and cigars. He knew that he was not well liked by the Americans and this may have been an attempt to court the Supreme Commander. When their discussion turned to how long the war was likely to last, Monty offered a bet that it would not be won before the end of 1944. Ike took care what he wagered. Monty recorded that 'He said he would accept the bet and it was written out for a level £5.' That was far less than the cost of a Fortress or Dakota, and Ike had it in writing.

Monty took this opportunity to register his doubts about the invasion of Italy. The landings had been well planned but the campaign that would develop from them had not: 'In fact, there was no master plan.' He had good reason to complain, having pestered Alexander, to whom Eisenhower had delegated overall command of the operation, for a full campaign plan setting clear objectives and receiving in reply only handwritten instructions on a single sheet of paper. Monty's obstreperous behaviour towards his seniors is easy to denigrate and was certainly wrong, but in explaining its cause – in this case Alexander's inability to develop and manage not just a master plan but any plan at all – Monty was undoubtedly right. Unfortunately in Italy, in the days ahead, he would exact a heavy price for going along with an operation in which he had no confidence – and it would be American troops who paid with their lives.

10. Monty's Crawl to Salerno

Napoleon was reported as saying that because Italy was shaped like a boot the way into it was from the top. At 0430 on 3 September 1943 the British Eighth Army invaded across the Strait of Messina, attempting an entry at the toe. In Monty's 'Personal Message to be Read Out to All Troops' he said, 'Forward to Victory! Let us knock Italy out of the war!'

A number of forts covered the landing zone and Monty ordered an artillery bombardment of the coast and the town of Reggio di Calabria. More than six hundred artillery guns fired four hundred tons of shells across the strait. It was later discovered that the forts were Great War defences and had contained no guns since 1915. A flotilla of boats carried 5th Division and 1st Canadian Division across. Italian soldiers emerged from slit trenches to surrender without a fight and were put to work on the beaches helping to unload supplies. Reggio had been occupied by 0930.

Monty went ashore an hour later: 'It was a great thrill to set foot on the continent from which we were pushed off three years ago at Dunkirk.' He took with him a supply of cigarettes to hand out to the men. Elated by the ease of the landing he allowed himself a rare glimpse of the surreal side of war:

The Germans evacuated Reggio before we got into the town, so we had no opposition there from soldiers, but there is a zoo in the town and our shelling broke open some cages; a puma and a monkey escaped and attacked some men of the 3rd Canadian Brigade, and heavy fire was opened by the Canadians; the puma got away; it is believed the monkey was wounded but it escaped in the confusion!

An intelligence report timed at 1900 confirmed that 'no effective resistance was encountered'.

Operation Baytown, the 'hop' from Messina to Reggio, was not as tactically misguided as Napoleon might have thought. It was to be followed a few days later by Operation Avalanche, the US Fifth Army landings at Salerno further to the north. It was hoped that Baytown might attract German forces into the toe and away from Salerno. Monty had argued from the outset that this would not work – that the Germans would anticipate the Italians changing sides and refuse to be drawn into the toe for fear of being trapped there – but Alexander and Eisenhower insisted on it.

Napoleon and Montgomery were proven correct: 29th Panzer Division, which had been in the Reggio area, had immediately moved north. All German forces evacuated the toe. At least Alexander had allowed for this in the handwritten objectives he set for Eighth Army: 'In the event of the enemy withdrawing from the 'toe', you will follow him up with such force as you can make available, bearing in mind the greater the extent to which you can engage enemy forces in the southern tip . . . the more assistance you will be giving to Avalanche.'

Monty reacted with an almost 'I told you so' disdain. Reconnaissance patrols sent out the next morning found no blocking force on the coastal road to Salerno, but instead of an immediate advance northward he insisted on building up his supply base before moving on. A vast force of men, tanks and other vehicles, ammunition, fuel and supplies had to be ferried across the three-mile strait and he complained about a shortage of shipping. Eisenhower's impatient criticism that 'we could do it in a rowboat' was ridiculous and indicated that he had (surprisingly) not yet learned that any operation undertaken by Monty would be meticulously 'managed' and would take time.

Perhaps Monty had reason to be cautious. He wrote in his diary on 5 September: 'It was known from agents and from the Italians that Rommel was in command in the north. I look forward to taking on my old opponent again.'

*

Immediately news of the landings reached Hitler, he asked to see Rommel. The two men dined together on the evening of 4 September and agreed that the Italians could not be trusted and would soon change sides. Rommel wrote proudly that 'The Führer agrees to my Italian campaign plan, which envisages a defence along the coastline.'

He returned to Munich to arrange the transfer of his head-quarters to Lake Garda in northern Italy. On the afternoon of 8 September a radio station in Rome announced that an armistice had been agreed between Italy and the Allies. At 1950 the German High Command telephoned Rommel with the code-word '*Achse*'. This was confirmed at 2020: 'Marshal Badoglio agrees accuracy of Allied radio broadcasts about Italian surrender. Achse takes immediate effect.'

For the last five weeks he had been infiltrating troops into northern Italy and by now he had eight divisions under his com-mand. His Achse orders were to disarm all Italian troops and transfer them to Germany as prisoners-of-war. Resistance in Milan and Turin was quickly put down. In Florence there was a short exchange between German and Italian tanks. Rome was seized and placed under the command of General Stahel.

From Lake Garda Rommel commanded the defence of the northern half of Italy's western coastline: 'Everything available is to be inserted along the coast itself; no reserves are to be held.' He attempted to prevent the looting of Italian homes by the SS and complained to Sepp Dietrich, the SS Corps commander; Dietrich knew of his interest in stamps and responded by sending him a looted collection. German troops under Kesselring, who com-manded the southern half of Italy, had seized control of crucial defence installations along the western coast where secondary Allied landings were expected, and six German divisions stood ready to press the enemy back into the sea.

The second landing came at 0330 on 9 September when General Mark Clark's Fifth Army stormed ashore in the Gulf of Salerno, midway between Rome and Monty's Eighth Army landings at the toe. The plan was to establish a beachhead, then press immediately north to take Naples. But, unlike Baytown, Operation Avalanche was met by fierce German resistance. Clark's troops struggled to get off the beaches as repeated enemy counter-attacks kept them pinned down.

Eighth Army was advancing northward from the toe to link up with them, following the retreating German forces but making slow progress. Monty explained this in his *Memoirs*: 'We marched and fought in good delaying country against an enemy whose use of demolition caused us bridging problems of the first magnitude.' There were bridging problems, but not 'of the first magnitude'. The commanding officer of the Royal Engineers reported 'no bridges blown on the road Gioia–Cittanuova' and 'two bridges blown at Corali are passable and bulldozers will take one hour to repair another'. The blown bridges were passable because it was September and many of the riverbeds were dry or at low water.

It had been planned that Monty would link up with the Americans, and Alexander signalled him 'to get forward as quickly as possible to threaten the southern flank of the Germans opposing the Salerno bridgehead'. Monty's response on 9 September came as a shock to both Alexander and Clark: 'My troops have now marched and fought over a distance of 100 miles in seven days. They are strung out and the infantry are tired. They must be rested.'

The following day Alexander signalled again: 'It is of the utmost importance that you maintain pressure on the Germans so that they cannot remove forces from your front and concentrate them against Avalanche . . . the next few days are critical.' Monty replied on 11 September: 'Am very stretched at present but will push on as soon as admin situation allows.' He noted in his diary that he 'would not move until the situation was clearer'. He halted his army even though he knew that German pressure on the Salerno bridgehead must increase. It is perhaps unfair to suggest that he was punishing Alexander and Eisenhower for following 'their' ill-defined plan and ignoring his own advice. But a note he made on 13 September suggests that he now anticipated the Americans being beaten at Salerno just as they had been at Kasserine, and that he would *then* come to the rescue: 'It was rather similar to the situation in North Africa when Rommel attacked the Americans and I received a cry for help from Alexander; I was then south of the Mareth line, but I drew Rommel off and beat him at Medenine.'

Alexander did all he could to keep the Germans from further reinforcing their troops around Salerno. Clark's PR officer received 'censorship instructions' that included the requirement to 'play up the Eighth Army progress'. More insensitive by far was: 'Americans may be mentioned.' Such propaganda was aimed, of course, at the BBC as much as the Germans, whose intelligence officers listened in. There is no evidence that such propaganda affected German deployments but it certainly affected American morale. The only English–language service the American troops at Salerno could pick up on their radios was the BBC World Service, and as Clark pointed out the British news reports gave a false view: 'The BBC gave the impression that Montgomery's army was dashing up the Italian boot to our rescue. This proved pretty irritating at times, particularly as the Eighth Army was making a slow advance towards Salerno, despite Alexander's almost daily efforts to prod it into greater speed.'

Clark soon realized that he would not be able to break through German forces between Fifth Army and Naples. By 14 September it even seemed possible the Germans might collapse the beachhead and plans for an evacuation were drawn up. Monty's army was still 150 miles to the south and not moving. Captain Butcher, Eisenhower's ADC, heard him wonder aloud what the situation might have been if Patton had landed in the toe of Italy instead of Montgomery.

The most astonishing meeting took place at Reggio on 15 September between Monty and Alexander's senior staff officers. In effect they were negotiating the terms on which Monty would agree to break his three-day 'halt' and advance to Salerno. Monty recorded what happened: 'The attitude was that Eighth Army could have everything it wanted if it could pull the show out of the fire and save Avalanche. We asked for everything we wanted, and got it. I could now drive hard ahead, and by very energetic action I could save a possible disaster to the American Army on the Salerno front.'

What 'we wanted' were adjustments to the invasion plan that, while minor in themselves, indicated to all concerned that Montgomery could and would get his way. His belief that Alexander and Eisenhower were not following the best strategy in Italy was in all probability correct. That he should in response inflict upon them the military equivalent of a tantrum is reprehensible, not least because the delay was measured in American lives.

Eisenhower asked Butcher whether he thought Monty would now make a belated dash north: 'I told him I thought Montgomery would be inspired by the opportunity to 'out Patton' Patton by reaching Clark under heroic acclaim. My guess was that Monty will move as he has never moved before . . . to add lustre to his name.'

Indeed by dusk the next day Eighth Army was only sixty miles from Fifth Army. Yet still Monty was preceded not by his own advance units but by the British journalists attached to his army. Under the loose 'command' of Evelyn Montague of the *Manchester Guardian* and travelling by car, they were the first Englishmen to appear in towns and villages along the coastal road and were welcomed by cheering Italians. A second vehicle, containing five journalists, reached Policastro to find the bridge blown, but within minutes two young boys had guided them to a ford where the gravel riverbed offered an easy crossing. There can be no straight comparison between the movement of two carloads of journalists and the progress of a large army, but the pressmen made contact with American forces outside Salerno at 1100 on 15 September, twenty-four hours before Monty's most advanced reconnaissance patrol reached them.

By then the panzer divisions were falling back in preference to fighting an enemy arriving on their flank. Although Monty would claim the credit, Alexander felt that by holding the beachhead for a week Clark had 'won the battle before Eighth Army got up'.

Having successfully 'dictated terms' for his advance, Monty now acted as if he would henceforth run the campaign. He declared that while Fifth Army 'held' Salerno, 'I would advance with my Army'. The unwritten codicil 'when I am ready' was now recognized by all. Clark informed him that *his* army would now break out of its bridgehead and take Naples. Monty's reaction revealed more about his own *modus operandi* and the assumptions behind it than it did about Clark's intention.

General Clark intended to advance on Naples on 20 September. This seemed to me very curious; his maintenance was still over open beaches, his troops had suffered heavy casualties, and there was great disorganization. I thought he would have made his position secure, re-organized, and built up his strength. The Americans do not understand the vital need to have your administration behind on a scale commensurate with what you want to achieve in front. The real point was to have no more disasters. If Fifth Army rushed off to Naples and got into trouble again, I would not be able to help them. I had already saved their bacon this time.

As for his own Eighth Army, Monty signalled Alexander that he would first pause 'to re-organize'. Butcher wrote, with barely concealed disgust, that 'Montgomery is resting his army and winding-up its tail, a process that delays his forward movement for a week or ten days.'

On 15 September Rommel suffered severe abdominal pains and was rushed to hospital with appendicitis. After an operation he remained bed-bound for ten days, and during this time he wrote to ask Lucie to look for a new home. Previously the northern cities of Germany had been most susceptible to bombing by planes taking off from airfields in the east of England. Wiener Neustadt had been bombed only once, during August, to little effect. But Rommel reasoned that the Allies would soon control all of southern Italy, including the airfields at Foggia, and from there they could more easily bomb targets in Austria. Those targets would be certain to include Wiener Neustadt, the site of the Messerschmitt aircraft factory, and his nearby bungalow might be hit. He urged Lucie to begin the search for a new home quickly before others realized the same and a scramble began for accommodation in safer areas.

He left hospital on 27 September and three days later attended a meeting with Hitler and Kesselring at the Rastenburg Wolfsschanze. A total of 800,000 Italian troops and 450 tanks had been captured. Rommel was furious to learn that 1,650,000 gallons of oil had been taken from the Italian Navy, which had said previously it could not escort supply convoys to his army in North Africa because of the fuel shortage. When Hitler stressed that German forces must 'hold on' in Italy, Rommel proposed they hold a line ninety miles north of Rome. Kesselring argued that this was too pessimistic and he could hold a line ninety miles south of Rome. Hitler approved Kesselring's plan and Rommel took it badly. The Führer was taking Kesselring's advice in preference to his own, and had not yet made Rommel the Supreme Commander in Italy as promised: Kesselring still had command of all forces in the south. From that point Rommel and Kesselring squabbled regularly. Field Marshal von Richthofen (a cousin of the 'Red Baron' of Great War fame) felt that Rommel was 'pig-headed and worn-out'.

Left behind in Sicily, Patton lived 'in exile' in the royal grandeur of Palermo Palace. Famous visitors who came to entertain the troops still on the island included Al Jolson and Jack Benny, the latter finding Patton 'very emotional'. When Eisenhower dropped in to see him on 17 September *en route* to visit Fifth Army at Salerno, Patton pleaded for 'any' operational position, but Ike refused him. General Richardson described him as 'looking very old and desiccated, keeping a low profile after his public disgrace'.

The slapping incidents were to blame but Patton felt the British had to have had something to do with it. A hint of paranoia appeared in his diary entries: 'Sometimes I think there is a deliberate campaign to hurt me. Certainly it is hard to be victimized for winning a campaign. Hap [Hobart Gay, his chief of staff] thinks the cousins [the British] are back of it because I made a fool out of Monty.'

On 29 September he felt that the British wanted to build up Monty's image as a war hero to Napoleonic proportions and that he [Patton] represented unwelcome competition. 'That is why they are not too fond of me. One British general said, "George is such a pushing fellow that if we don't stop him he will have Monty surrounded." I know I can outfight the little fart anytime.' When he accompanied a fellow officer (who was leaving for Italy) to the quay and a British brigadier pulled rank to take the stateroom allotted the American, Patton wrote that 'The act is typical of the race. They have no shame and will take all they can get.'

He had no great liking for Jews or black Americans either. He enjoyed a prime item of gossip passed to him by General Arthur Wilson, that Clark (commanding Fifth Army in Italy in what was, Patton believed, *his* rightful command) 'only got his job due to the efforts of Jews in the US'. When three black troops accused of rape were tried by a military court that included two black officers he wrote that 'Although the men were guilty as hell, the colored officers would not vote death – a useless race.' Hidden behind the racist outburst there was 'another' Patton: *he* had deliberately appointed the two black officers to the court so that the accused men would not face the death penalty: 'In the first five rape cases we had, one was white and four were negroes. Through a miscarriage of justice the white man only got life and the niggers were hung. When three more negroes from the same battalion were arrested for rape I put two negro officers on the court.'

He invited Kay Summersby, Eisenhower's driver, to visit Palermo and gave her an escorted tour of the city. She reported to Ike that in a medieval church Patton had dropped to his knees and 'prayed aloud for his troops, his family and a safe flight back for me'. Eisenhower told her: 'George is one of the best generals I have, but he's like a time bomb. You can never be sure when he's going to go off. All you can be sure of is that it will probably be at the wrong place at the wrong time.'

By 17 September Kesselring had begun the withdrawal of all German forces to hold at the Gustav line centred on Cassino. Monty took the vital airfields at Foggia on 27 September and followed the enemy north along the east coast. Clark took Naples on 1 October and followed along the west coast. Both were now heading for Rome. On 11 October Monty informed Eisenhower he could take the city but not with the four divisions he had, and requested that additional forces be transferred to him from Clark. The American refused to give up a single man and the idea was dropped.

No substitute plan for taking Rome was forthcoming and Monty became even more frustrated, writing on 27 October that 'the indecision and lack of grip is bad – it is a scandal. Alexander is the nicest chap I have ever known, but he does not understand the conduct of war.' The problem, he felt, was that it was 'a major campaign lacking a predetermined master plan'. In Italy, as so often elsewhere, Monty's behaviour – which was undoubtedly wrong – made it more difficult for others to hear and accept those aspects of his complaints that were undoubtedly right: the campaign plan *was* unclear and *would* have benefited from something more akin to 'a master plan'. That cannot excuse his acting like a small boy who had not got his way with toy soldiers; real lives were lost while Montgomery made his point.

Rommel had as much disdain for his own High Command. General Kurt Dittmer, who was by then a broadcaster with German radio, visited him on 8 November and pencilled in his diary those comments that were firmly 'off the record': 'Rommel's damning verdict on OKW operations staff: Jodl, Warlimont and rest. Says the Führer is far-sighted, but the officers around him – from Keitel through Schmundt – dismiss any decision that takes real situations into account.'

Patton was so convinced that Eisenhower had got it wrong that he occupied his time at Palermo by drawing up plans for an amphibious landing on the north-western coast of Italy near Pisa and Leghorn (Livorno), in expectation that the whole operation would founder and he would be called on. Indeed he was, although not in quite the way he had hoped. Eisenhower and Alexander wanted to convince the Germans that an Allied landing in that area was soon to occur, so that enemy troops were retained in northern Italy and not sent south to reinforce the opposition to Clark and Monty.

On 29 October he flew to Corsica, located west of Leghorn and the obvious site from which to mount such an operation. While intelligence 'misinformation' suggested to the Germans that an Anglo-French force including Patton's Seventh Army was gathering on the island, he and his staff spent two days inspecting the harbour facilities and the small number of French troops actually stationed there. He also found time to visit Napoleon's birthplace. When he was shown the sofa on which the emperor had been born he ran his fingertips lightly over the covering, then blew on them, saying, 'Just for good luck.'

He would rather have been involved in the fight in Italy. But he quite liked the idea of 'pulling a fast one' on Rommel, who commanded Axis forces in the north and who might otherwise transfer men south, if not for 'the Patton effect'.

On 11 November, Patton's birthday, he wrote: 'One year ago today we took Casablanca. Now I command little more than my self-respect.' His mess sergeants presented him with a huge birthday cake.

His fifty-eighth year began badly. News of the slapping incidents was leaked to Washington columnist Drew Pearson, who ran the story on his weekly syndicated radio programme on 21 November. It caused a sensation across the US. Ordinary Americans were divided – some were disgusted, others approved – but there were strong calls from Capitol Hill for his dismissal and Eisenhower had to act. He relieved Patton of combat duty and could give no indication of what the US Army intended to do with him. Patton received a great deal of personal mail supporting him, and he shrugged off all of his critics except one: the old soldier he idolized, Jack Pershing, 'disowned' him. Patton wrote to Pershing admitting he had been 'a damned fool' but that 'something burst in me'. He received no reply. In Berlin it was thought the whole matter was a propaganda trick to convince them Patton had no present role in the Allied war effort, and they watched him all the closer.

On Rommel's fifty-second birthday, 15 November, he was full of enthusiasm for his new job. Hitler had appointed him to inspect the Normandy defences in anticipation of a cross-Channel invasion. The Allied landings in Sicily and at Salerno had proven their ability, and the 'Atlantic Wall' of German fortifications along the coastline facing England no longer appeared invulnerable. The Führer asked Rommel to apply his experience of fighting the British to the western front. Führer Directive 51 defined his task: to 'inspect and improve the Atlantic Wall defences' using the Todt Organization (construction unit) and report directly to Hitler with recommendations on how best to defeat an invasion force. It seemed to Rommel that his reinstatement in the Führer's affections was now complete. He exclaimed to Lucie: 'What power he radiates!'

He had a new home, too, and on 21 November he took a week's leave to inspect it. At his urging, Lucie and Manfred had left the bungalow at Wiener Neustadt for fear of enemy bombing and rented a large country house at Herrlingen near the city of Ulm. It had been a Jewish home for the elderly until in 1942 it was 'confiscated' for use by the mayor of Ulm. Local Nazi Party officials had condemned him for this extravagance and it was empty. The city of Ulm was honoured to rent the house and its grounds to Rommel, and during his first visit several local dignitaries called to pay their respects to Herrlingen's most distinguished new resident.

He was taking no chances with the lives of his wife and son, and Russian prisoners-of-war were drafted in to excavate a seven-metre-deep pit in preparation for local builders to construct an air-raid shelter. He put his ADC, Captain Aldinger, in charge of this and ordered that the Russians also work in the gardens.

Monty spent his fifty-sixth birthday, 17 November, preparing an attack on Rome. He wrote to tell Brooke about it the next day: 'I have gone all out for surprise and have concentrated such strength . . . I will hit the Bosch a crack that will be heard all over Italy. I have lined up three divisions, with 400 tanks and the whole of my air power.' His attack across the Sangro would take German forces defending the city in the flank, forcing them across the face of Clark's Fifth Army where their destruction would be completed. That assumed the Americans were up to it. Monty explained to Brooke that he was not too sure: 'Fifth Army is absolutely whacked. So long as you fight somebody all the time, then you don't get far. My observations lead me to the conclusion that Clark would be delighted to be given quiet advice as to how to fight his army; I think he is a decent chap. If he received good and clear guidance he would do very well.'

At the same time he wrote to General Leese in England and confided that 'When we have captured Rome I shall want some leave. I shall probably write a book entitled *Alamein to Rome*.' He declared that 'Given fine weather, nothing can stop us' – and the next day the rains began. Several days later the Sangro, normally eighty feet wide, had swollen to three hundred. The surrounding valley was completely waterlogged and offensive operations using tanks became impossible. Monty realized that Rome would not be taken until early the following year: 'I understand Caesar used to go into winter quarters about this time, when he commanded an army in these parts. And very wise too!'

On 30 November, Churchill's sixty-ninth birthday, the Prime Minister invited President Roosevelt and Josef Stalin to a cele-bratory dinner. The British, American and Russian leaders had met for a summit conference in Teheran to discuss their future war plans. Churchill pressed for the main effort to remain in Italy. Roosevelt doubted his motives, commenting privately that 'The British look on the Mediterranean as an area under their domination . . . the Prime Minister is thinking too much of the post-war and where England will be.' Stalin argued forcefully that no further delay in the Anglo-American invasion of occupied France (codenamed Operation Overlord) was possible; it must take place no later than May 1944. He insisted that Russian forces could not hold on longer than that, and implied that he might then have to make terms with Hitler. Churchill felt he was bluffing but dare not call him on it. Roosevelt later told Henry Stimson that he and Stalin had combined their efforts to force Churchill to accept moving the focus of all future operations from the Mediterranean to the English Channel.

The Prime Minister's birthday meal almost ended in a brawl between Alan Brooke and Stalin. Brooke spoke as if the British were taking the lead in winning the war for the Allies. Stalin (perhaps grateful for American support in securing priority for Overlord) angrily replied that it was a war of machines – *Materialschlacht* on tank tracks and in the air – and as most of these machines were being manufactured by the United States, it was the Americans who were winning the war for them all.

In Sicily when Monty had realized he could not take Messina, he had handed that task to Patton and concentrated on the invasion of Italy. So in Italy when he realized he would not take Rome, he pressed to take over the planning of the Normandy invasion. Johnny Henderson confirmed it later: 'As the Eighth Army got bogged down in Italy it became clear to Monty that this campaign was a stalemate, and his mind wandered to the Second Front, where he might get command.'

On 23 December he wrote to General Simpson, director of Military Operations in the War Office. 'I consider you must at once begin to prepare [the invasion] and get the plan properly shaped. That means you must transfer from here some good chaps who really understand the business.' Two days later he wrote again: 'We must get the Army in England in good shape, and tee up the cross-Channel venture. I am not certain that the Army in England *is* in good shape. Some fresh air seems to be needed.'

The hint was hardly subtle, but Churchill was not sure Monty was one of the 'good chaps' he needed. Brooke, always a Monty advocate, convinced the Prime Minister that this troublesome general might well trouble the German defenders of Normandy. Brooke informed Monty that he would command the 21st Army Group in the invasion. It became obvious at once that with Monty came Monty's style of command. He asked permission to take with him to London his 'team': his chief of staff, chief of intelligence and chief of administration. When told that the last must remain in Italy, Monty decided to take him anyway.

In the old theatre at Vasto on 30 December he said farewell to the officers and men of Eighth Army whom he had commanded in North Africa, Sicily and Italy: 'I am leaving officers and men who have been my comrades during hard and victorious fighting . . . I do not know if you will miss me, but I will miss you more than I can say . . . You have made this Army a household word all over the world.' De Guingand wrote that 'We cheered him and then he walked out to his car; I had tears on my cheeks . . . I could not help thinking of Napoleon and his Marshals.'

A more objective appraisal was given by the BBC correspondent Denis Johnston who heard his farewell address, and despite a great deal of later analyses by soldiers and historians it remains the most balanced summing up of Montgomery:

As I listened to him, I thought to myself, what a headache, what a bore, what a bounder he must be. And at the same time what a great man he is as a leader of troops, and how right he is to wear funny hats so that the soldiers along the roads will know their general and answer his wave. Maybe he is not as great as he thinks he is, but he out-foxed Rommel, and turned the men of the Desert Army from the shoulder-shrugging cynics they used to be into the confident, self-advertising crowd they are now.

Through the first two weeks of December Rommel and his staff toured the part of the Atlantic Wall that followed the Danish coastline. Although it consisted of no more than occasional gun batteries, he concluded there was no possibility of the Allies invading there: the beaches were more readily reached by the *Luftwaffe* than by the RAF, and he knew the Allies would want air supremacy over their landing sites. His main concern, he told one of his new staff officers, Lieutenant Hammermann, was that 'the main battle-line is drawn too far back from the coast'. Not only tanks and troops but the fuel and ammunition reserves they required had to be located near the beaches. He knew from his experience in North Africa that when the invasion began German supply lines would be unable to function properly (if at all) because of enemy air attacks.

On 7 December President Roosevelt, on his way back to the US from Teheran, called in at Algiers to see Eisenhower. It was in the general's Cadillac that he said, 'Well, Ike, you are going to command Overlord.'

The next day he and Eisenhower 'dropped in' at Palermo to see Patton, who remained in residence in the royal palace but was in effect unemployed. The President's personal adviser, Harry Hopkins, told Patton 'not to let anything that son-of-a-bitch Pearson said bother you'. Ike took him aside and said privately that he 'would soon get orders to go to the UK and command an Army'. In the meantime he must be on his best behaviour. Patton wrote in his diary that 'If I am sent there to simply train troops, I shall resign.' Four days later he began a tour of Allied bases throughout the Mediterranean, intended to suggest to German intelligence that another major operation in that area was being prepared. He lectured to British and American officers on 'Landing Operations' and when asked by Sir Henry Maitland-Wilson for his views on General Montgomery, 'I was very careful on what I said and refused to be drawn out.'

On 14 December during Rommel's flight by Heinkel from Denmark for Christmas leave in Herrlingen, he talked to General

Wilhelm Meise about 'the coming invasion'. He had decided that 'Our only possible defence will be at the beaches where the enemy is always weakest.' He would construct a deep 'wall' of minefields and bunkers along the whole invasion coast, and use his influence with Hitler to see that panzer divisions were so positioned that they could destroy the British and American forces on the beaches.

Finally, on 27 December, Monty flew to Algiers for a meeting with Eisenhower. Churchill had appointed him to command British troops in Overlord, under Ike as Supreme Commander. Ike now said that he wanted Monty to 'analyse and revise the invasion plan' and to 'take complete charge of the initial land battle'. Monty was cock-a-hoop when Ike said he would 'place the American armies in England under my command for the landing and subsequent operations'.

The most colourful members of the cast – the 'big names' from the North Africa and Sicily campaigns – had been selected for the biggest scene of all: the cross–Channel invasion of occupied France.

11. The Battering Ram and the Atlantic Wall

Montgomery spent New Year's Day 1944 in Marrakesh with Winston Churchill and his wife Clementine. He arrived to find the Prime Minister (who was recovering from pneumonia) 'in bed reading a copy of the Overlord invasion plan drafted by General Frederick Morgan. He told me I was to read it and give him my opinion about it.'

Eisenhower had said in Algiers that the plan as it stood would not work, and called it 'a mess' with 'not enough wallop in the initial attack'. Monty agreed: it had to be a battering ram. He wrote down what was wrong with it, which was essentially everything. The invasion force was too small, the landing site was too narrow and too many divisions were being funnelled inland across the same beach instead of landing abreast along a wider front. The only certain result of Morgan's plan would be 'the most appalling confusion'. He had his notes typed up and took them to Churchill. Monty's idea of a much stronger force assaulting several beaches simultaneously struck the Prime Minister as 'right' and from that moment he was virtually given *carte blanche* to write the Montgomery Master Plan.

Later that day Churchill was well enough to take a drive into the Atlas Mountains with Clementine and they invited Monty along. During the two-hour journey their guest lectured them on the techniques of amphibious assault and they may have been relieved when, according to the Prime Minister, 'Monty got out and walked straight up the hill "to keep himself in training" as he put it. The General was in the highest spirits; he leapt about the rocks like an antelope, and I felt a strong reassurance that all would be well.' He had previously doubted whether Montgomery was right for the job; now after this demonstration of self-confidence he felt that the general would 'grip the show'.

Eisenhower, as Supreme Commander, would oversee the invasion from Supreme Headquarters Allied Expeditionary Force (SHAEF) established in St James's Square, London. Monty was to command 21st Army Group but with operational control of all land forces during the assault phase of the invasion. Once this had been completed and an American Army Group had been formed under the command of General Bradley, Eisenhower would assume direct control, with Montgomery and Bradley having equal status under him.

On 2 January Monty flew from Marrakesh to London to begin planning the invasion. He stayed at Claridge's Hotel while a flat was made ready for him at Latymer Court. De Guingand and the rest of his 'team' had already established the headquarters of 21st Army Group in St Paul's School in West Kensington, which Monty had attended as a boy. The school had been evacuated to Berkshire at the beginning of the war and its London buildings were requisitioned by the Army. Monty chose as his office the headmaster's study (without the slightest sense of self-parody, although his junior staff officers had long known him as 'Master'). 'I had been a school prefect, captain of the 1st XV [rugby team], in the cricket XI and in the swimming team, but I had never entered that room before. I had to become a Commander-in-Chief to do so.'

He held his first meetings of British and American commanders at St Paul's on 7, 10 and 21 January. Bradley was the most senior American present (as Eisenhower was in Washington) and felt that Monty stressed a little too much that *he* would command US troops for the invasion. He had a 'quiet word' and henceforth Monty spoke of commanding the British and Canadian troops and 'suggesting' to Bradley the orders for American troops. It was a small matter, but indicative of sensitivities still raw from earlier wounds.

The objectives of Overlord as set out by Monty on 21 January, based on the minutes of the meeting and his own account, can be summarized as:

Two armies must be committed. The American Army will be on the right. Their task will be the capture of Cherbourg and the clearing of the Cherbourg Peninsula. They will subsequently develop operations to the south and west.

The British Army will be on the left. Their task will be to operate to the south to prevent any interference with the American Army from the east.

I hope eventually to get a firm lodgement from Caen to Nantes with the British being built up through Cherbourg and the Americans through Brittany.

The town of Caen is an important road centre and must be secured. The object of Second Army, therefore, is to seize Caen and the airfield area to the south-east, subsequently exploiting to the south to cover more effectively the First US Army.

Monty had his portrait painted in North Africa but in February he decided that he wanted to be painted again. His ADC Kit Dawnay felt that 'Monty was getting a bit vain'. He was asked to find 'the best portrait painter in England' but to offer only 'a modest fee'. Dawnay arranged for him to sit for Augustus John. When it was done Monty took an instant dislike to the portrait and Dawnay agreed: 'There was no attempt to define the general on the verge of decisive military achievement. So I went back to John. I said, "General Montgomery doesn't like it." John was delighted. "Good! Now I can sell it to someone else for *much* more."'

Rommel's new headquarters at Fontainebleau, forty miles south of Paris, was a small but opulent château once occupied by Madame de Pompadour, the famous mistress of King Louis XV. The locals knew it as Maison Pompadour and after two centuries they still revelled in her salacious reputation. Rommel, however, was 'not too enthusiastic about drafting his plans in the house of this notorious lady'.

Among senior officers manning the Atlantic Wall defences this distant connection between the field marshal and the courtesan was the source of some *risqué* wit. All but one knew better than to share the joke with the Führer's inspector. Admiral Ruge, who was Rommel's naval adviser and accompanied him on his inspection tours, noted that 'He was no prude, but so-called humour of a certain kind was not tolerated in his presence. Once, during a tour, a commander made an attempt in that direction, but stopped cold when he saw the expression on Rommel's face.'

He was unhappy with his headquarters at Fontainebleau from the outset, not primarily because of its fame as the home of a courtesan but because of its great distance from the likely site of an Allied invasion on the north coast. He made the best of it, personally taking the wheel of a powerful Horch car and bragging to Manfred in a letter home about 'raising plenty of dust wherever I go'. Ruge, driving a smaller Mercury, had trouble keeping up with him on the long, straight stretches of road north of Paris.

What Rommel found on arriving at the north coast shocked him. Like all German soldiers and civilians he had seen the Atlantic Wall on the newsreels and had been reassured by how formidable it appeared. He now discovered that this was only a short section. For most of its length, senior officers stationed in France referred to it privately as 'the Propaganda Wall'.

Like it or not, Rommel was now part of the façade. A propaganda unit followed him everywhere. Photographers asked him to pose looking out to sea, his unrelenting stare aimed at the Allies. To those at home he was the one German who could deliver battlefield victories over the British and Americans, and they felt that his deployment to the Atlantic Wall added immeasurably to its strength. Hitler was fully aware of the 'Rommel effect', which may have played a part in the field marshal's appointment: he said that against the Allies Rommel's name alone was worth several divisions.

He provided the newsreel cameras with good footage and his every pronouncement was reported in the national press. The 10 May 1944 edition of *Die Zeit* quoted him at length: 'I am convinced that every German soldier will assist in dealing a great blow against the British and Americans in retaliation for their criminal and uncivilized bombing campaign against our cities.' Rommel helped to change the mood of the nation, calming the fear of invasion and fostering a sense that it provided an opportunity to defeat the Allies and turn the course of the war.

His reports to Hitler were more realistic. In the first, written after only ten days inspecting the Atlantic Wall, he concluded that 'the enemy will probably succeed in creating bridgeheads at several different points and in achieving a major penetration of our coastal defences.' As a remedy he designed his own range of anti-tank beach obstacles and proposed a six-mile-deep minefield overlooked by bunkers to run the full length of the wall. At the top of his list of supplies urgently required were mines: 'I want anti-personnel mines and anti-tank mines. I want mines to sink ships and mines to sink landing craft. I want mines that detonate when a wire is tripped, mines that explode when a wire is cut, mines that can be remote controlled, and mines encased in non-ferrous metals so that mine-detectors will not register them.' But most of all he needed time. In a letter to Lucie on 19 January he felt that 'We will win the defensive battle in the west, providing that the Allies allow us more time to prepare.'

As there was as yet no definite sign of an Anglo-American invasion, he relaxed a little and sent his soldier-servant Corporal Herbert Gunther to collect his brown civilian suit, coat and hat because 'I want to go out without a Marshal's baton for once.' He was overjoyed when the Todt Organization presented him with two dachshunds and wrote immediately to tell Lucie: 'One is a year old, long-haired. The other is only three months. The younger one was very affectionate immediately, but the older one is not so forthcoming. The two of them are now lying beneath my writing-desk.' He walked the dogs each day in the woods around

Fontainebleau – what Ruge called his 'armed promenades' because he always took a hunting rifle with him and occasionally brought down a wild boar. But his conversation was dominated by North Africa, and when Colonel Hesse, field commander at Saint-Germaine and author of *Der Feldherr Psychologos* (Military Psychology) came to visit, they discussed the influence of Great War military literature (including Rommel's own *Infanterie greift an*) on the present generation of young soldiers.

Churchill had given Montgomery command of the invasion and Hitler had given Rommel charge of the Atlantic coastal defences. While Eisenhower wanted to do as much for Patton, who had proven himself to an equal extent in battle, any high-profile appointment had been made impossible by the slapping incidents in Sicily. Nevertheless an American force fighting its way across Europe was unthinkable, and perhaps impracticable, without Patton somewhere in its vanguard.

A signal reached him in Palermo on 22 January ordering him to London. Four days later on a cold and foggy morning he flew into the city to a similarly icy reception from Eisenhower. He had been recalled but not forgiven. Patton did his best to appear repentant. At that time he had a white spot on his lip that he worried might be cancerous but his doctor assured him it was a harmless sore caused by too much sun in North Africa and Sicily. Patton disagreed: 'After all the ass-kissing I have to do here, no wonder I have a sore lip.'

He was disappointed to learn that despite 'all the ass-kissing' he was to play no part in Overlord. Ike had selected Bradley to lead the US Army in the initial invasion. Once a sufficiently large bridgehead had been established US Third Army would cross the Channel with Patton in command, and would become part of Bradley's Army Group.

Ike detailed Kay Summersby to give Patton the 'sixty-four-dollar tour' of London. As she drove him through streets badly damaged by the *Luftwaffe* and the flying bombs (Hitler's *Vergeltungswaffe* or 'flying revenge weapon') he muttered several times, 'Those sons-of-bitches.' She described him as 'the most glamorous, dramatic general I'd ever met', but found this macho image contradicted by 'the world's most unfortunate voice, a high-pitched womanish squeak'.

Patton was provided with a flat on Mount Street, near the American Embassy in Grosvenor Square, but he disliked it from the first, and particularly the bedroom: the walls and curtains were of pink brocade, the bed was covered with embroidered silk and there was an ornate dressing-table to match. He told General Sandy Patch that 'I'd rather be shot than spend the evening sitting around this Anglican bordello.' The two men went to the Haymarket Theatre to see a performance of *There Shall Be No Night* during which a flying bomb hit the adjacent building; plaster showered the stage and the audience alike.

Patton delighted in two items of Grosvenor Square gossip. General Spaatz, who commanded the US Air Force in Europe, told Patton he was 'fed up with the British and said we have paid a hell of a price for the Supreme Command'. General Hull, who had recently arrived from Washington, 'told me that General Marshall said, "I wish Patton was commanding at the beachhead." He also said that I had captured Sicily in spite of Monty.'

On 29 January he travelled north to Knutsford and Peover, twenty miles south of Manchester, where Third Army was to be stationed, although most of the men had yet to arrive. His headquarters was a further five miles south at Peover Hall, an eleventh-century manor house. The English knew it correctly as 'Pea-ver'; to the American troops it was at first 'Pee-over' and from that mispronunciation it quickly became 'Piss-over'. Patton wrote to tell Beatrice that it had 'last been repaired in 1627 or thereabouts', but it had 'history' and he loved it.

On 11 February he was back in London, having been summoned to a meeting with Montgomery at St Paul's School. Bradley went with him. Monty appeared to hope that even Patton would be in awe of the master plan:

MONTY: You've seen the plan, Patton?

PATTON: I've seen it but not had a chance to study it yet.

MONTY: Did you like what you saw of it?

PATTON: No. It doesn't give me anything to do.

MONTY: It was not good at all when the P.M. showed it to me at Marrakesh.

Monty explained that instead of the original single force assaulting a single beach, 'the initial landing must be made on the widest possible front, the British and American areas of landing must be kept separate, and the air battle must be won before the operation is launched'.

Bradley then commented that 'Ike had come to the same conclusion independently as soon as he saw the [Morgan] plan.' Monty did not rise to the challenge. Eisenhower had seen the Morgan plan and expressed his opinion about it to Monty before the latter saw it in Marrakesh and reached the same conclusion. Each was said to be now privately accusing the other of plagiarism.

When Monty emphasized that 'We must aim at success in the land battle by the speed and violence of our operations', Patton could only agree, although the knowing glance he exchanged with Bradley carried his doubt that the speaker had anything more than the words for it. His own army would not be involved in the landings and not operative in France until three to four weeks later. After the meeting Patton told Bradley that 'The plan was redesigned along Monty's suggestions, and his suggestions move along the lines that are best for him, and all others be damned.'

Later he visited SHAEF to be decorated with the British Companion of the Order of the Bath by Brooke, representing the King. As Brooke pinned on the ribbon he said, 'Don't wince, Patton, I shan't kiss you.' Brooke told him that he believed he deserved the honour more than anyone else. Patton wrote that he probably said that to everyone he decorated.

★

Hitler was convinced by Rommel's plans to bolster the coastal defences and promoted him from inspector to commander of the Atlantic Wall, under Field Marshal Gerd von Rundstedt, Commander-in-Chief, West. A crucial difference soon developed between these two about how best to defeat the invasion forces.

Von Rundstedt claimed to know where the Allies would land. In 1940 when drawing up plans for the German invasion of England he had chosen a frontal assault across the narrowest part of the English Channel: from the Pas de Calais to Dover, a mere twenty miles. He now argued that the Allies would do the same in reverse. He was elderly and drinking too much, but there is no evidence that either of these diminished his skills as a strategist and the assumption had much to recommend it. Calais was three hundred miles nearer to Germany than the likely site of any Normandy landings, nearer to the *Vergeltungswaffe* launch sites that the Allies would surely want to disable at the earliest opportunity, and offered them their most direct route into Germany.

Only the precise direction of their thrust inland from the Pas de Calais then remained uncertain, and von Rundstedt's remedy was to hold a strong armoured reserve force some distance inland (he suggested north of Paris). When the enemy had established his bridgeheads and moved out from them, thereby revealing his 'line of thrust', a counter-attack could be launched to turn him back.

Rommel's disagreement with von Rundstedt involved not so much the assumed invasion site as the conclusion drawn from it. Both men had a great deal of battle experience. The difference between them stemmed from where they had fought. In North Africa Rommel had experienced the impossibility of moving armoured divisions across country when the enemy held air superiority. Von Rundstedt had seen none of that in his advance across southern Russia. Rommel argued that it was crucial to defeat the enemy on the beaches before he could establish bridgeheads, and that therefore *all* the panzer divisions had to be deployed along the coast or close behind it, not held inland. The invasion could only be defeated on the coast; any attempted counter-attack after the Allies had moved inland would come too late.

Rommel also based his assessment on personal experience of the man the Allies had put in command of the invasion: Montgomery. He knew that Monty would build up his armies, equipment and supplies until they were able to attack with overwhelming odds in his favour. But this was not El Alamein and he had to get his men and matériel across the Channel and established in secure beachheads before he would dare to press inland. *That*, then, was where Rommel believed Monty could and must be defeated – on the beaches. Once his armies and their supplies were ashore, the greater resources of the Allies and the law of attrition meant that they must eventually win.

This difference between Rommel and von Rundstedt caused no personal antagonism. Von Rundstedt appeared content to be a figurehead, delegating much of his responsibility, and it had been at his suggestion that Rommel was promoted from inspector of the coastal defences to their commander. Although the concept of holding reserves inland was von Rundstedt's, its most vigorous proponent was General Leo Geyr von Schweppenburg, his commander of Panzer Troops West. Rommel's 'difference' was with von Rundstedt but the heated arguments were with von Schweppenburg.

Rommel felt so strongly that a crucial mistake was being made that he asked Hitler to intervene and countermand von Rundstedt's decision to hold many of his panzer divisions inland. The Führer refused to do so and, as in Italy, Rommel felt personally hurt that another's advice had been preferred to his own. However, Hitler's refusal had more to do with the eastern front: divisions held just north of Paris could more quickly be moved eastward than if they were deployed along the north-west coast of France.

Having adopted the role of 'Lord of the Manor' Patton began reading Freeman's *The Norman Conquest*, 'paying particular attention to the roads William the Conqueror used in his operations in Normandy. The roads used in those days had to be on ground which was always practicable.' He also decided that he must have a dog. Kay Summersby elicited the aid of Lady Leese (wife of General Leese) and they found for him a fifteen-month-old English bull terrier called Punch. Patton changed the dog's name to William (after 'the Conqueror'), which quickly became shortened to Willie. Willie was pure white with pink-rimmed eyes and a permanent scowl. Fittingly he not only looked fierce but had real battle experience: he had belonged to an RAF Lancaster pilot and had accompanied his master on six bombing raids over Berlin before machine-gun fire from a Messerschmitt claimed the man and spared the dog.

In early February Patton had his officers and men – over a thousand in total – assemble on the terrace in front of the hall and treated them to one of his trademark pep talks. His grand entrance was described by one of his staff officers:

General Patton walked out on the terrace, three steps above us. Most of us had never seen him in the flesh, and when the drumbeat and bugles sounded the General's march, we stood transfixed, with not one square inch of flesh that was not covered in goose pimples. It was one of the greatest thrills I shall ever know. That towering figure, impeccably attired, froze you in place and electrified the air.

Wearing an expensively tailored uniform, cavalry boots polished to a high sheen, and spurs, he faced them with the medieval manor house as a backdrop and the bull terrier Willie held tightly on a leash at his side.

At ease, gentlemen. I can assure you that the Third United States Army will be the greatest army in American history. We shall be in Berlin ahead of everyone and to gain that end I shall drive you until hell won't have it. We are going to kill German bastards. I would prefer to skin them alive, but some of our people at home would accuse me of being too rough.

Some crazy German bastards decided they were supermen and it was their mission to rule the world. They've been pushing people around and they were getting set to do the same to us. We are fighting to wipe out the Nazis who started it all. But it will take more than guts to win this war. A man with guts and no brains is only half a soldier. We licked the Germans in Africa and Sicily because we had brains as well as guts. We're going to lick them in Europe for the same reason. That's all, and good luck.

Patton's coarse language failed to impress the officers – it was not what they expected of a commanding officer – although one wrote home to say that 'Here was a man for whom you *would* go to hell and back.' The men took to him at once and eighteen-year-old David Terry, who had arrived at Knutsford that same day, explained why: 'This was the first time I had ever heard anyone talk as I thought a warrior would. I thought, this man Patton is a warrior, and I'm glad we're on the same side.'

On 14 March Basil Liddell Hart visited Patton at Peover Hall. In the pre-war years he had been the main (and sometimes only) proponent of the use of the tank as an attacking weapon in its own right. Patton had read his books and was honoured to meet him. The two men were in total agreement on tank tactics and became friends. Patton wrote to tell Beatrice that Liddell Hart was 'a funny-looking man, tall and skinny' who had 'developed a great love for me'. Liddell Hart wrote to tell Patton how stimulating he found their meeting: 'The forbidding photographs (in a steel helmet) that have appeared in the newspapers certainly don't do you justice.'

On 9 March Rommel moved from Fontainebleau to his new Army Group B headquarters at the twelfth-century château of La Roche-Guyon, which stood on the north bank of the Seine between the towns of Mantes-la-Jolie and Vernon, thirty-five miles north-west of Paris. Rommel was enchanted by the sixty-year-old Duc de la Rochefoucauld, and Ruge confirmed that 'In keeping with the old customs of chivalry in war, the inhabitants were not forced to leave their ancestral home.' Rommel occupied the ground floor. The great hall, its stone walls covered with oil paintings and tapestries, became his study. He worked at a Renaissance desk that looked out over a rose garden.

Here he would be closer to the invasion front but also within easy reach of enemy planes. Anti-aircraft batteries were established on the surrounding high ground, and his engineers constructed underground tunnels large enough to house his twenty staff officers and eighty administrative staff members, so their work would not be disrupted by air raids.

The *Kriegstagebuch* (war diary) of the *Wehrmacht* habitually referred to the American forces by army designations, not by the names of their commanders. That changed in the entry for 20 March 1944: 'It has now been ascertained that General Patton, who is well regarded for his ability, is in England.' The Germans were certain he would play a central, perhaps leading role in what they already called the *Grossinvasion*. The intelligence services were charged with finding both him and his troops, which with surprising largesse they termed Armeegruppe Patton.

This pre-invasion German interest in Patton has often been overstated, with the American represented as the Allied general they most feared. Although that later became the case, it is unlikely they rated him so highly on the evidence of North Africa and Sicily alone, particularly as in the latter theatre he had fought mostly Italian troops. More significant for Berlin at this point was that the *American* press made such a fuss of his achievements and rarely mentioned others by name. The Germans focused on Patton not because they believed he was the best Allied general but because the available evidence suggested the Americans believed it, and as an American (Eisenhower) was now running the show they expected Armeegruppe Patton to figure large in the *Grossinvasion*.

Similarly it has sometimes been suggested that Operation Fortitude, designed to mislead the Germans as to where the invasion would take place, was developed with Patton in mind and moulded around his ability to attract their attention. In fact Morgan's original plan for Overlord, drawn up much earlier, had included a plan to misrepresent the likely site of the invasion as the Pas de Calais and thereby cause the Germans to misplace at least part of their defence forces. When Montgomery rewrote Morgan's draft he retained Fortitude. Patton, denied a role in the initial assault phase and yet expected by the enemy to figure prominently, was the obvious candidate to play the lead role in a deception already scripted.

The Germans had to know that the Allies planned to invade France from England; the build-up of tanks, supplies and American troops along the south coast clearly indicated as much. But they could be misled about its timing and location. Enigma transcripts had revealed that von Rundstedt was convinced the invasion would be launched across the Pas de Calais. Fortitude was meant to convince Hitler and the German High Command that he was right.

Operation Fortitude involved the creation of a fictitious First United States Army Group (FUSAG). This was no mere paper ruse. Designers from the Shepperton Studios in London were brought in to oversee the construction of a 'set' that occupied large areas in East Anglia and around the port of Dover. German air reconnaissance could not fail to spot the build-up of tanks, artillery and trucks (made of rubber, plywood and canvas), the (fake) ammunition dumps and oil storage tanks. Massive encampments of (empty) tents would assist German intelligence in calculating the strength of the army. The landing craft (made of canvas and wood and floating on empty oil drums) that filled Dover harbour indicated the embarkation point, directly across the English Channel from the Pas de Calais.

The operation required not just a set but an elaborate script. FUSAG had its own (real) signals unit to transmit (fake) radio traffic around the clock. It also had its lead actor, complete with a Hollywood sense for the melodramatic: George S. Patton. In establishing the First Army Group in East Anglia and Dover, commanded by Patton, Fortitude gave the Germans both the crossing *and* the commander they expected.

The German High Command was convinced. Even Hitler was deceived. On 28 May the Allies intercepted a telegram sent by General Oshima Hiroshi (Japanese ambassador to Germany) reporting to Tokyo on a meeting with Hitler held the previous day. Hiroshi quoted the Führer's assessment of the coming invasion: 'Diversionary actions will take place in a number of places. After that – when they have established bridgeheads in Normandy and Brittany – they will come forward with an all-out second front along the Pas de Calais.'

On 7 April Monty briefed British and American commanders on the invasion and Dawnay made notes of what was said. By then he knew a great deal from Enigma transcripts about the state of German defences, and was even aware of the Rommel/von Rundstedt debate over the best use and deployment of panzer divisions in France. He still personified the enemy as 'Rommel' and clearly relished this rematch with his old advocate:

Since Rommel toured the Atlantic Wall the enemy has stiffened up his coastal crust but our target area is not yet known to the enemy. Rommel will hold his mobile divisions back from the coast until he is certain where our main effort is being made. By D plus 5 the enemy will have brought in six Panzer type divisions. By then we will have fifteen divisions ashore . . . the armoured divisions are being kept directly under Rundstedt, and delay may be caused before they are released to Rommel. This may help us and quarrels may arise between the two of them.

After outlining the landing operation Monty concluded: 'I do not see how the enemy can stop us from developing the operation as we plan – if only we can get a good lodgement in the first four days.' In effect he acknowledged that Rommel was correct in his disagreement with von Rundstedt. The Allied landings would succeed if bridgeheads could be established in the first four days:

four days because Rommel would not have his panzer divisions on the coast until D-plus-5.

Monty was provided with a train (named *Rapier*) comprising four coaches (headquarters, mess, kitchen and sleeping accommodation) and a flat-car to carry his Rolls-Royce (so that he could journey out to visit soldiers in their camps and workers in their factories). He toured England, and then his son David accompanied him on the trip to Scotland (as it took place during the April school holiday):

The train went to Glasgow, where we visited factories and I remember we went to see an international football match at Hampden Park. On all occasions my father was treated as a great hero. Coming out of factories he was virtually mobbed by people and indeed when we went to Hampden Park there was a roar from the crowd for him as he entered the Royal Box.

Monty estimated that by the middle of May he had visited every formation in the United Kingdom and inspected ('and been inspected by') more than a million men. He had prepared both the army and the people of Great Britain for the invasion soon to come.

On 25 April Patton attended the opening of a 'Warm Welcome Club' organized by the Knutsford WVS (Women's Voluntary Service) to entertain GIs based at Peover and Toft camps. Mrs Constantine Smith introduced the general to the sixty women present and repeated what he had stressed to her: 'Nothing he will say must be quoted.' Patton thanked the ladies for the warmth of their welcome, and his short speech comprised the three parts that so typified his public persona. First came the bluntness bordering on the indiscreet: 'The only welcoming I have done for some time has been welcoming Germans and Italians into Hell. I have done that quite successfully and my troops have got about 170,000 of our enemies there.' Then the charm that saved him: 'As soon as our soldiers meet and know the English ladies and write home and tell our ladies how truly lovely you are, the sooner the American ladies will get jealous and force this war to a quick conclusion.' And finally the thoughtless blunder: 'Since it is the evident destiny of the British and Americans to rule the world, the better we know each other, the better job we will do.' The ceremony closed with a few words from Mr Mould of the British Ministry of Information, then a band played 'God Save The King' and 'The Star Spangled Banner'. Patton left, unaware that he had said anything controversial and in any case assured by Mrs Smith that even his presence would not be reported.

The next morning the *Washington Post* headlined Patton's claim that Britain and America would run the post-war world. An editorial suggested that while such a pronouncement was merely indiscreet, to omit Russia was a blatant insult to an ally whose war effort was crucial to ultimate victory. 'General Patton has progressed from simple assaults on individuals to collective assaults on entire nationalities.'

Patton's issue of an official transcript of the speech in which his words became 'the British and Americans, and, of course, the Russians' was too transparent a sop. The evidence of some who heard the speech that he had indeed included reference to the Russians is of doubtful weight, given his quip to his brother-in-law Fred Ayer after the furore had died down: 'Anybody, Freddy, who wants the Russians to rule any part of this world is a God-damned fool.'

On 29 April Eisenhower wrote to tell him: 'I am thoroughly weary of your failure to control your tongue and have begun to doubt your all-round judgement . . . I want to tell you officially and definitely that if you are again guilty of any indiscretion in speech or action, I will relieve you instantly from command.' At a meeting two days later Ike repeated 'Keep your goddamned mouth shut.' On 3 May Henry Stimson wrote from Washington: 'Keep your mouth absolutely shut until you have reached the beachhead, and then by successful drive and successful fighting, win back your reputation as a soldier who can contain himself as well as conquer the enemy.'

Stimson chose his words well, and the incident certainly strengthened Patton's resolve to prove himself as a great field commander. It also added to another, less constructive obsession: the feeling that some in high places, almost certainly Englishmen, were conspiring against him. He told of his suspicion in a letter to Everett Hughes: 'My last alleged escapade smells strongly of having been a frame-up in view of the fact that I was told nothing would be said, and that the thing was under the auspices of the Minister of Information who was present.'

In fact, a local reporter had been present too and filed the story, although the British papers made little of it. The *Manchester Evening News* ignored the speech altogether and published a photograph of the general and his dog over the caption: 'General Patton proudly presents Willie, his pet bull terrier, after the opening of a Warm Welcome Club'. The American press picked up the full story. Another blunder by Patton made first-rate copy and editors wrote the inflammatory headlines to match.

With only three weeks to go before the invasion, so many British, American and Canadian troops, so many tanks and trucks, and such massive stockpiles of munitions and other supplies were made ready along the south coast of England that Eisenhower joked it was only the vast number of barrage balloons floating in the skies and tied firmly to the soil that kept the island from sinking under the sea.

Montgomery gave a final briefing at St Paul's School on 15 May, attended by King George VI, the Prime Minister, Eisenhower and all the senior commanders. The conference was to begin at 0900 and he ordered the door to be closed and locked at precisely that time. Whether this was merely an expression of his penchant for punctuality or because he had noticed that the only invited officer not yet present was General Patton, it is impossible to say. Minutes later, as Monty began speaking, Patton arrived and hammered loudly on the door. Monty carried on regardless, but when it became apparent that the hammering was not going to stop he ordered the door unlocked. Patton marched in and took his seat. Monty made no comment.

The main theme of the presentation was the need for rapid movement inland between D-Day and D-plus-4. Monty stressed that Rommel would attempt to destroy the invasion force on the beaches before bridgeheads could be established, and that he expected a 'full-blooded counter-attack' on or soon after D-plus-5. Monty's own contribution to this swift push inland would be the taking of Caen: 'Rommel is best at spoiling attacks as he is too impulsive for a set-piece battle, and he will do his best to Dunkirk us ... by preventing our tanks landing. Our armoured columns must move deep inland on D-Day; this will upset the enemy plans and hold him off while we build up strength.'

After Monty's address the King, Churchill and Eisenhower each spoke briefly, but a Canadian general went on at a greater length than seemed appropriate to his standing. Everyone listened politely except Patton. Admiral Hall was sitting in front of him: 'George whispered loud enough so I could hear, "If that so-and-so doesn't shut up, I'm going to so-and-so," both terms very vulgar. Montgomery's Chief-of-Staff heard and got up and whispered in the ear of the Canadian and he shut up.'

Monty issued a message to all invasion troops:

The time has come to deal the enemy a terrific blow in Western Europe. To us is given the honour of striking a blow for freedom which will live in history. Let us pray that 'The Lord Mighty in Battle' will go forth with our armies, and that His special providence will aid us in the struggle. Good luck to each one of you. And good hunting on the mainland.

Rommel set his objective in a Letter of Instructions to army commanders:

In the short time left before the enemy's great offensive starts, we must strengthen all our defences so that they will hold up against the strongest attacks. Never in history was there a defence of such an extent. The enemy must be annihilated on the beaches before he can press inland.

The Germans knew that the invasion must come soon, although many felt that the Allies had delayed too long, giving Rommel the time he needed to turn 'the Propaganda Wall' into a formidable defence. On 16 May he spoke to Hitler by telephone to report further progress. 'The Führer was in an excellent mood and was full of praise for our work in the west.' Goebbels confirmed that 'The Führer is continually praising Rommel's work. The Field Marshal has an old score to settle with the British and Americans, he is burning with rage and has used all his cunning to perfect the defences. He is his old self again.'

Indeed he was. Rommel felt confident. All his old pessimism had gone. His relationship with Hitler had recovered and he told Lang: 'The Führer trusts me and that is all I ask.' He wrote to tell Lucie that 'I am now fully confident of our success in the west. We are getting stronger day by day. I am looking forward to the great battle.'

But by 31 May there was still no sign of the expected invasion, and the weather forecast for the first week of June was so bad that he was assured no Channel crossing would be possible then. Rommel wanted to take this opportunity to report personally to Hitler and to ask (again) that additional panzer divisions be deployed along the coast instead of retained inland. Such a visit was not strictly necessary – neither the report nor the request required his presence – and it is likely that he wished to cement his renewed relationship with Hitler face to face. A meeting with the Führer was scheduled for 6 June.

On 1 June Monty called Dempsey, Bradley and Crerar (the British, American and Canadian commanders of his invasion force) to a final conference at Southwick House, his Portsmouth head-quarters and, surprisingly, he invited Patton too. At dinner that evening the atmosphere was convivial and a bet was laid and written down: 'General Patton bets General Montgomery that the first Grand National run after the present war will be won by an American-owned horse – an even £10.' Monty toasted his four guests and it was Patton who returned the toast. Later that night the American confided to his diary his surprise that in expressing his satisfaction in serving under Monty 'the lightning did not strike me'.

As Bradley knew, their amiable comradeship over the dinner table was one thing and the truth quite another. It had been to him that Patton had explained his chief reservations about Monty's invasion plan:

Monty told us, you and me, that he wants to operate swiftly and boldly in the beachhead, but I think he is incapable of that, congenitally incapable. He is a good general, Brad. I think he's the best general the Limeys have. But he is not a man for fast and bold action. He is a master of the set battle, more concerned with not losing the battle than with winning one.

Patton had then made another, far more serious, prediction, on which no money was bet but on which many thousands of lives and perhaps the outcome of the war depended:

Monty is supposed to take Caen on D–Day and it is essential that he takes it. Well, Brad, he won't take it. He'll need some time to get established in his beachhead and then to prepare the next move slowly and carefully. He'll take his time and in the meantime the Germans will get ready for the counter-attack.

Such talk was more than just spite. Patton proposed (on paper, and in the hope that his plan would reach Eisenhower) that if Monty got 'boxed in' Third Army should immediately be sent over to regain the momentum. There is no evidence that the plan was seen by Eisenhower and if it was then the Supreme Commander ignored it.

On 3 June, with only two days to go, the meteorologists reported that bad weather over Iceland was moving quickly southward. Monty wrote that it was 'the worst forecast we have had', while Kay Summersby heard Eisenhower grumbling that the weather was 'always on the side of the Germans'. Eisenhower called a conference of Allied commanders at Southwick House at 0430 the next day. After the meteorologist reported worsening conditions over the Channel, only Montgomery advised keeping to schedule and Ike ordered a twenty-four-hour postponement.

The group reconvened at 0430 on 5 June. A decision was required within thirty minutes if the invasion was to be launched the next day and all eyes turned to the meteorologist as he delivered a brief forecast: 'A clear spell is possible for twenty-four hours from late today.' Eisenhower decided that Overlord would 'go' on Tuesday, 6 June.

Later that morning Rommel left La Roche-Guyon in his open-topped Horch staff car for the long drive into Germany. Hitler had ordered that senior commanders must not use air transport because of the danger of attack by Allied aircraft. That suited Rommel: it meant that he could break his journey at Herrlingen to spend the night at home with Lucie. It was her fiftieth birthday on 6 June and he would keep her gift – a pair of French shoes – concealed as a surprise for the morning.

Part Two Appendix
Original Patton: The Invasion Speech to Third Army

Patton's speech to the men of Third Army prior to the D-Day landings was the greatest motivational speech of the war and perhaps of all time, exceeding (in its morale-boosting effect if not as literature) the words Shakespeare gave King Henry V at Agincourt. What has become known as 'the speech' was delivered at least four and possibly six times in late May and early June 1944. Some historians have named it Patton's 'eve of D-Day speech', but that was not so: he had no prior knowledge of the invasion date and in any case his Third Army was not part of the 6 June invasion force.

On each occasion Patton arrived in a black Mercedes driven by Sergeant Mims. He wore his helmet, dark-green uniform, highly polished cavalry boots, and the deep-set scowl that he called his 'war face'. He carried a riding crop that he occasionally snapped for effect while he spoke.

Patton addressed the fears that were in the mind of every man. Most were inexperienced troops. His task, using language alone, was to make them fight like hardened veterans. A few of the officers balked at the vulgarity, but it was the language of the barracks and the men loved it. Some, however, felt that his words were too strong for their folks back home. One man changed every occurrence of 'bullshit' to 'baloney', while another rendered 'We're going to hold on to him [the enemy] by his balls' as 'hold on to him by the nose'.

Patton's nephew queried his overuse of obscene language. The general replied:

When I want my men to remember something important, to really make it stick, I give it to them double dirty. It may not sound nice to some bunch of little old ladies at an afternoon tea party, but it helps my soldiers to remember. You can't run an army without profanity, and it has to be eloquent profanity. An army without profanity couldn't fight its way out of a piss-soaked paper bag.

The speech, made without notes, was substantially the same at each delivery, although the order of its several parts varied. A number of the troops wrote it down from memory and their versions differ slightly in the precise words used while agreeing on the content and tone. This reconstruction is derived from the written recollections of Gilbert R. Cook (Cook Papers, Dwight D. Eisenhower Library, Abilene, Kansas), Hobart R. Gay (Hobart R. Gay Papers, United States Military Academy Library, West Point), Lynn A. Hoppe (Patton Papers), Colonel Theodore J. Krokus (Patton Papers), Lieutenant Joshua Miner (Oral History Research Office, Columbia University Library, New York City), and Neil H. Shreve (XII Corps Third Army unit history).

Be seated!

Men, all this stuff you hear about America not wanting to fight, wanting to stay out of the war, is a lot of bullshit. Americans love to fight! All real Americans love the sting and clash of battle. When you were kids you all admired the champion marble shooter, the fastest runner, the big-league ball players and the toughest boxers. Americans love a winner and will not tolerate a loser. Americans play to win all the time. That's why Americans have never lost and never will lose a war. The very thought of losing is hateful to Americans. Battle is the most significant competition in which a man can indulge. It brings out all that is best and it removes all that is base.

You are not all going to die. Only two per cent of you right here today would be killed in a major battle. Every man is scared in his first action. If he says he's not, he's a Goddamn liar. But the real hero is the man who fights even though he's scared. Some men get over their fright in a minute under fire, some take an hour, and for some it takes days. But the real man never lets his fear of death overpower his honour, his sense of duty to his country, and his innate manhood.

All through your army career you men have bitched about what you call 'this chicken-shit drilling'. That is all for a purpose – to ensure instant obedience to orders and to create constant alertness. This must be bred into every soldier. I don't give a fuck for a man who is not always on his toes. But the drilling has made veterans of all you men. You are ready! A man has to be alert all the time if he expects to keep on breathing. If not, some German son-of-a-bitch will sneak up behind him and beat him to death with a sock full of shit. There are four hundred neatly marked graves in Sicily, all because one man went to sleep on the job – but they are German graves, because we caught the bastard asleep before his officer did!

An army is a team. It lives, eats, sleeps and fights as a team. This individual hero stuff is bullshit. The bilious bastards who write that stuff for the Saturday Evening Post *don't know any more about real battle than they do about fucking. And we have the best team – we have the finest food and equipment, the best spirit, and the best men in the world. Why, by God, I actually pity these poor bastards we're going up against.*

All the real heroes are not storybook combat fighters. Every single man in the army plays a vital role. So don't ever let up. Don't ever think that your job is unimportant. What if every truck driver decided that he didn't like the whine of the shells and turned yellow and jumped headlong into a ditch? The cowardly bastard could say to himself, 'Hell, they won't miss me, just one man in thousands.' What if every man said that? Where in the hell would we be then? No, thank God, Americans don't say that. Every man does his job. Every job is important. The ordnance men are needed to supply the guns, the quartermaster is needed to bring up the food and clothes for us because where we are going there isn't a hell of a lot to steal. Every last damn man in the mess hall, even the one who boils the water to keep us from getting the GI shits, has a job to do.

Each man must think not only of himself, but think of his buddy fighting alongside him. We don't want yellow cowards in the army. They should be killed off like flies. If not, they will go back home after the war, Goddamn cowards, and breed more cowards. The brave men will breed more brave men. Kill off the Goddamn cowards and we'll have a nation of brave men.

One of the bravest men I saw in the African campaign was on a telegraph pole in the midst of furious fire while we were moving towards Tunis. I stopped and asked him what the hell he was doing up there. He answered, 'Fixing the wire, sir.' 'Isn't it a little unhealthy up there right now?' I asked. 'Yes, sir, but this Goddamn wire has got to be fixed.' I asked, 'Don't those planes strafing the road bother you?' And he answered, 'No, sir, but you sure as hell do.' Now there was a real soldier. A real man. A man who devoted all he had to his duty, no matter how great the odds, no matter how seemingly insignificant his duty appeared at the time.

And you should have seen the trucks on the road to Gabès. Those drivers were magnificent. All day and all night they crawled along those son-of-a-bitch roads, never stopping, never deviating from their course with shells bursting all around them. Many of the men drove for over forty consecutive hours. We got through on good old American guts. These were not combat men. But they were soldiers with a job to do. They were part of a team. Without them the fight would have been lost.

Sure, we all want to go home. We want to get this war over with. But you can't win a war lying down. The quickest way to get it over with is to get the bastards who started it. We want to get the hell over there and clean the Goddamn thing up, and then get at those purple-pissing Japs. The quicker they are whipped, the quicker we go home. The shortest way home is through Berlin and Tokyo. So keep moving. And when we get to Berlin, I am personally going to shoot that paper-hanging son-of-a-bitch Hitler.

When a man is lying in a shell hole, if he just stays there all day, a Boche will get him eventually. The hell with that. My men don't dig foxholes. Foxholes only slow up an offensive. Keep moving. We'll win this war but we'll win it only by fighting and by showing the Germans we've got more guts than they have or ever will have. We're not just going to shoot the bastards, we're going to rip out their living Goddamned guts and use them to grease the treads of our tanks. We're going to murder those lousy Hun cock-suckers by the bushel-fucking-basket.

Some of you men are wondering whether or not you'll chicken out under fire. Don't worry about it. I can assure you that you'll all do your duty. War is a bloody business, a killing business. The Nazis are the enemy. Wade into them. Spill their blood or they will spill yours. Shoot them in the guts. Rip open their belly. When shells are hitting all around you and you wipe the dirt from your face and realize that it's not dirt, it's the blood and guts of what was once your best friend, you'll know what to do.

I don't want any messages saying, 'I'm holding my position.' We're not holding a Goddamned thing. We're advancing constantly and we're not interested in holding on to anything except the enemy's balls. We're going to hold on to him by his balls and we're going to kick him in the ass; twist his balls and kick the living shit out of him all the time. Our plan of operation is to advance and to keep on advancing. We're going to go through the enemy like shit through a tinhorn.

There will be some complaints that we are pushing our people too hard. I don't give a damn about such complaints. I believe that an ounce of sweat will save a gallon of blood. The harder we push, the more Germans we kill. The more Germans we kill, the fewer of our men will be killed. Pushing hard means fewer casualties. I want you all to remember that.

My men don't surrender. I don't want to hear of any soldier under my command being captured unless he is hit. Even if you are hit, you can still fight. That's not just bullshit either. I want men like the lieutenant in Libya who, with a Luger against his chest, swept aside the gun with his hand, jerked his helmet off with the other and busted hell out of the Boche with the helmet. Then he picked up the gun and killed another German. All the time this man had a bullet through his lung. That's a man for you!

Don't forget, you don't know I'm here at all. No word of that fact is to be mentioned in any letters. The world is not supposed to know what the hell they did with me. I'm not supposed to be commanding this army. I'm not even supposed to be in England. Let the first bastards to find out be the Goddamn Germans. Some day I want them to rise up on their piss-soaked hind legs and howl, 'Ach! It's the Goddamn Third Army and that son-of-a-bitch Patton again!'

Then there's one thing you men will be able to say when this war is over and you get back home. Thirty years from now when you're sitting by your fireside with your grandson on your knee and he asks, 'What did you do in the great World War Two?', you won't have to cough and say, 'Well, your granddaddy shovelled shit in Louisiana.' No, sir, you can look him straight in the eye and say, 'Son, your granddaddy rode with the great Third Army and a son-of-a-Goddamned bitch named George Patton!'

All right, you sons of bitches, you know how I feel. I'll be proud to lead you wonderful guys into battle anywhere, anytime. That's all.

PART THREE

12. The Master Plan and the *Grossinvasion*

At 0630 on 6 June 1944 the telephone rang in the drawing room of Rommel's home in Herrlingen. He had just come downstairs in his dressing gown to retrieve Lucie's birthday present from its hiding place, and answered immediately. It was General Speidel, his chief of staff, calling from La Roche-Guyon. Enemy paratroops had landed in Normandy. An unconfirmed report of Allied naval guns bombarding the coast had just come in. Speidel had ordered all units to their battle stations but it was unclear whether this was the *Grossinvasion* they expected. Rommel asked him to ring back immediately the situation became clearer. Meanwhile he had breakfast with Lucie and she opened her gift. She told him that she was delighted with the shoes from Paris, but 'they did not fit well'.

At the same moment that Rommel took the first call from Speidel, American troops went ashore at the Omaha and Utah beaches, followed twenty minutes later by British and Canadian landings at Gold, Juno and Sword, all meticulously choreographed by Monty's master plan. Each landing was preceded by a naval and air bombardment intended to destroy the 'wall' of bunkers and enemy forces stationed on the coast, and damage the roads and railways by which they could be reinforced. The first wave of landing craft carried in tanks to knock out the surviving beach defences and demolition squads to clear paths through the beach obstacles. The second wave landed artillery, trucks and the infantry, who stormed through the 'lanes' to overcome surviving enemy troops and take the roads immediately above the beach.

The British landings went as planned, as did the American landing at Utah. Opposition was weaker than expected (although

Monty never acknowledged as much) because German infantry divisions stationed near the coast were mostly non-mechanized and the slightest delay in reaching the assault beaches assisted the Allies. Von Rundstedt later confirmed that 'twenty of our divisions had no mobility'. It was Rommel's 'underwater obstacles' (many with mines attached) that caused the landing craft most difficulty, and his mines that most effectively slowed the move off the beaches.

The situation at Omaha was quite different. The Americans landed to find themselves facing a crack German division stationed on the coast, which had somehow been overlooked by Allied intelligence. The close fire of 352nd Infantry kept them pinned down on the beach, preventing 'lanes' being cleared through that tanks, artillery and trucks could move up, and an evacuation was considered. Prolonged naval and air support eventually enabled American troops racing single file along narrow paths through the mines, under constant fire and suffering heavy losses, to push inland. The difficulty experienced at Omaha, compared with the relative ease by which troops moved up elsewhere, suggests that Rommel was correct: the invaders were at their most vulnerable on the beach and, if attacked there instantly and in force, they might be defeated.

At 0930 SHAEF issued the first communiqué to the press: 'Under the command of General Eisenhower, Allied naval forces, supported by strong air forces, began landing Allied armies this morning on the northern coast of France.' The BBC broke the news by quoting that statement word for word.

Patton was at Peover Hall. He had been excluded from the final two planning meetings and he learned of the invasion from the BBC. He wrote to his son George: 'I wish I was there now as it is a lovely sunny day for a battle, and I am fed up with just sitting.' It was no consolation that his 'just sitting' had convinced the German High Command that the Normandy landings were a feint and *his* army was poised to make the true *Grossinvasion* across the Pas de Calais.

Rommel cancelled his meeting with Hitler and left Herrlingen in the Horch at 1030 with an aide, Captain Hellmuth Lang. Several times during their high-speed race back to the headquarters at La Roche-Guyon his driver Corporal Daniel had to pull into the cover of a wooded area or copse because of enemy aircraft overhead, continuing only when the skies were clear.

While he was dashing towards his headquarters his worst fears for the defence were realized. Von Rundstedt, despite his differences with Rommel, saw at once that the panzer divisions held inland as a reserve force should be moved towards the Normandy coast to mount an immediate counter-attack. He ordered 12th SS Panzer Division and Panzer Lehr forward and only then sought permission from the High Command. Jodl told him they could not be moved to the coast without an explicit order from the Führer and must hold their present positions. Hitler was asleep; Jodl refused to wake him and von Rundstedt dared not disobey the order to hold. He finally received Hitler's consent at 1600.

Rommel reached La Roche-Guyon at 2000. The panzer reserves were moving but the Allies had already established deep bridgeheads and their fighter-bombers patrolled every route to the coast. Lang saw him punch one gloved fist into the other as he exclaimed, 'I was right all the time. I should have had the panzers near the beaches. If Montgomery only knew the mess we are in he would have no worries tonight.'

Monty remained at Southwick House in Portsmouth throughout the day, taking regular walks in the garden between reading reports coming in from the beaches, and recording a message to be broadcast by the BBC. At 2200 he boarded HMS *Faulknor* and made a short entry in his pocket diary: 'Invaded Normandy; left Portsmouth 10.30 p.m.'

He knew by then that the Allied invasion force had advanced up to six miles inland and casualties had been lower than anticipated, except at Omaha beach where they had been considerably more. More than 156,000 Allied troops had been landed in

Normandy with fewer than 10,000 killed or wounded. But not all of the D-Day objectives he had set had been taken and chief among these was Caen. The town was twelve miles inland and despite a successful drop by 6th Airborne near the Caen canal, ground troops of 3rd Division had been held up by traffic congestion on the roads leading from the beaches, and German reinforcements arrived first. General Nigel Poett commanded 5th Parachute Brigade and believed that the objective was too far inland to take in the time allowed: 'Monty could have taken Caen on the first day, but 3rd Division were late up from the beach. At first light one of my companies could have taken Caen; in the afternoon it would have needed a whole battalion, and by night a division.'

Despite a massive naval and air bombardment and the continuous presence of Allied fighter-bombers overhead, von Rundstedt was able to move sufficient forces (including 21st Panzer Division, which Allied intelligence *did* know was stationed nearby) into Caen to hold it, and 3rd Division had to halt three miles away. The difference between the town, which the Allies failed to take, and objectives that *were* occupied on D-Day by the Allies, was one of a few miles and a few hours. Had the panzer divisions stationed inland been placed nearer the coast, according to Rommel's rejected plan – and if not for the invasion the field marshal would at that time have been meeting Hitler to plead for their release – the invasion would have proven far more costly for the British and Americans and might conceivably have failed.

In Monty's master plan the capture of Caen on D-Day was crucial to the 'breakout' phase that would follow, and this failure would affect the campaign for weeks to come. The 3rd Division cannot be blamed, and it is remarkable that a battle manager of Monty's expertise planned the taking of an objective twelve miles inland on the first day of the invasion in the knowledge that a prime panzer division was stationed nearby. It appeared to be a critical mistake of the first magnitude. He spent the rest of the campaign (and, indeed, the rest of his life) arguing that it was not.

He arrived in Normandy the next morning on HMS *Faulknor*.

He had ordered that he was to be woken at 0600 by which time the ship was expected to be lying off the assault beaches. Johnny Henderson let that time slip because the captain told him they were lost and had hove to, 'waiting for daylight to see where we are'. By 0630 Henderson could wait no longer. 'I went and told Monty. He seemed very unperturbed. Got dressed, came up on the bridge. And said, "I hear we're lost."' Minutes later the lookouts spotted a battleship. 'Action Stations' was called and the guns were manned but it turned out to be an American vessel patrolling the western extreme of the landing zone. The *Faulknor* had veered so far off-course that it reached the Cherbourg peninsula, and now the captain made all speed back towards the assault beaches.

Monty's tactical headquarters were established at the Château de Creullet eight miles east of Bayeux, home of the Marquis de Druval. His command caravans were sited in the gardens where the marquise fed him sandwiches and provided a *pot-de-chambre* for his personal use. His Dakota was flown in to a nearby airfield and used to ferry daily reports between his Tac HQ and SHAEF in London, and brought post too. The first message Monty received from England was from his mother, congratulatory but with a sting in the tail: 'You landed at the same spot from which William the Conqueror sailed in 1066. Our ancestor Roger Montgomery was his 2nd in Command – like you.'

During the first days of the campaign it was essential that the forces established at each beachhead move outward as well as forward, to join up with those to each side and create a single, firm lodgement sixty miles long and three to twelve miles deep. This was achieved by 9 June. Equipment and supplies were still coming in, according to plan, to build up Allied strength prior to a breakout. Monty was satisfied with the general situation but had one concern: Caen had still not been taken. When he improvised a high-risk plan to capture the town, involving a drop by 1st Airborne Division, the air commander, Trafford Leigh-Mallory, vetoed the idea. Monty wrote to tell de Guingand that 'L. M. is a gutless bugger, who refuses to take a chance.'

Nevertheless he added that 'I am enjoying life greatly and it is great fun fighting battles again after five months in England', and wrote to tell the Reynoldses that 'The operation is going to plan.' His comments masked the anxieties he felt during this early stage of the campaign. Despite the length of the Allied incursion, it was still shallow and vulnerable to an enemy counter-attack in force.

At dinner one evening Montgomery told his staff officers that he wanted a dog. A few days later Kit Dawnay brought him a King Charles spaniel puppy. That same day the BBC's war correspondent Frank Gillard arrived with a Jack Russell terrier pup. Monty took to them both at once, naming the spaniel 'Rommel' and the terrier 'Hitler'. According to Johnny Henderson, 'He adored the new younger members of our Tac HQ and played with the two small dogs whenever he had a free moment.' Monty joked that 'They both get beaten regularly.'

On the morning of 10 June Rommel drove to the front to consult with General von Schweppenburg. Although using secondary roads his car had to take cover thirty times when enemy planes were spotted. Von Schweppenburg's command centre was hidden between the trees of an orchard near Le Caine twenty miles south of Caen. The Panzer Lehr Division had arrived but lost five tanks and 123 trucks *en route* and was now positioned near Tilly, ten miles west of Caen. A counter-attack in force was impossible because of enemy air superiority.

Soon after Rommel left the command post it was hit by an air attack and several junior officers killed. He had intended to travel forward to visit Sepp Dietrich's 1st SS Panzer Corps, but abandoned that. In any case, von Schweppenburg had already confirmed that the enemy air force was preventing adequate supplies getting through, and armour and artillery could not be moved without considerable losses.

Early the next day he drove to see von Rundstedt. Bombing and strafing by Allied planes had disrupted the rail network and they discussed emergency measures to alleviate the supply problem. Back at La Roche-Guyon, Rommel invited Ruge to accompany him for a walk up the hill behind the château, and the reason soon became clear. The field marshal told him that the Allies could not be stopped and that this was von Rundstedt's opinion too. More significantly, and for the first time, he expressed a view that he knew contradicted the intent of the Führer: 'In Rommel's opinion the best solution to the immediate situation was to stop the war while Germany still held some territory for bargaining. Hitler however had no intention of negotiating. *He* wanted to fight to the last house.' When Rommel spoke about the post-war reconstruction of their country, Ruge told him that 'Of all the political leaders you are the only suitable man for the job.'

He was not yet prepared to act or even speak behind Hitler's back and on 12 June he sent a report to Berlin, expressing himself most frankly, with a request that it be presented to the Führer.

The strength of the enemy is increasing more quickly than our reserves can reach the front . . . Our operations in Normandy are made exceptionally difficult by the overwhelming superiority of the Allied Air Force. The enemy has total control of the battle area and up to sixty miles behind the front. Virtually all transport and movement of troops is prevented by day by strong fighter-bomber formations . . . It is difficult to bring up ammunition and food. Artillery and tank deployments are immediately bombarded with annihilating effect . . . Our position is becoming increasingly difficult.

The next day he wrote home to Lucie: 'The battle is not going well for us. I reported to the Führer yesterday. It's time for politics now. The strength of two world powers is having its effect. It will all be over quite quickly.'

Monty's campaign was marked by his VIP visitors as much as his battles. He urged Sir James Grigg, secretary of state for war, to keep all visitors away from his Tac HQ, but his own optimistic reports back to SHAEF encouraged them to come anyway, led by the Prime Minister on 12 June. Churchill described their lunch: 'We eat in a tent looking towards the enemy. I asked the General how far away was the front. He said about three miles. I asked him if he had a continuous line. He said, "No." "What is there then to prevent an incursion of German armour breaking up our luncheon?" He said he did not think they would come.' The Prime Minister found him 'in the highest spirits' because the Americans had that morning taken Caumont, which enabled the British 7th Armoured Division to outflank the Panzer Lehr Division blocking the direct route to Caen. That night Monty told de Guingand that this could be 'a turning point in the battle'.

By the time his next visitor, Charles de Gaulle, arrived on 14 June he knew that the advance had failed and was not in such good humour. The two men lunched together, but when de Gaulle left to make a speech in Bayeux and it was reported back that this had caused traffic jams and military convoys were delayed, he ordered that the general be returned to London immediately.

King George VI arrived two days later. Monty told him that he did not think the people of Normandy had any wish to be liberated. He explained what he meant later in a letter to Phyllis Reynolds:

When you read in the papers that another town has been 'liberated' it really means that very heavy fighting has taken place around it and that it has been destroyed – that not one house is left standing – and that a good many of the inhabitants have been killed. Such is the price the French are now paying. When they chucked their hand in in 1940 they thought they could avoid all this – but they cannot.

Neither could Monty. The correspondents travelling with the King included in their stories the approximate area of Monty's headquarters and referred to it as 'a château'. Two nights later a nearby château was bombed by the *Luftwaffe*. 'I decided to move my HQ and I am having no more visitors.'

★

Von Rundstedt and Rommel had asked Hitler to visit the front to see the situation for himself. Late on 16 June a call from Berlin ordered them to report to the 'Battle Headquarters at Margival' the following day. This was the underground bunker near Soissons built in 1940 and from which Hitler had intended to command the invasion of England. Rommel had hoped he would visit him at La Roche-Guyon. Even if one could count his headquarters as 'the front' – which no soldier would do – Margival was still 140 miles to the rear. Speidel went with them and was present throughout: 'Hitler greeted us curtly and coldly, and then in a loud voice berated us bitterly for our failure to halt the Allied landings.' Von Rundstedt spoke briefly and handed over to Rommel to brief the Führer.

Rommel with a ruthless honesty pointed out that the struggle was hopeless against the overpowering Allied superiority in the air, at sea and on land, ignoring several interruptions by Hitler. The Führer doubted the shocking picture that he painted of the enemy's destructive power and this irritated Rommel . . . After more accusations of failures in generalship and more refusals to believe in the power of the enemy, Rommel sharply replied that he formed his impression from daily visits to the front, pointedly remarking that neither the Führer nor anyone from the High Command had come to the front to see for himself.

Rommel suggested that offensives in the Caen sector would be turned back by the enemy and result only in weakening the panzer divisions. It would be better to withdraw them, preserving them for a possible strike into the flank of Montgomery's Second Army. Hitler appeared to have no understanding of tactical withdrawal. All withdrawal was for him an indication of the commanding officer's defeatism. According to Speidel, Hitler told them that 'The enemy cannot last longer than through summer.'

Although this tense meeting revealed how differently Hitler and Rommel now viewed the situation on the western front, some-thing of their special relationship remained and he agreed to visit

La Roche-Guyon the following day so that field commanders could report to him personally. That evening a *Vergeltungswaffe* veered off target and landed within the bunker compound. The underground facility, separated from the blast by twenty feet of concrete, was undamaged and no one was hurt, but that was near enough for Hitler. He cancelled his visit to Rommel's headquarters and returned to Germany.

Despite that, Rommel was once again (albeit for a short time) brought under what Ruge called the Führer's 'uncanny magnetism'. After returning to La Roche-Guyon, Rommel invited Ruge for a walk in the gardens: 'He told me about his meeting with Hitler, who had big plans for a counter-offensive. Rommel had relentlessly described the seriousness of the situation. Hitler was very optimistic and calm, judged the situation differently, and apparently influenced Rommel somewhat.

Ruge was correct, for on 18 June Rommel wrote to tell Lucie: 'I saw the Führer yesterday . . . and I am looking forward to the future with less anxiety . . . we've got a lot of stuff coming up. The Führer was very cordial and in good humour. He realizes the gravity of the situation.'

Whether or not Hitler understood 'the gravity of the situation', Montgomery was about to make matters much worse. The Americans under Bradley were held up in 'the battle of the hedgerows', fighting their way through the *bocage* south of Omaha beach, and had not taken Cherbourg as planned. The British had not taken Caen. Casualties that had been fewer than 10,000 on the beaches now exceeded 30,000 as the Allies consolidated their gains but failed to take all of their objectives. In a briefing document written on 18 June Monty outlined the plan for Operation Epsom, to begin in three days' time. It is important to note his summary of its objectives in the light of his later claims that he never intended to capture Caen but merely to threaten to take it and thereby draw German forces to its defence, reducing pressure on the American sector and enabling them to take Cherbourg.

Once we can capture Caen and Cherbourg . . . the enemy problem becomes enormous . . . It is then that we have a mighty chance to make the German army come to our threat and to defeat it . . . Caen is really the key to Cherbourg: its capture will release forces which are now locked up in ensuring that our left flank holds secure.

A three-day storm that began on 19 June caused (in Monty's own words) 'delays in Divisions arriving here and marrying up with their vehicles and becoming operative'. Enigma transcripts showed that German reinforcements *were* moving towards Caen during that period, taking advantage of the bad weather that reduced Allied air patrols. Monty assured Eisenhower that 'On 23 June I shall put VIII Corps through in a blitz attack.' That date slipped again and the attack took place on 25 June.

Monty's intention was clear. General Crerar recorded that in briefing his officers he spoke of 'the capture of Caen', and in summing up he said: 'The enemy is firming up in front of the Second Army . . . He obviously means to hold us up in the Caen sector . . . We have now reached the showdown stage.'

For Epsom Monty deployed the whole of Second Army for the first time in Normandy. O'Connor's 8 Corps led the attack, supported by 30 Corps to the west and 1 Corps to the east: a total of 600 tanks supported by 675 artillery guns and 60,000 infantry. To take the town they had to cross the Odon and Orne rivers. Resistance on the first day was stronger than anticipated and Second Army failed to reach the Odon, but Montgomery signalled Eisenhower: 'I will continue battling until one of us cracks; it will not be us.'

SS General Dietrich, commanding the Odon line, was equally certain it would not be him and sent an urgent appeal to Rommel for reinforcements. Rommel immediately ordered up battle groups from 2nd SS Panzer Division and others in the area. These allowed a German counter-attack on 27 June but it was turned back, and by that evening advance units of 8 Corps had reached the Odon and taken one bridge. The tanks of 11th Armoured crossed early the next morning, but in view of the possibility of a second German counter-attack no attempt to cross the Orne was made.

That same day Bradley's First Army took Cherbourg. General 'Lightning Joe' Collins's VII Corps was first into the town. When Bradley joined him he commented that Monty had said Caen was the key to Cherbourg. Collins laughed and suggested they 'wire Monty to ask for the key'.

Rommel's conviction that Hitler 'understood' the situation soon faded, and on 28 June he and von Rundstedt drove together to Berlin, accompanied by Lang and Major Wolfram (one of his staff officers), to present the facts again. Wolfram overheard much of the conversation in the car. Rommel told von Rundstedt that, because there was now no possibility of an *Endsieg* (final victory) in the west, 'the war must be ended immediately. I shall tell the Führer so, clearly and unequivocally.' According to Wolfram, von Rundstedt appeared to be in agreement with all that was said.

In Berlin the next morning, Rommel spoke to both Goebbels and Himmler before he was called in to see Hitler. He told them that he was going to put the full facts before the Führer and ask that appropriate action be taken, and hoped for their support. There is no evidence that he received any. Later Rommel told Admiral Ruge how the meeting with Hitler had begun.

ROMMEL: *Mein Führer*, I think this is perhaps my last opportunity to inform you of the situation in the west. I would like to begin with our political situation. The whole world stands together against Germany, and this disproportionate balance of strength . . .
HITLER: *Herr Feldmarschall* will restrict himself to the military situation.
ROMMEL: History demands of me that I should speak first about our overall situation.
HITLER: No! You will stick to military matters.

Rommel delivered his report, restricting himself to the military situation in France. Hitler could not dispute the facts but they hardly mattered in comparison with his fanciful 'directive'. The enemy must be contained in his bridgehead where he will be destroyed by the *Luftwaffe* (using a thousand new aircraft soon to be provided) and the navy (using torpedo boats and submarines soon to appear in the Channel). Rommel told Speidel that he had been particularly hurt by one comment Hitler flung at him: 'Everything will be all right if you will only fight better.'

As the meeting came to an end he tried again: 'I must speak bluntly, *Mein Führer*. I cannot leave here without speaking about Germany.' Hitler glared at him: '*Herr Feldmarschall, ich glaube Sie verlassen besser das Zimmer!*' (Field Marshal, I think you had better leave the room!) Rommel was correct: it was to be the last time he met Hitler.

This time the Führer's 'uncanny magnetism' failed to work on Rommel. The spell had been broken. And that set him free to follow a quite different course.

On the day after Rommel attempted to tell Hitler that the war in the west was lost, German forces launched a counter-attack near Caen with more than two hundred tanks. British forces stood firm but losses were heavy on both sides and the following day Monty cancelled Operation Epsom. British and American newspapers labelled it a failure.

Monty explained that it was a success. He had attracted additional German divisions into the British sector centred on Caen, thus easing the pressure on the American sector and making it more likely that Bradley's First Army could break out. That much was undeniable. The problem was Monty's claim that this, and only this, was what he had intended all along: he had not failed to take Caen because that had never been his plan and anyone who thought it had been his plan – the Supreme Commander and the British Prime Minister among them – was mistaken.

Eisenhower was not so easily bamboozled. He knew that the real objective of the British attacks had been the capture of Caen because Monty had told him so, and this had not been achieved.

A breakout in the British sector opened up the most direct route to Paris; only the failure of that breakout passed the responsibility along the front to the Americans. Monty was interpreting a major change of plan as something he had intended all along: 'All I wanted to do was to keep German armour tied down on this flank so that my breakout with the Americans could go more easily.'

Brigadier Kenneth McLean, who had been one of the Overlord planners, called Montgomery a 'big cheat' for making such a claim. 'For him to say that he was holding the Germans so Bradley could break out was absolute rubbish and a complete fabrication that only developed after he was stopped outside Caen.'

Nevertheless an early breakout was only possible now from the American sector. Bradley made it clear to Eisenhower that he was in no position to do that. Monty was kept fully briefed on First Army. It is unlikely that he would have launched an operation whose main objective was to facilitate an American breakthrough at a time when the Americans were not ready.

On 4 July Patton travelled to London to meet Jean Gordon, the niece with whom he had had an affair and who had enlisted as a Red Cross 'doughnut girl' to be near 'Uncle Georgie'. He told Everett Hughes that her presence in London must be kept secret from Beatrice. Patton noted that with Monty held outside Caen and Bradley making no progress in the American sector, 'we will die of old age before we finish'.

He still commanded the fictitious FUSAG and it still had believers in Berlin. On 5 July the Japanese ambassador there based this assessment on information from a highly placed informant in the Reich: 'Germany is still waiting for the forces under Patton to engage in a second landing across the Channel. There is no great danger in the Normandy area as long as we are facing only Montgomery's forces. There will be time for the Germans to work out counter-measures after it becomes clear what Patton's forces intend to do.' The ambassador's informant had clearly not visited the front or spoken to Field Marshal Rommel.

After their meeting with Hitler both Rommel and von Rundstedt had expected to be sacked, but only the latter was replaced,
by Field Marshal Günther von Kluge. He arrived at La Roche-
Guyon on 5 July, convinced by his briefing in Berlin that Rommel
was a defeatist and the situation could not be as bad as he had
suggested. He would set out immediately for the front to see for
himself. Ruge overheard Rommel's warning:

ROMMEL: Do so, but be careful. Enemy planes patrol the roads continuously.
KLUGE: Oh, they'll not bother me. I won't even get out of the car.
ROMMEL: I warn you, be careful. Whenever I go up forward I keep my
hand on the door release, ready to jump out. I have to dive into a
ditch ten or fifteen times, and I don't permit the presence of my driver
or the accompanying officers to embarrass me.

On his return Kluge had the grace to admit that he had 'left the
car' many times. He also acknowledged that Rommel's appreciation of the military situation was correct.

Almost one month after the invasion when Rommel and Kluge
were certain they were losing the war, Eisenhower began to doubt
whether Montgomery was winning it. On 5 July while he was in
London to see Churchill, he called Patton in to see him. Patton
later wrote that 'Ike seemed cheerful but a little fed up with
Monty's lack of drive' and felt that he ought to take personal
command, but 'he cannot bring himself to take the plunge'. Ike
told him that there would eventually be four American armies
operating in France, three under Bradley and one under Montgomery. 'Why an American army has to go with Monty I do not
see, except to save the face of the little monkey.' The good news
for Patton was that Fortitude had played out its role and he could
now travel to Normandy to take command of Third Army under
Bradley, although this would not become active until the end of
the month.

At Eisenhower's meeting with the Prime Minister that same day, Montgomery was the main topic. There is no record of what was said, but Alan Brooke wrote in his diary on the evening of 6 July: 'A frightful meeting [at 10 Downing Street] with Winston after he had too much to drink. He was maudlin, bad-tempered, drunken and highly vindictive against the USA because he was annoyed at being told by Eisenhower that Monty was over-cautious. Winston started abusing Monty because operations were not going faster.' Brooke wrote to warn his friend Monty that Eisenhower had called him 'over-cautious' and believed him to be 'bogged down', and that Churchill was furious.

Perhaps Eisenhower had hoped that the Prime Minister would intervene and deal with 'the Monty problem', but for now it was left to the Supreme Commander. While Hitler and Rommel aired their differences face to face, the 'showdown' between Eisenhower and Monty took place by letter.

Montgomery had already taken the offensive. In a letter dated 3 July, when he might have been expected to explain his *failure* to take Caen, he again claimed that by drawing Rommel's panzer divisions towards Caen in the British sector he had relieved pressure on the American sector and thus enabled Bradley to take Cherbourg. He had therefore scored a great *success*. He then asked for further bombing by Allied aircraft as a prerequisite of any fresh advance on Caen. He concluded, 'All this is good.'

Eisenhower showed the letter to Butcher, who noted the change of emphasis from taking Caen to threatening Caen: 'In Monty's previous directive he seemed to be all out to capture Caen, which he still doesn't have.' Eisenhower replied angrily on 7 July after his meeting with Churchill:

Dear Monty. When we began this operation we demanded from the air forces that they obtain air superiority and delay the arrival of enemy reinforcements. Both of these have been done. Our ground build-up on the British side is approaching the limit of our available resources. Also we are approaching the limit in the capacity of the ports to receive and maintain American troops. Thereafter it is possible for the enemy to increase his relative strength. These things make it necessary . . . to use our forces before the enemy can obtain substantial equality in infantry, tanks and artillery. We must use all possible energy in a determined effort to prevent a stalemate. We have not yet attempted a major full-dress attack on the left flank [the British sector] supported by everything we could bring to bear.

Monty replied immediately with the news that the British Second Army had just begun a fresh offensive with the objective of taking Caen: 'You can be quite sure there will be no stalemate.' On 8 July, after Allied aircraft had dropped almost six thousand tons of bombs on German positions around the town, General Dempsey ordered the attack. By the end of the next day he had again reached the Orne and occupied part of the town north of the river. But the Germans held the greater part of the town south of the river and no further British advance could be made.

Montgomery signalled Eisenhower on 10 July: 'Operations on eastern flank proceeding entirely according to plan.' A signal to Bradley sent at the same time included the words, 'Now that Caen has been captured . . .' Neither statement was true.

Eisenhower became increasingly concerned, not merely at the failure to take Caen but at the regular discrepancy between the objectives Monty claimed before and after an operation. The British were growing restless too. Sir James Grigg, secretary of state for war and an admirer of Montgomery, sent him a copy of a fifteenth-century document indicating that the British siege of Caen in 1417 had taken less than a month. Montgomery had been held at Caen for more than a month. The document was sent in jest, and perhaps also to alert his friend to more powerful criticisms in London.

Patton arrived in Normandy on 6 July in a C-47 that landed near Omaha Beach. He as yet had no army to command, but stopped to address the first troops he met: 'I'm proud to be here to fight beside you. Now let's cut the guts out of those Krauts and get the hell on to Berlin. And when we get to Berlin, I am going to personally shoot that paper-hanging Goddamned son-of-a-bitch, Adolf Hitler.'

The next morning he drove to Third Army Headquarters, code-named Lucky Forward and established in the grounds of the Château Bricquebec at Néhou ten miles behind the front line. He immediately went to see Montgomery at Blay and arrived to find him decorating First Army men. The medal ceremony was being filmed by George Stevens, a Hollywood director whose films included *Gunga Din*. Patton observed that 'There were at least twenty-five cameramen of various types, also a microphone on a pole was held over Montgomery's head so his priceless words would not be lost.' Stevens asked permission to film Patton, 'but I told him I was still a secret'. He and Bradley had lunch with Montgomery. That evening Patton wrote in his diary: 'Montgomery went to great lengths explaining why the British have done nothing. Caen is still their D-Day objective and they have not taken it yet.'

Patton had brought with him to France all six volumes of Edward Freeman's *History of the Norman Conquest*. His biographers explain that, as the essential topography of the area remained unchanged, the ancient routes along which troops were moved *then* must remain the most effective routes for armoured divisions to move *now*. That may well have been so. But Patton also brought a Michelin road map marked, he said, with 'railroads, road networks and rivers, all that you have to know about the terrain'.

Montgomery's master plan defined Patton's role: 'All operations in Brittany will come under the direction and control of Third U.S. Army, which Army will have the task of clearing the whole of the Brittany peninsula.' It seemed to Patton that an army whose operations were restricted to Brittany, distant from the most crucial battleground where a breakout might open a route to Paris, was effectively sidelined. The implication was clear: Monty was making certain the American could not 'do a Sicily' and reach the French capital first. He blamed Eisenhower for allowing Monty to get away with it: 'Ike is bound hand and foot by the British and does not know it. Poor fool. We actually have no Supreme Commander – no one who can take hold and say that this shall be done.' His opinion of his fellow generals attempting to make a breakout in the American sector was no better.

Brad and Hodges are such nothings. I could break through in three days if I commanded. They try to push all along the front and have no power anywhere . . . All that is necessary now is to take chances by leading with armoured divisions and covering their advance with air bursts. Such an attack would have to be made on a narrow sector, whereas at present we are trying to attack all along the line.

Although Patton would hardly have guessed it, Monty agreed and had told the Americans as much. In a 10 July meeting when Bradley admitted that they had made 'small and slow progress', Monty said, 'If I were you I think I should concentrate my forces a little more.'

Hidden within the German forces in France, the anti-war camp comprised a small inner group surrounded by a much larger circle. *Luftwaffe* Colonel Caesar von Hofacker, a staff officer at the Paris headquarters of the Commander in Chief West, was a cousin of Colonel Klaus von Stauffenberg and a member of the inner group of those German officers plotting to overthrow Hitler. While those in the outer ring were aware of the intent to replace the Führer – the implication being that he was to be arrested – only members of the inner group knew of the plan to kill him, Göring and Himmler.

Hofacker visited Rommel at La Roche-Guyon on 9 July. His official task was to make a presentation to the field marshal and his staff on the strategic situation. The true purpose of his visit, revealed during a private discussion between the two men, was to win

Rommel's support for the plot. There are two contradictory versions of what Rommel was told and what he agreed to, and there is a crucial difference between them: was he being recruited to the outer circle or the inner group?

According to Speidel, to whom Rommel later spoke about the meeting, the two men agreed that the war could not be won by Germany and that a political solution was necessary. Hofacker asked Rommel if he was willing to participate in peace negotiations with the Allies, and he said he would. The implication was that some kind of *putsch* would previously have taken place in Berlin to allow such negotiations to begin. Nothing was said about an attempt on Hitler's life.

In Paris, von Hofacker, according to his fellow conspirators (von Falkenhausen, von Stülpnagel and von Teuchert), told them he had won Rommel over. It was this that Hofacker reported when he met with von Stauffenberg and others in Berlin on 11 July. Some of the conspirators claimed later that Hofacker told them Rommel had given his support to the bomb plot.

The difference is crucial, but it is possible that Hofacker deliberately left the matter ambiguous. At least one conspirator reported that he told them Rommel agreed to 'the removal of the Führer', a phrase that can imply an arrest or a killing. That allowed the conspirators to claim him as one of their own, and Rommel (well known for his personal allegiance to Hitler) to be won over but remain ignorant of their intent to kill. This ambiguity is supported in evidence given at the Nuremberg trials by the Gestapo's chief inquisitor, Dr Georg Kiessel, who 'interviewed' several of the conspirators after their arrest: 'Hofacker spoke about half an hour with Rommel, explained that the situation called for swift action and that if the Führer refused to act then he must be coerced. There was no talk of an assassination.'

At his 9 July meeting Rommel unwittingly set the timetable for the bomb plot. He told Hofacker that he could hold the western front for another three weeks, but no more than that. The conspirators decided to act immediately.

For his part Rommel began sounding out his senior officers. If he gave orders clearly contrary to the wishes of the Führer, would they obey him? The day after the meeting with Hofacker he told Kluge, 'We have lost the war in the west. It must be brought to an end.' Kluge agreed. Later he spoke with Colonel Hans Lattmann:

ROMMEL: What do you think about the end of the war?
LATTMANN: *Herr Feldmarschall*, that we can't win is evident. I hope we can keep enough strength to end it well.
ROMMEL: I will try to use my reputation with the Allies to make a truce, even against Hitler's wishes.

The plotters expected to begin and conclude their action on 11 July when von Stauffenberg attended a conference at which Hitler was present. He carried a briefcase bomb with him, but because neither Göring nor Himmler was present he decided not to explode it.

On 13 July Lady Tedder, wife of the British air chief marshal, told General Hughes that 'He is going to Churchill to complain about Montgomery . . . things are not moving fast enough . . . he is sick of the whole situation.'

Patton was frustrated by the waiting too. His army was gathering in France but it was not yet being used. 'Brad says he will put me in as soon as he can. He could do it now if he had any backbone. Of course, Monty does not want me as he fears I will steal the show, which I will.'

That same day Monty began letting others know that the show was his and it would be decisive. In a letter to Brooke dated 14 July he gave his most succinct description of a new plan, Operation Goodwood: 'The time has come for a real "showdown" on the eastern flank, and to loose a corps of three armoured divisions into the open country about the Caen–Falaise road. The possibilities are immense with 700 tanks loosed . . . anything may happen.'

He informed Tedder that Goodwood, 'if successful, promises to be decisive'. Tedder replied: 'All the air forces will be full out to support your far-reaching decisive plan.' Finally Monty wrote to Ike, promising that 'my whole eastern flank will burst into flames' and 'the operations may have far-reaching effects'. Ike responded: 'I would not be at all surprised to see you gaining a victory that will make some of the "Old Classics" look like a skirmish between patrols.' Ike appreciated that the open country south of Caen was perfect tank country; the *bocage* that faced the Americans was definitely not.

Monty knew there was a growing dissatisfaction with him and that it centred on his failure to take Caen, a D-Day objective he himself had set. Only the success of Goodwood could answer all of his critics and ensure the continuation of his command. The operation would be launched in the early hours of 18 July. It might bring about a swift conclusion to the war.

On 15 July, while visiting the Caen sector of the western front, Rommel met with Colonel Warning of 17th Luftwaffe Division. Warning had been on Rommel's staff in North Africa, and when the two of them were alone, he spoke freely to this trusted colleague. When interviewed after the war Warning recalled what was said:

WARNING: *Herr Feldmarschall*, what's really going to happen here? Twelve German divisions are trying to contain the whole front. We can count the days off on our tunic buttons before the breakthrough comes.

ROMMEL: I'll tell you something. I am sending the Führer an ultimatum. Militarily the war cannot be won and he must make a political decision.

WARNING: And what if the Führer refuses?

ROMMEL: Then I open the west front. There would only be one important matter left – that the Anglo-Americans reach Berlin before the Russians.

Rommel spoke frankly to Admiral von Ruge in the back seat of the staff car as they drove back to La Roche-Guyon. He said that he expected Montgomery to launch a large-scale British attack in this area in the next few days.

Then the discussion became more serious. It dealt with how life would be in the event of a sudden peace or armistice and the absurdity of continuing an aimless war. Rommel remarked that it would be better to continue as a western dominion instead of letting Germany be destroyed completely. Over and over he put the question: 'When will a decision come?'

On reaching the château Rommel dictated to Speidel an urgent report to be sent to Hitler. He began: *'Die Lage an der Front der Normandie wird von Tag zu Tag schwieriger, sie nähert sich einer schweren Krise'* (The situation on the Normandy front is growing worse day by day and is now approaching a serious crisis).

This 'report' was in effect Rommel's ultimatum to the Führer. We can use that term because Rommel himself did so earlier that same day when talking to Warning. First he listed the facts. His army had lost 97,000 men and only 6000 replacements had reached the front. Of 225 tanks destroyed only 17 replacements had arrived. The enemy air force was now operating up to 150 kilometres (90 miles) behind the front, disrupting all rail and road transport. All of the troops in Normandy were exhausted and he urgently required two fresh divisions. Then he went further than he or anyone had done before in spelling out to Hitler the full consequences:

Due to the intensity of the fighting, the enemy's enormous strength in artillery and tanks, and his total command of the air, our losses are so high that the fighting capacity of our divisions is quickly diminishing. The enemy's use of his material strength will smash our army to pieces . . . In the near future he will successfully break through our thin front and then thrust deep into France. We have no mobile reserves to defend against such a breakthrough . . . Our troops are fighting heroically, but this unequal struggle is nearing its end. I feel it is my duty as *Oberbefehls-haber* [commander-in-chief] of the Army Group to state this clearly.

Rommel read Speidel's typed draft the next day and then scribbled a concluding sentence himself. '*Ich muß Sie bitten, die politischen Folgerungen aus dieser Lage unverzüglich zu ziehen*' (I must ask you to draw the political consequences from this situation immediately). But before it was sent off they thought it best to delete the word '*politischen*'. Speidel advised that 'This would have been a red flag to Hitler. We decided "consequences" could be read to include both military and political matters. At this point Rommel said to me, "I am giving Hitler this last chance before we negotiate ourselves."'

The threat behind the report could not be conveyed along with it to the Führer: that if he failed to 'take the consequences' (in the original German this phrase implies that some action must follow) Rommel himself would do so without permission. The field marshal had two options. He could either negotiate with the Allies or, if that proved difficult given the speed at which events were moving, he could order Army Group B to cease all resistance and open the front to Montgomery, so that the British and Americans might reach Berlin before the Russians. His headquarters staff had already established radio communications with the Allies (to arrange exchanges of seriously wounded men).

Rommel signed and submitted the final version to Field Marshal Kluge on 16 July to be forwarded to Hitler. Although the report is proof enough that he had not yet 'given up' on the Führer, neither had he any great expectation that the necessary action would be taken, for that same evening he discussed with Speidel what conditions the Allies might impose if and when he negotiated for peace. He revealed that he had already drawn up a commission to take part in the discussions, comprising Speidel, Ruge, Stülpna-gel, Hofacker, von Schweppenburg and Count von Schwerin (one of his divisional commanders).

★

Ironically, at about the time this conversation took place, Johnny Henderson was recording in his diary a conversation that had taken place earlier that day: 'General Browning (commander of all the airborne divs) asked General Monty if he would like to have Rommel bumped off. They have found his HQ and know where he goes down to shoot pigeons. General Monty said 'Yes', so the party is to be arranged.'

In England Brigadier McLeod of the Special Air Service had already drawn up a detailed plan for the operation. He submitted it immediately, dated 16 July. Local informants, including a member of the French Resistance who owned land near the château used by Rommel as his headquarters, had provided details of the grounds and the timing of his daily walks. Two officers and four soldiers would be dropped by parachute close to La Roche-Guyon during the night in order to 'kill or kidnap and remove to England Field Marshal Rommel'. The first option was likely to prove the most practicable. The SAS was ready to go immediately the plan was approved. That would take several days.

Meanwhile von Kluge agreed that Rommel's report properly represented the situation and attached a note to the Führer stating as much. His covering letter is dated 21 July. It is not known why he delayed forwarding it, but perhaps the fearful consequences of telling Hitler that the war on the western front was lost, and that something must be done, gave him pause. The report reached Hitler on 23 July, seven days after leaving Rommel's hand. A short period, but during which everything changed, both for Rommel and for the Führer.

13. The Caen Controversy

The war ended for Rommel on Monday, 17 July. That morning he visited 277th and then 276th Infantry Divisions, both of which had come under massive enemy bombardment during the last two days. Air reconnaissance had shown that the British were massing tanks and artillery to the east of Caen and he expected Montgomery to attempt a breakthrough there within the next few days. He wanted to be certain that his defence was fully prepared for the enemy.

In the afternoon he visited the headquarters of 1st and 2nd SS Panzer Corps and spoke with their respective commanders, SS-Gruppenführer Bittrich and SS-Gruppenführer Sepp Dietrich. Both units had suffered heavy casualties. Their ability to resist the expected Allied assault had to be doubted. It was here that Captain Hellmuth Lang overheard a crucial exchange between Rommel and Dietrich that suggested the field marshal's schedule for the day had another, quite different purpose.

ROMMEL: Would you always execute my orders, even if they contradicted the Führer's orders?
DIETRICH: You're the boss, *Herr Feldmarschall*. I obey only you, whatever it is you're planning.

According to Lang, the two men then shook hands, and more was said, but he was unable to hear this.

Before leaving to return to La Roche-Guyon, Rommel took a call from Speidel to say that Allied aircraft were active throughout the area. Dietrich offered him the use of a Volkswagen 'Beetle', which would be much less conspicuous than his open-topped Horch staff car, but Rommel declined.

They left at 1600. Rommel travelled in the front beside his driver, Sergeant Daniel. With him were two staff officers, Lang and Major Neuhaus, and Corporal Holke, who acted as aircraft lookout. Lang wrote that 'All along the roads we could see burning vehicles, and from time to time enemy aircraft forced us to take secondary roads.' They were also slowed down by the large number of French families fleeing Normandy with their belongings on horse-drawn carts. As they reached Livarot at about 1800, intending to take the main road to Vimoutiers, Holke spotted two enemy aircraft circling the town. Rommel told Daniel to take a secondary road that ran through a wooded area they could see some distance away.

In the lead Spitfire, Squadron Leader Le Roux of the South African OC 602 Squadron patrolled the roads around Livarot where burning vehicles marked his earlier strikes. The flatlands of Normandy offered perfect hunting country for ground-attack aircraft; the open fields and the wind kept the roads dirty and ensured that every vehicle threw up a dustcloud, easily spotted from above, while the long, straight stretches of tarmac allowed an easy end run. Only the woods and copses that dotted the landscape offered cover for a German truck or vehicle, once in his sights. On spotting a particularly large dustcloud, he knew it must have been put up by a powerful car – speed rather than the greater weight of a vehicle threw up the most dust – and began his approach.

Crouched on the back seat of the Horch, Corporal Holke shouted a warning as the two enemy aircraft swooped down, the Spitfires one behind the other, following the line of the road, flying low and fast towards the staff car. Rommel, his hand already on the door handle, yelled to Daniel to make a dash for the copse now only three hundred yards ahead and turn in among the trees. But reaching that camouflage was beyond them. Lang saw Rommel turn to stare at the lead aircraft, now five hundred yards behind them and only yards above the tarmac; both men saw its wings spark rapidly with cannon-fire as it strafed the open-topped car.

The first burst hit the left side. Daniel yelped as a shell shattered his left shoulder and arm, and he lost control. Rommel's face was peppered with fragments of glass. The car veered off the right side of the road, struck the stump of a tree and was thrown back across the road at an acute angle. Rommel's door swung open and he was flung head first on to the tarmac as the car rolled over and came to rest upside down in a ditch. Rommel lay on the road, twenty yards back, unconscious.

Lang and Corporal Holke scrambled out as the second aircraft flew low overhead, but its burst of fire missed both the vehicle and the men. As he picked himself up, Lang spotted Rommel on the road and raced back to him. He 'lay unconscious and was covered with blood, which flowed from many wounds on his face, particularly from his left eye and mouth'.

After receiving first aid from the pharmacist in Livarot and a French doctor, he was driven in a commandeered Mercedes to the *Luftwaffe* hospital at Bernay, twenty-eight miles away. The doctor who examined him diagnosed 'severe injuries to the skull – a fracture at the base, two additional fractures, severe injury to the cheekbone, a wound in the left eye, glass splinters in the face, and concussion'. Sergeant Daniel had lost so much blood that he could not be saved. Major Neuhaus had suffered a fracture of the pelvis. Ironically Rommel's staff car had been attacked, and his active participation in the defence of Normandy brought to an end, near the village of Sainte-Foy-de-Montgommery.

The following day as Rommel lay in a hospital bed, Montgomery launched a massive Second Army attack east of Caen. Operation Goodwood was to be the last battle between the two men and Rommel's absence from the battlefield hardly mattered: he had prepared the defences in advance.

The Allies now had more than 1,300,000 men in France (about seventy per cent of them American) and casualties exceeded

100,000. The Americans were still stuck in the *bocage* and the British still blocked by German forces holding Caen. Some spoke of a stalemate and blamed Monty. His answer was Goodwood.

The belief at SHAEF was that this operation would deliver a decisive victory: an armoured breakthrough from Caen south-wards, most likely as far as Falaise, severing the German defensive line and opening the way to Paris. Goodwood began at 0530 with the heaviest aerial bombardment of the war so far, targeting the villages in the path of the British advance and the known locations of German forces. In three hours 1599 heavy bombers dropped 7700 tons of high explosives. Shells from field guns and the warships anchored within range of Caen contributed too. This unprece-dented saturation bombing was expected to be so effective that Eisenhower believed 'the infantry could then practically walk through'. More realistically it was aimed at destroying artillery, anti-tank guns and the tanks harboured inside stone buildings. Rommel's strongest forces were grouped around Caen and the British advance would have to cross Bailey bridges over the Orne and the Caen canal in full view and in range of his guns. Despite Monty's superiority in tank numbers, this difficult approach meant that the odds on the ground were against him and he counted on the air bombardment swinging those odds in his favour (as they had at El Alamein).

It did not. As the tank attack began at 0900 the Germans manned their surviving panzers – some had been positioned in woods outside the bombing area, others simply pulled out of the stone buildings that had collapsed around them – to meet the advance of all three British armoured divisions: Monty's armoured battering ram that would force a way through Caen. He had 720 tanks against Rommel's 230, but an average of three British tanks was lost for each German tank destroyed.

By this time the Panzer Mark IV was being replaced by the Tiger (of which there were eighty in the Caen area) and the Panther (of which there were 200). The guns of British and Ameri-can tanks could not penetrate their thicker armour (except the one in four Shermans equipped with a 17-pounder gun). Reports reached Monty from junior officers indicating the impossibility of breaking through such a defence. He told his army commander, General Dempsey, what he thought of that: 'At a time like this, with large forces employed and great issues at stake, we must be careful that morale and confidence are maintained at the highest level. Alarmist reports, written by officers with no responsibility and little battle experience, could do a great deal of harm. There will therefore be no reports.'

By noon the advance had been brought to a halt. Not only did Monty deny that the German armour was one cause (among others) of Goodwood's failure, he denied that there had been a failure at all. At 1620 (more than four hours after the advance had been stopped) he signalled SHAEF: 'Operations a complete success. Few enemy tanks met and no mines.' In a press communiqué issued twenty minutes later he confirmed that the planned British breakout had occurred: 'Early this morning British and Canadian troops of the Second Army attacked and broke through into the area east of the Orne and south-east of Caen. Heavy fighting continues. General Montgomery is well satisfied with the progress made in the first day's fighting of this battle.' He followed that with a message to Ike timed at 1940 in which he said his three armoured divisions were 'now operating in the open country to the south and south-west of Caen'. In fact Second Army had lost 186 tanks and 1500 men in extending British-held territory (outward rather than forward) by about seven miles into an area east of the Orne. That ground had been gained by 1100 and was not extended after that time. Rommel's defences had held.

The following morning the headline in *The Times* broke the news of a breakthrough and made clear its most authoritative source: 'SECOND ARMY BREAKS THROUGH – ARMOURED FORCES REACH OPEN COUNTRY – GENERAL MONTGOMERY WELL SATISFIED'.

None of it was true, and by then Ike knew as much. Captain

Butcher recorded his reaction: 'Ike said that with 7000 tons of bombs dropped in the most elaborate bombing of enemy front line positions ever accomplished, only seven miles were gained – can we afford a thousand tons of bombs per mile?'

That day Second Army renewed the attack. Dempsey reported that the area still 'bristled with Tigers' and the three divisions lost a further sixty-five tanks. Yet Monty now wrote to tell James Grigg, 'We have got off to a very good start.' His letter contains a clue to his apparent ability to see a success where there was none: 'The great thing is to "write off" enemy personnel and equipment so as to weaken his war potential, and this we will do.' Although in a wider context the Normandy campaign can be seen as a battle of attrition, Operation Goodwood was not and its specific objectives had not been attained.

Eisenhower was now seriously concerned about Monty, and Butcher was concerned about *him*: 'The slowness of the battle, his inward but generally unspoken criticism of Monty for being so cautious: all these pump up his system. It ain't good.' Ike had written to Monty on 7 July about his previous failure to take Caen and now on 19 July he wrote again to say much the same:

The recent advances near Caen have partially eliminated the necessity for a defensive attitude, so I feel you should insist that Dempsey keep up the strength of his attack. Right now we have the ground and air strength and the stores to support major assaults . . . The enemy has no immediately available resources. We do not need to fear, at this moment, a great counter-offensive.

Some among Eisenhower's staff were convinced he was about to sack Monty but there is no evidence he considered that. He was, however, deeply frustrated by the general. Kay Summersby heard him exclaim that 'Monty seems quite satisfied regarding his progress!'

Monty's statements to SHAEF and to the press had clearly exceeded the facts, but Patton was never going to be beaten on hyperbole. Liddell Hart had a prearranged meeting with him at Peover Hall on 19 July:

He remarked that the American forces had penetrated much deeper than the British, at every stage, and that round Caen the British had failed to gain any of their objectives, while the Americans were overrunning the Cherbourg peninsula ... He also said that there were more German divisions facing the Americans than facing the British ... This was quite contrary to the facts.

On 20 July Montgomery called Goodwood off. Had Rommel in his ultimatum to Hitler three days earlier underestimated the strength of his own defences, or had Monty either overestimated the strength of his attack or not persisted with it long enough? Major Hans von Luck, who had fought Montgomery in North Africa and was now commanding Panzer Grenadier Regiment 125, part of General Feuchtinger's 21st Panzer Division placed near Caen specifically to prevent a British breakout, believed the British stopped too soon:

Things became critical, but the British attack was not continued. It was astonishing that the attack in my sector should be so hesitant. The shock from our 88-mm anti-tank guns, the few Tigers and Becker's assault-guns seemed to have struck deep among the British. With only about 400 grenadiers left we had to hold a long front. That was too few to withstand a vigorous attack.

The irony was that Monty, in assuming he would win by *Materialschlacht* – that his superiority in matériel would win the battle for him – was right, but his hesitancy caused him to cancel the operation when, had he continued the fight one more day, the German defence would have broken and a British breakout become possible. In Monty's defence it is likely that such hesitancy was fed by the high level of casualties among the infantry. That was not for him so much a humanitarian as a military concern: he

302

had been warned from London that replacement tanks were available but infantry losses could no longer be replenished. In halting Goodwood he was conserving his scarcest resource.

That same day Air Marshal Tedder (who was no friend of Monty, and was angry that the airfields he needed around Caen had still not been taken) wrote to the general suggesting that he had deceived his colleagues: 'An overwhelming air bombardment opened the door, but there was no immediate deep penetration whilst the door remained open and we are now little beyond the immediate bomb craters. It is clear that there was no intention of making this operation the decisive one which you so clearly indicated.'

The Times felt that it had been misled by Monty, too, and now commented drily that there had been 'too much booming' in the early stages and that 'it is better to do the booming after success has been secured'. More seriously, both Eisenhower and Churchill felt that he had misled *them*. Monty explained that he had not done so: *they* had misunderstood him. From those differing viewpoints two contradictory accounts of Goodwood arose (even before the operation was over). That is not surprising: war is ever so. The remarkable thing about Goodwood is that *both* accounts are derived from Montgomery. They largely repeat the previous 'misunderstandings' that had arisen about his intentions towards Caen.

Monty claimed that Goodwood was a success. The operation could not be said to have failed to take Caen because that was not *and had never been* its objective. His written order to Dempsey before the offensive had begun clearly stated its limited nature. It was intended to draw German forces into the British sector to prevent what he deliberately presented to the enemy as an attempted breakout, thus removing pressure from the American front and allowing Bradley to attempt a genuine breakout. He said that he explicitly communicated this to Eisenhower and Churchill, and there are indeed references to that intention in the plans both leaders must have seen. The problem then is why both men believed they had been told by Monty that he intended to take Caen. It is not good enough, as one of Monty's biographers claims, to say that he explained his true intentions to Eisenhower but that the Supreme Commander 'kept forgetting'.

Churchill, Eisenhower and a good many others, including the British and American press, believed that Goodwood had failed, and they took this belief not from an alternative source but from Montgomery himself. Before the operation began he had said it would 'be decisive', 'have far-reaching effects', and that the front would 'burst into flames'. He would 'loose three armoured divisions into the open country' and then 'anything could happen'. During the operation he informed Eisenhower, SHAEF and the press that it had been a 'complete success' and that Second Army 'attacked and broke through'. He had planned Overlord and knew better than anyone how heavily loaded the term 'breakthrough' was, and the sense that others must take from his statements.

It is interesting to consider how the Germans perceived Goodwood. Major von Luck had no doubt that they had prevented 'a breakthrough in the direction of Falaise'. Writing after the war, he dismissed Monty's claim that it was intended to do no more than tie down German divisions: 'Captured Canadians told us that shortly before the attack Monty had called out to them: "To Falaise, boys, we're going to march on Paris." Anyone who knew Monty and his ambition would have taken it for granted that he would not have been content with a mere "tying down of German divisions".'

The two contradictory accounts of Goodwood first divide at 1640 on 19 July when Monty announced that it had been 'a complete success'. The Germans knew that the British thrust had been stopped, and had known it by midday, more than four hours earlier. Von Luck wrote: 'At my command post I made contact with Feuchtinger. I described the situation as it appeared to me at about midday: "General, I believe that the whole British attack has come to a standstill."' Lieutenant Freiherr von Rosen, commanding a Panzer VI (Tiger) company, agreed: 'The British offensive had come to a halt. The gain in territory was not very great.'

If von Luck had known by 1200 that the British had been fought to a standstill, Monty must have known it by 1640. He stressed the importance of a commander knowing the battlefield situation, and had established a system by which young officers visited the front and reported back personally to him. It is inconceivable that he did not have the facts before him.

His announcement far exceeded the customary 'brave face' applied to a setback: it was blatantly untrue. Monty was no fool and would hardly tell a lie that was certain to be revealed as such within a short period. Therefore it is likely that he knew his statement was not at that point true, but was convinced it would soon become so – the battle situation would catch up with him in time to substantiate tomorrow's headlines. He had perhaps so convinced himself of matériel as the deciding factor in battle, that the facts that mattered most were his greater numbers of tanks and artillery, and his air superiority, and from these pre-existing factors he could deduce the outcome of the day's fighting more accurately than reports coming in from the field. If the pure mathematics of attrition allowed him to reject reports from field officers about the unplanned effects of Tiger and Panther tanks, it might also have convinced him that the battle would eventually catch up with his 'results'.

Monty argued consistently that Goodwood and his previous operations around Caen were intended to draw German divisions into that area and away from the American sector, thus weakening German resistance in front of Bradley and facilitating his breakout. Evidence for this can be found in both his original plans for Overlord and his later plans for Epsom and Goodwood. Supporting that, Bradley did plan to launch his First Army push, Operation Cobra, two days after Goodwood.

One difficulty with this argument is that the Germans were already convinced that the British front presented the greatest threat, and already had the bulk of their forces there. Enigma transcripts dated between 7 and 17 July – just before Goodwood began – indicated that German forces facing the Americans were already so sparse that they could not have held against a First Army attack. A successful breakout from the American sector was made even more likely by their 'secret weapon'. It was virtually impossible for tanks to move through the thick *bocage*; the Germans believed, not unreasonably, that their relatively few tanks, which had to keep to hard surfaces, could turn back an assault by enemy armour whose operation was similarly restrained. The Americans, however, had developed the Rhinoceros, heavy steel prongs welded to the front of their Shermans that made short work of hedgerows and banks and allowed them to operate off-road. Shermans could cut across country and take enemy Panthers and Tigers, restricted to hard surfaces, in the flank.

Why, then, was Goodwood necessary at all, if the effect on the German forces it was intended to bring about was already the case? Montgomery needed a success to answer accusations that he had not made the promised gains in the area of Caen, to support his ambition to retain command of Allied land forces (despite a previous agreement that it would eventually be transferred to Ike), and to confirm and capitalize on the public adulation he had experienced while touring Britain (which specifically compared him with Wellington).

Just as significantly, Eisenhower needed a British success too. He was under pressure from Washington and the President's concerns were regularly conveyed by Henry Stimson, who shared them. The American public had gained an impression that the US was fighting for primarily British interests (instead of concentrating its war effort against the Japanese, who had attacked *them*) and that Americans were doing a disproportionate amount of the fighting (and suffering disproportionate losses). And 1944 was a presidential election year. Stimson made it clear: 'I told Eisenhower that this was no Anglophobia criticism of the British but a real problem arising out of British limitation of strength . . . and the public relations problem in the US which might arise from that in a presidential year.' The success of Goodwood would ease that problem. After its failure Eisenhower put his disappointment in writing to Monty: 'I thought that at last we had the enemy and were going to roll him up . . .' He thought *that* because of Monty's own optimistic statements about its likely outcome.

There can be no certain and final judgement on Monty's objectives for Goodwood, and that may be precisely the outcome he intended. Typical of his handling of the operation was his written order to Dempsey before the offensive began, which stated the limited nature of the assault. Monty normally gave only oral instructions: it was not his way to give written orders. This sudden break in *modus operandi* can be interpreted in two ways. He may have realized that Eisenhower and others might be expecting more from Goodwood than was planned, and thought it necessary to put his limited aims in writing. Or his motive might have been altogether more Machiavellian: the order provided an indelible justification, if no breakout occurred, for the negligible *territorial* gains made by Goodwood, but its very existence could remain unknown to others if his divisions broke free.

Monty took the lessons of history seriously and none more so than those of his immediate personal history: El Alamein and Sicily. He expected Goodwood to be a second El Alamein which, calculating by his matériel superiority and the mathematics of attrition, he must win. But if it turned out to be a second Sicily – if he was once again held at Caen while Bradley (*à la* Patton) burst out of the American sector – he could claim credit as the grand architect of an overall Allied success (with the added gloss of his own army bearing the full attention of the panzers in order to ease Bradley's passage through the front).

Given his several aspirations in mid-July 1944, it is this writer's conclusion that on the balance of probability Monty *did* intend Goodwood to break through German defences and pass beyond Falaise. The primary evidence comes from Monty himself. Yet so skilfully did he establish an 'alibi' that the opposite interpretation can be argued with as much recourse to the oral and written statements of the man himself.

In his *Memoirs* he confirms that Goodwood was meant to be a deception, not a breakthrough. It is not unreasonable to suggest that, had German opposition crumbled and Monty's army broken through, his *Memoirs* would have quoted this latter outcome as that which he had always intended. Not only was that Monty's way, it would by then have been his unalterable belief.

Whatever the verdict on Montgomery, for Rommel this final clash with his old enemy has to be counted a success. Major von Luck, about to be promoted to lieutenant colonel for his own part in stopping the British, praised Monty's battle management but felt he had been defeated by Rommel's meticulous preparations:

Goodwood was a masterpiece of preparation and logistics. Yet we were able to prevent the enemy from making a breakthrough. Only then did we hear that Field Marshal Rommel had been severely wounded in a fighter-bomber attack on his car. We could hardly take it in; to us he had seemed invulnerable. Nevertheless, the defensive front set up by Rommel had held against Montgomery's attack. As late as 15 and 17 July he had checked this defence in the area of our corps. It is probably true to say that with this Rommel had denied his constant adversary the way to Paris, his last military success.

While Patton told the brutal truth about war to the men of his Third Army, and Rommel risked everything by telling the truth about the western front to Hitler, a beleaguered Montgomery now survived by telling a blatant lie to Winston Churchill. On 18 July, the first day of Goodwood, the Prime Minister had sent a signal saying that he intended to visit the headquarters of Second Army. The operation was not going well and Monty immediately signalled back, via Eisenhower, to say he could not accept visitors at that time. Ike suggested Churchill visit the invasion beaches instead, but the Prime Minister was furious. He told General Brooke: 'With hundreds of war correspondents moving about freely this cannot be considered an unreasonable request. If however General

Montgomery disputes about it in any way the matter will be taken up officially, because I have both a right and a duty to acquaint myself with the facts on the spot.'

Brooke had a meeting with Monty scheduled for the following day and informed him of the Prime Minister's reaction. He made it clear that the situation was critical and that Monty's job was in question. For once, Monty's confidence was shaken and he backed down. At Brooke's suggestion he went straight to his caravan and wrote to Churchill.

My Dear Prime Minister
I have just heard from the C.I.G.S. that you are proposing to come over this way shortly; this is the first I have heard of your visit. I hope that you will come over here whenever you like; I have recently been trying to keep visitors away as we have much on hand.

But you are quite different, and in your capacity as Minister of Defence you naturally are above all these rules. So far as I personally am concerned I hope you will visit Normandy whenever you like; and if I myself am too busy to be with you I will always send a staff officer. And if you ever feel you would like to stop the night and stay in one of my caravans – which will be held ready for you at any time.
Yours sincerely
B. L. Montgomery

In other circumstances Monty might have asked Brooke to read the letter and confirm that it 'would do', but he dared not: Brooke would know that it began with an outright lie. Churchill's message about the proposed visit had reached Monty and he had personally rejected the idea. By stating that 'this is the first I have heard of it', he blamed others by implication for a serious lapse, communications between a prime minister and his senior commander in the field being of the utmost importance. The lie provides an insight into the man, not because it was particularly serious – it was not – but because it was unnecessary. Another man might have apologized for attempting to put off the Prime Minister's visit, quoting the pressures that occupied a headquarters and its commander in mid-operation, asking for (and undoubtedly receiving) understanding. Monty was incapable of that apology. He could not be wrong. What he could be, however, was *misunderstood*. If others had interpreted his wish to keep visitors away at this time to include the Prime Minister, they were mistaken.

Churchill arranged to visit Monty at his headquarters the very next day. Monty did not know whether his letter had placated the Prime Minister or whether he was still in trouble. His staff officers believed it to be the latter. Johnny Henderson wrote afterwards:

It was common knowledge at Tac HQ that Churchill had come to sack Monty. He came in his blue coat and in his pocket he had the order. There was quite an 'atmosphere'. However Monty showed not the least nervousness. He shook hands, took him into the Operations caravan – and when they came out Churchill was beaming. But we all knew how near 'Master' had come to being sacked.

There is no record of what was said between the two men in Monty's caravan, although he admitted to giving Churchill a bottle of brandy 'as a peace offering'. Henderson felt that Monty had escaped by turning on the charm. By then Churchill had awkward questions to ask about Goodwood, and Monty's defence of himself might well have included the assertion that his intentions had been misunderstood, just as his veto on visitors to Tac HQ had been misunderstood to include his present guest. However, there was a further reason for the Prime Minister to have been 'beaming' as he left. The first news was just coming in of an astounding event in Rastenburg and both men must have savoured the possibility of an early German surrender.

At approximately 1235 on Thursday, 20 July, during Hitler's conference at his *Führerhauptquartier* in East Prussia, Colonel Heinz Brandt crossed his legs beneath the long, oak table. In doing so he unwittingly kicked and moved the briefcase of his neighbour,

Colonel Count Klaus von Stauffenberg, and might have quietly apologized for it, had Stauffenberg not just left the room to take an urgent telephone call. Across the table Lieutenant General Adolf Heusinger was briefing the Führer, and the twenty-one senior officers assembled there, on plans to counter a Russian advance on the eastern front west of Duna. Field Marshal Keitel, who had asked Stauffenberg to attend the meeting to report on the raising of new divisions, glanced worriedly towards the door: Heusinger was nearing the end of his report and Stauffenberg was scheduled to speak next.

The conference should have taken place in Hitler's underground bunker but because engineers were strengthening it against a possible direct hit from the air, the meeting took place in a hut. Stauffenberg was waiting nervously outside. There was no telephone call. Before the conference began he had reached carefully into his distinctive yellow-leather briefcase packed with explosives, triggered a timing mechanism with his one good hand – the other had been smashed beyond saving by an enemy shell in Tunisia – then placed the case at his feet beneath the table. He had been seated only two places from Hitler; the Führer had even welcomed him with a nod of recognition. Now his escape was planned and a plane was waiting, but he wanted to be certain before leaving.

The bomb exploded at 1240. Some of those present were hurt by the force of the explosion: General Schmundt, Hitler's military secretary, lost both legs and bled to death before he could be helped. Others were impaled by fragments of oak as the table disintegrated: General Korten died immediately with a stake through his chest. But Colonel Brandt had unwittingly saved the Führer's life. In crossing his legs he had moved the briefcase-bomb behind one of the table's heavy wooden legs, which shielded Hitler from the worst of the blast. Brandt survived too, although ironically he lost a foot.

Keitel, who was unhurt, helped Hitler from the acrid smoke that hung over the wounded and the rubble. The Führer's face was cut and blood streaked his cheeks, his right arm was temporarily paralysed, his trousers had been virtually ripped away and his left leg was badly scorched, but he had survived. The doctor who treated his wounds removed more than a hundred small splinters of wood from his left thigh. 'Someone must have thrown a grenade,' Hitler said. But fragments of a yellow-leather briefcase were clearly visible at all points of the demolished hut, and a search for Stauffenberg established that he had passed through the compound guard post immediately after the explosion.

That afternoon Admiral Ruge was attending the funeral of Daniel, Rommel's driver. Afterwards he heard about the bomb. The first reports – *Attentat! Führer ist tot!* (Assassination attempt! Führer is dead!) – proved incorrect. Even military men engage in 'what if' suppositions and his first thoughts were that if Rommel had been less seriously injured by the RAF, the field marshal would by now be speaking with Eisenhower:

Rommel in our talks agreed that to end the war a political solution was imperative. In my opinion he was the only man in Germany strong enough to bring about a change, even after the revelation that Hitler was still alive. I believe he would have mastered the situation in the west if on 20 July he had been able to act. I think it possible that the offer of an armistice proposal to the Allies as planned by him would have been accepted. The moment was favourable after the heavy British losses at Caen, where the unsuccessful breakthrough attempt had cost as many British lives as they had estimated it would cost them to get to Berlin. The 20-mm shell, which fatally injured Daniel, greatly influenced the course of the war.

Bradley's plan for an American offensive to follow Goodwood was twice postponed because bad weather prevented the preliminary air bombardment. Operation Cobra was finally launched on 25 July and immediately blighted by two Blue-on-Blue incidents in which American planes bombed their own troops. Bradley attacked on a narrow front west of St Lô, German forces stood firm and by dusk he believed the offensive had failed. The next day General Collins led VII Corps on an advance of six miles, breaking the German front, and when Bradley sent First Army divisions swarming through the gap, enemy resistance collapsed. Two days later they took Coutances and Bradley realized the way was open for a rapid advance south of the *bocage* to Le Mans. It was an opportunity for what Liddell Hart called the 'expanding torrent' and the man most suited to it was Patton. That day Bradley placed VIII Corps under Patton's command.

The slapping incidents still hung over him and some wondered whether he was really up to the job. His close friend Everett Hughes confessed as much: 'I was fearful that he had been cowed by the fools who didn't realize that a fighter couldn't be a saint or a psychiatrist when the job was to kill Germans. Or cowed by those who didn't like pearl-handled pistols or fancy uniforms or all the little idiosyncrasies that are George.'

Hughes need not have worried. Patton told Collins: 'You know, Collins, you and I are the only people around here who seem to be enjoying this Goddamned war! But I'm in the doghouse, I'm in the doghouse! I got to do something spectacular!'

The Allies' secret weapon had been launched. He wrote to tell Beatrice: '*L'audace, l'audace, tout jour [sic] l'audace!* When I emerge it will be quite an explosion.'

14. Patton's Blitzkrieg and the 'Montgomery Thrust'

Patton's first and most inspired act in command of US Third Army was to have standard Michelin road maps issued to every unit. There were few Germans left to fight in Brittany – following Bradley's breakthrough most enemy forces had been pulled back to consolidate a line further inland and prevent a general push out from Normandy – but his advance was blocked by one great Allied traffic jam. Bradley's rapid advance meant that Third Army was left far behind the front and its first task was not to fight but to reach the enemy.

All transport moved slowly, if at all. The sparse road network was blocked by trucks ferrying troops and supplies to the front, and there was no space for Patton's seven divisions. As there was no system for giving one unit precedence over another he sent staff officers out to every bottleneck and crossroads, and told them to 'ignore the battle plans and sort out the Goddamn traffic'. If they had to become 'traffic cops' they must do so. Bradley was impressed: 'Every manual on road movement was ground into the dust. Patton and his staff did what couldn't be done: it was flat impossible to put a whole army out on a narrow two-lane road and move it at high speed. Everything was going to come to a halt.'

Within seven days he had caught up with the front and had his army through the Avranches–Pontaubault gap, beyond which there was open country and a retreating enemy: 'I am going in there to kick someone's ass.'

Patton's Third Army had become operational at 1200 on 1 August 1944 as part of a larger reorganization of the Allied command structure. General Courtney Hodges took command of

First Army. Together these two armies formed the US 12th Army Group under General Bradley. Montgomery, as overall land-forces commander, had tactical command of both the British 21st Army Group and the US 12th Army Group, and answered only to the Supreme Commander, Eisenhower. Patton was far from pleased to have Monty two levels above him in the command structure, but after the indignity of FUSAG he now had a 'real' army, and a powerful need to prove himself that would drive it across France.

His orders to his commanders virtually repeat Rommel's words prior to the 1940 invasion of France. Patton told them to keep moving forward and to ignore their flanks because 'flanks are something for the enemy to worry about, not us'. He did not want to receive reports informing him that positions were being held. 'We are advancing constantly and we're not interested in holding on to anything except the enemy. We're going to kick the hell out of him.'

He had one regret. He had known that Rommel was in charge of the German defences and still looked forward to meeting him in combat. But on the day that Third Army became active the British (informed by German prisoners) announced that Rommel had been seriously injured by an RAF attack on his car and was believed to be dead.

He should have been. Moved from Normandy to the Le Vesinet Hospital outside Paris, X-rays showed that he had sustained a quadruple skull fracture. His doctors told him that according to the textbooks such an injury invariably results in death. He suffered frequent, severe headaches. Ruge visited every day to read to him. Often in a semi-delirious state Rommel asked to see Hitler so that he could tell him Germany's only chance was to make peace in the west and transfer all forces to the east, to save themselves from the Russians. It was several days before he was told of the assassination attempt and Lang claimed that he visibly blanched and called out, 'Madness! Against the Führer! No one wanted that!' He dictated a letter to Lucie, assuring her that he would recover and condemning the bomb plot: 'Coming straight after my injury, the attempt on the Führer's life has shocked me deeply. Thank the Lord that the worst did not happen.'

In reply to the British announcement he was helped to pull his uniform jacket over the shoulders of his hospital pyjama top and had himself photographed from the uninjured side. He even came up with a caption: 'I'm not gone yet.'

Meanwhile in Berlin Ernst Kaltenbrunner, chief of the Gestapo, was compiling a report for Hitler on the interrogation of Colonel Hofacker, who had been part of the plot and had implicated several others, most prominent and surprising among them being Field Marshal Rommel.

The Allies' strategic plan for the period following the establishment of a secure bridgehead in Normandy required the occupation of the north-west of France up to the Seine. It was expected that there would then be a pause while the army was resupplied, before the Seine was crossed and the advance towards Germany begun. The task of Patton's Third Army was first to move south-west to secure Brittany, then head east towards the Seine.

Within two days Patton had advanced fifty miles and taken Rennes, and by the end of the next day he had sealed off the whole Brittany peninsula. Eisenhower considered it a remarkable achievement but not everyone was impressed. The Germans were retreating and he had not yet met determined resistance. Andy Rooney of the US Army newspaper *Stars and Stripes* wrote: 'Patton's Third Army went west, away from the main German forces, so naturally they made great distances every day. They weren't fighting anyone, that was the thing.' But that hardly mattered to the press in the US. After weeks of reporting US forces stalled in the *bocage*, ground was being taken at a phenomenal rate. Eisenhower had still not released the name of the Third Army commander and there was no mention of Patton in the British or American papers.

By the end of the first week Patton's blitzkrieg had carried him to St Malo, Dinan and Brest. Instead of stopping to engage the enemy he deliberately bypassed them in order to speed up his advance – an enemy cut off was in any case an enemy neutralized – in order to take his final objective by surprise. When 4th Armored Division, commanded by Robert Grow, was halfway to Brest it was ordered to halt by the VIII Corps commander, Troy Middleton, in order to engage an isolated German unit. Patton was furious and told Grow: 'Don't take any notice of this order, or any other order telling you to halt, unless it comes from me. Get going and keep going until you get to Brest.'

Now he was turning a large part of his army to move east towards the distant Seine. He told Major Edward Hamilton that 'I'll be in Paris in two weeks.' Thunderbolt strike planes flew constant missions in support of his advance – his leading tanks carried bright fluorescent panels to distinguish friend from foe, but in a new development the command tank had radio contact with the air support and could direct their fire as effectively as the tanks. He covered fifty miles a day and destroyed four enemy tanks for each one he lost (although a good number of these 'hits' must be credited to the Thunderbolts).

From a German perspective the Allied breakout they had feared had happened and Field Marshal von Kluge immediately began a general withdrawal to the Seine. He was astonished to receive a quite different order from Hitler: he must make a counter-attack 'to annihilate the enemy'. Von Kluge ordered Seventh Army and Fifth Panzer Armee to attack westward. He was perhaps too candid with his army commanders, telling them that he could 'foresee the failure of this attack' but '*es ist ein Führer Befehl*' (they are the Führer's orders).

Patton's progress in the field had been remarkable but it was Bradley, staring at Allied and enemy positions on a headquarters map desk on 7 August, who suddenly realized they had an opportunity to trap and destroy virtually the whole of the German Army in Normandy, amounting to about 100,000 men and their equipment. The Germans were heading west into the gap between

Montgomery's Army Group to the north and Patton's Third Army to the south. Patton was moving towards Le Mans, and Bradley saw that if he swung north to Argentan the enemy would be trapped between him and the Canadian First Army, which was moving south towards Falaise. He put the idea to Eisenhower who immediately rang Montgomery, and they agreed to encircle the Germans in the Falaise–Mortain sector. Writing after the war Montgomery described what he called 'my plan' and how 'I decided' to bottle up the enemy.

While part of Patton's army moved north to spring the Falaise trap, the rest moved on to Le Mans and took the city the following day. Patton's focus was shaken momentarily by a letter from Beatrice. She had discovered that Jean Gordon had joined the Red Cross as a volunteer and was in England, and was clearly worried that he might be seeing something of her. He replied straight away to assure her that 'The first I knew about Jean's being over here was in your letter. We are in the middle of a battle so don't meet people. So don't worry.' He told Everett Hughes about this exchange of letters. They both knew more than Beatrice. Hughes wrote in his diary: 'Jean Gordon to France. Will please Uncle George.'

On 8 August Rommel insisted that he be moved from Paris to his home in Herrlingen, against the urgent advice of his doctors. To convince him of the seriousness of his injuries a human skull was procured from the pathology department and hit with a hammer for him to witness the results. He merely repeated his demand. He was determined not to fall into the hands of the enemy. Ruge felt that he had more to fear from his own people than the Allies and that becoming a prisoner of war might be the best for him, but 'I never plucked up the courage to suggest it.' Rommel agreed to the condition of a midway overnight stop to rest, but once his ambulance was on the road he cancelled that arrangement. Some of Germany's top eye and brain specialists were near Herrlingen at Tübingen University and he was placed under their care.

In Berlin the evidence against him was growing. The Gestapo interrogated Colonel Georg Hansen, who had attended the meeting of conspirators on the occasion that Hofacker claimed he had recruited the field marshal. Hansen could not confirm that Rommel had known about the assassination attempt, but it was reported by Hofacker that he had said the front would collapse in three weeks and that he would take the necessary action after the Führer had been 'removed'. Hitler resolved to question Rommel as soon as he was well enough. He did not want to believe the worst.

Although by 12 August Third Army's XV Corps was approaching Argentan and was in position to spring the trap, the Canadians were still some distance from Falaise. Patton told his commanders to 'pay no attention to Monty's Goddamned boundaries. Be prepared to push even beyond Falaise if necessary.' At the same time he appealed to Bradley to let him continue north, beyond his agreed 'halt' position, to close the gap between himself and the Canadians and complete the encirclement of the enemy. Bradley refused to allow it: Patton would be moving north to occupy territory allotted to the Canadians, who were moving south, and these two armies might meet and take each other for the enemy. There had been too many Blue-on-Blue losses already. He ordered Patton to hold his present position. Patton was indignant: 'The XV could easily have entered Falaise and completely closed the gap, but we were ordered not to do so. The halt was a great mistake. We had reconnaissance parties near the town when we were ordered to pull back.'

He believed that 'The order emanated from Montgomery and was either due to jealousy of the Americans or utter ignorance of the situation.' Bradley insisted that it was his decision and not Monty's to halt Patton at Argentan, but nothing could persuade Patton that the order, from whomever it came, was not intended to allow Monty's Canadians to catch up. He was mad at the British (and for Patton that term included the colonial nations) and pleaded with Bradley: 'Let me go on to Falaise and we'll drive the British back into the sea for another Dunkirk.'

The Falaise encirclement had been Bradley's idea. But when by 14 August the Canadians had still not closed the gap and German troops were escaping in their tens of thousands, he had to choose between talking to Monty with a suggestion that Patton be permitted to move up and close the gap himself, or allowing Patton to continue his advance east towards the Seine. He decided that securing a bridgehead on the Seine was of greater tactical value than the operation at Falaise: 'If Montgomery wants help in closing the gap, I thought, then let him ask for it. Since there was little likelihood of his asking, we would push on to the east.'

The British press had given the credit for the rapid dash across France to Montgomery. He was in command of all land forces and it was 'his' army. Patton's name had still not been released to the press. Ike's son John wrote that 'This exaggerated canonization of Monty was exacerbated by the fact that the British press, radio, and newsreels emphasized British accomplishments to everyone else's expense.'

Ike had the remedy. On 14 August he released the name of the Third Army commander and now it was the turn of the American papers to proclaim a national hero. There was much boasting of 'an American blitzkrieg'. Patton's name was in the headlines and his photo on the front pages. After Ike's announcement, Patton quipped to correspondents, 'That's the first I've heard of it.' Asked how he made such rapid progress he joked that 'I never worried about my flanks, which was probably due to my long-felt masculine virility.' It was also due to his awareness that the fighter-bombers of the XIX Tactical Air Command were constantly overhead.

Bradley had told Patton to continue his eastward advance and the following forty-eight hours saw the most astonishing blitzkrieg of all. By 16 August he had reached Orléans, Chartres and Dreux, and was overtaking enemy forces that were falling back. However, those among his biographers who credit him with 'out-Rommeling' Rommel make too straight a comparison. It had certainly been the German Army that set the standard for a blitz-krieg advance, but that same army was slow in retreat. In its infantry divisions the main form of transport was the horse. Remarkably the German forces in France in August 1944 had 14,500 trucks and 60,000 horses. The Tiger tank, latest star of the panzer divisions, was made all but invulnerable in battle by its unprecedented 100-mm-thick armour, and was, by the same token, an extremely slow-moving vehicle. Patton had four-wheel-drive trucks and half-track troop carriers for his infantry and no use for horses; his tanks were faster than the panzers, all of his artillery was motorized, and he could rely on air superiority. He was making notes on the campaign as it proceeded, much as Rommel had done in 1940, and he recognized the reciprocal advantages of a combined air and ground advance:

Just east of Le Mans for about two miles the road was full of enemy motor transport and armor, which bore the calling card of a fighter-bomber – namely a group of fifty-caliber holes. Whenever armor and air can work together, the results are excellent. Armor can move fast enough to prevent the enemy having time to deploy off the roads, and so long as he stays on the roads the fighter-bomber is his most deadly opponent. To accomplish this teamwork, two things are necessary: intimate contact between air and ground, and incessant, ruthless driving on the part of the ground commander. A pint of sweat saves a gallon of blood.

Every assessment of Patton's rapid advance includes his claim that 'The Third Army had advanced farther and faster than any army in history', as the general's own comment on his armoured blitzkrieg. In fact that statement in his campaign notes, later to be published as *War As I Knew It*, followed from a much more astute observation, which indicates his own insight into the factors that enabled his progress. The movement of a large army presents con-siderable difficulties in its own right, quite distinct from those resulting from the presence of an enemy, which is why Montgomery had to halt so often to 'regroup' and allow his 'administration' to catch up. Patton wrote: 'The system of administration in the Third Army passed direct from divisions to army, leaving the corps as a tactical unit. Because of this arrangement we had perfect facility in shifting divisions without losing a moment's time. We never had to regroup, which seemed to be the chief form of amusement in the British armies.'

With Third Army in Orléans, Chartres and Dreux, Monty explained to his staff officers that 'The general picture in this part of the front is that Patton is heading for Paris and is determined to get there and will probably do so.' As he commanded all land forces, when addressing war correspondents he made a point of referring to Patton's troops as his own, which technically they were. He wrote to tell the Reynoldses that 'I now have troops in Orléans, Chartres and Dreux.'

That, Bradley now ordered Patton, was where they must for the moment hold. He was within striking distance of Paris and the Germans were hurriedly evacuating the city, but he must make no move towards it. For political reasons it had been decided that General Jacques Leclerc's French 2nd Armoured Division should enter first. That detail did not stop American correspondents who went in with the French describing the liberation of the capital by 'Patton's Third Army'.

In his frustration at having Paris 'taken' from him by the French, Patton accused Bradley of 'losing his nerve' (although, of course, the decision had been taken at the highest political level). It was, he felt, a missed opportunity. An Allied bridgehead along the Seine could feasibly trap the large number of German troops still west of the river, many of them falling back from the Falaise area. When he heard that Monty had proposed dropping a British airborne division between Paris and Orléans to cut off the German retreat, he exclaimed, 'If I find any Limeys in my way, I will shoot them down.'

Bradley had another reason for temporarily halting Patton's advance. It was feared that he might over-extend his supply line. Supply problems were increasing but it is perhaps more true to say that Patton had outrun SHAEF plans for the campaign, which were necessarily being developed in real-time. His advance had moved faster than the overall strategy could be adjusted to encompass it and project forward from it. Patton had to wait for SHAEF to catch up with him. In any case the overall strategic plan for the campaign east of the Seine was an advance on a broad front and for that to occur the rest of the Allied forces had to catch up with Patton too.

On 18 August Field Marshal von Kluge, who was otherwise quite busy attempting to salvage what he could of German forces in France, was asked to attend an interview in Berlin. He broke a cyanide capsule instead. He knew that to be a more certain escape than the recent bungled attempt of General von Stülpnagel to shoot himself, which had merely resulted in the loss of one eye but brought the Gestapo to his hospital bedside, where in his delirium he had repeatedly called the name 'Rommel'. Von Stülpnagel was tortured until he talked, then hanged. Von Kluge wished to go quietly.

The secret unit established in Berlin to investigate officers suspected of involvement in the bomb plot now made an interesting link. On the evening of the explosion von Kluge had been at La Roche-Guyon and von Stülpnagel had visited him there. Rommel himself was by then in hospital, but the number of links to him was increasing.

At Herrlingen Rommel was able to sit in the garden to take the sun. He still suffered from regular headaches but Professor Albrecht of Tübingen University felt his rate of recovery was quite remarkable. Rommel was puzzled that neither Hitler nor any of his staff had enquired after his health, but then, he told Lucie, there was a war on.

Patton's Third Army was halted but he sent out regular patrols and encouraged them to advance further than their reconnaissance tasks required. On 19 August a patrol moved forward thirty miles and discovered an intact bridge over the Seine at Mantes-la-Jolie, thirty-five miles south-east of Paris and half that distance from Rommel's headquarters at La Roche-Guyon (which had been hastily evacuated the previous day). Patton joined them shortly afterwards, then returned to his command truck to use the scrambler-radio. He asked Bradley's permission to cross the Seine and was told that he should not be anywhere near it. Patton replied that he 'in fact had pissed in the river that morning and just come back from there', and asked whether he was now to pull his advance units back. He was given permission to cross and establish a bridgehead, and by the next day he had his whole army over.

The German retreat was now frantic. Falaise had been occupied with 10,000 German troops killed and 50,000 taken prisoner. British and American accounts emphasize the high number of Germans to escape the encirclement because of the delay in closing the gap, suggesting that it failed in its primary objective, while German accounts bemoan its unexpected success. General Speidel, Rommel's chief of staff, complained that 'There were barely a hundred tanks left out of six panzer divisions.' Field Marshal Walther Model calculated that 'All that remained of eleven armoured divisions amounted to eleven regiments, each with five or six tanks.' Surviving forces were falling back to the *Westwall* (Western Wall), known to the Allies as the Siegfried Line, fortified by pillboxes and minefields, and as the last line of defence it was expected that those manning it would fight to the last.

On 21 August a group of senior Pentagon officers visited Eisenhower's command post in Normandy, then flew on to see Patton. This was an influential audience and Patton put on a one-man performance for them, its tone perfectly captured by one of the group, Lieutenant General Brehon Somervell, the US Army's Chief of Supply:

Patton's HQ near Brou. He is in trucks, parked in a forest. Full of life, banging away. Wants to go straight through to Germany without stopping. Wants to push as fast as he can, before Germans can organize in front of him. Certain that he will go fast unless higher command stops him. Has Germans on the run. Showed figures on his casualties and Germans – 12,000 to 120,000.

That convinced Somervell, who fully understood the implicit requirement that 'higher command' provide adequate supplies for the dash: 'Patton has the right idea – straight ahead.' It was not Somervell but Eisenhower who commanded the distribution of supplies in the battle zone, and two days later Montgomery staked his own claim for the lion's share of available resources.

Both Montgomery and Patton separately presented to Eisenhower their plans for extending the campaign across the Rhine and winning the war. These were the two 'heroes' of the European campaign so far (judged by the sentiments of the British and American press and public, if not the SHAEF strategists and the politicians) and their opposing plans met on Eisenhower's desk. The conflict between them – both the plans and their outspoken advocates – was to occupy much of Ike's time for the remainder of the war.

Montgomery argued that the war could be quickly won if his 21st Army Group now made a powerful thrust north through Belgium and Holland, then on into the Ruhr, and he asked that several American divisions be transferred to his command, along with additional ammunition and fuel, to enable him to do so. Patton's plan (although he called it 'a strategical idea') was that Third Army should advance from the Seine south of the Argonne and on into the Saar, and he, too, would require more than his proportional share of the available fuel and ammunition.

German defences were collapsing and either of these thrusts might have worked. It is sometimes said that Eisenhower was aware of the personal antipathy between the two men and, not wanting either to feel *his* plan had been rejected in favour of the other, he rejected both and chose to advance on a broad front. But it had always been the Allies' intention to advance from the Seine on a broad front. Ike merely applied the agreed strategy. Claims that if he had immediately thrown all Allied resources behind either Monty's or Patton's single-thrust plan – and the Germans were in such disarray that it probably did not matter which – then the war could have been shortened by several months may be correct but cannot be proven. As it was he included both plans within his broad-front advance, which denied each the preponderance of supplies it would have required as a single thrust.

To the north Monty was advancing into Belgium to take the deep-water port of Antwerp. Use of the port was vital in landing the equipment and supplies needed by the vast Allied force, and for that reason Ike did see that he received adequate fuel. To the south Patton's advance came to a halt south-east of Paris because Third Army ran out of fuel, and Patton was not ready to accept Ike's sound operational reasons for giving Monty's fuel needs priority over his own: 'The British have put it over again.' He complained to Beatrice: 'I have to battle for every yard but it is not the enemy who is trying to stop me, it is "They". If I could steal some gas I could win this war.'

Patton's near paranoia – it is difficult to know whether he believed his own most extreme comments or merely indulged a weakness for Brit-bashing – was misplaced. His own fuel and ammunition needs could be more easily met if the deep-water port of Antwerp was taken. But that was not how it seemed to him on 29 August as Third Army approached Rheims, fifty miles east of the Seine. Patrols reported no enemy forces ahead. He could make a rapid advance, cross the Meuse and Moselle rivers, and break through the Siegfried Line before the enemy had prepared the defences: 'There was no real threat against us . . . Suddenly it was reported to me that the 140,000 gallons of gas we were to get that day had not arrived . . . the delay was due to a change of plan by the High Command, implemented in my opinion by General Montgomery.' He asked for the fuel and for permission for 'a rapid advance to the east for the purpose of cutting the Siegfried Line before it could be manned'. He was denied both.

He disobeyed the order and found a way of extending his limited fuel, telling his commanders that when necessary they should drain three-quarters of their fuel tanks in order to fill the remaining quarter, and keep moving forward. By 1 September his most advanced troops had reached the east bank of the Meuse near St Mihiel and had taken Verdun.

Eisenhower assumed direct command of all Allied land forces on 1 September and Monty henceforth had command of the British 21st Army Group only, as agreed from the outset. To the British public – fiercely nationalistic in time of war, and expressing that not only in hatred of the German enemy but also in a growing antagonism towards the American ally – it appeared to be a demotion. At a London press conference Eisenhower said that anyone who thought that 'simply won't look facts in the face'. He called Monty 'one of the great soldiers of this war' (and did not know that Monty had written of him that 'Ike's ignorance of how to run a war is absolute and complete'). Monty could have helped by making an announcement of his own confirming the facts – this change in the command structure had been agreed before the invasion – but he kept quiet. In recognition of his achievements (and perhaps to placate both him and the British public) he was promoted to the rank of field marshal. King George VI sent him congratulations and noted his 'masterly plan' for Overlord.

Patton wrote to tell Beatrice that 'The field marshal thing made us sick, that is Bradley and me.' Bradley himself went further in a letter to Bedell Smith: 'Montgomery is a third rate general and he never did anything or won any battle that any other general could not have won as well or better.' And Bedell Smith himself, arguably more impartial than either Patton or Bradley, wrote that Montgomery was 'not worth a damn without Patton' and thought the baton only came to him after 'behind-the-scenes conniving to enhance his own prestige'.

Despite Monty's pre-invasion acceptance that Eisenhower would eventually take direct command of land forces, he now began a campaign aimed at having that command transferred back to himself. It began quietly, with the not unreasonable observation that Ike's headquarters at Granville were by now four hundred miles behind the Allied front line: 'This was a suitable place for a Supreme Commander; but it was useless for a land force commander who had to keep his finger on the pulse of his armies.'

The day after taking direct command Eisenhower flew to see Bradley in Versailles but his main message was for Patton. Kay Summersby noted in her diary that 'E. says he is going to give Patton hell because he is stretching his line too far and making supply difficulties.' He knew what Patton would never concede: that there is a point at which the benefits won by a blitzkrieg advance are outweighed by the extra logistical problems it creates. He may also have suspected what Patton later admitted: that he was deliberately pressing ahead and refusing orders to conserve fuel because the fuel thus saved would be transferred to Montgomery to sustain *his* drive.

Patton was called in to Bradley's headquarters to see Ike. His response to being given 'hell' for stretching his line was to ask permission for a further thrust at the Siegfried Line, and for the additional fuel and ammunition that required. Ike refused. Bradley and Patton were not natural friends, but they were brought together now by a common resentment that an American commander was starving them of fuel to allow a British commander to continue his advance. Bradley told Patton that he would turn a blind eye if he was able to secure a crossing over the Moselle and close on the Siegfried Line 'if I could get the fuel to move'.

Accounts of Patton's dash through France figure large in the history books and there is little mention of Monty's blitzkrieg which began on the very day he became a field marshal. American war correspondents had reported his promotion by comparing the prolonged difficulty in extracting his army from Caen to Patton's rapid advance across France. The next day British correspondents extolled Monty's advance: 'It was clear that the British left flank was advancing as fast as or faster than Patton. Amiens is ours. Last night we had three bridges over the Somme.' That was just the beginning. Second Army had hardly paused to take Amiens, and by 4 September, after an advance of 250 miles in four days, it reached Antwerp.

Monty felt this proved that all available resources should be transferred to him for a single thrust in preference to a broad-front advance. He signalled Ike to 'explain' the situation, and to ask that a choice be made between his own proposed thrust and Patton's:

One really powerful and full-blooded thrust towards Berlin is likely to get there and thus end the war. We have not enough resources for two full-blooded thrusts. The selected thrust must have all the resources it needs and any other operation must do the best it can with what is left over. There are two possible thrusts, one via the Ruhr and the other via Metz and the Saar. The thrust likely to give the best and quickest results is the northern one via the Ruhr.

That same day Monty wrote to the King. He was not confident his recommendation would be accepted: 'Our great Allies are very "nationally minded", it is election year in the States, the Supreme Commander is an American; I fear we may take the wrong decision. I am throwing all my weight into the contest.'

The reaction of the Americans at SHAEF was summed up by General Whiteley, who said Monty's plan required them to 'keep three-quarters of the entire Allied Expeditionary Force sitting on their fannies', starving them of ammunition and petrol in order to provide Montgomery with more than his share. Monty's fears were well placed, for a plan that (if it succeeded) would result in a predominantly British victory was unacceptable to the Americans, who were now providing the majority of the troops and matériel.

At least the supply shortage would ease once the port of Antwerp was opened up. Unfortunately, although Monty had taken the city, strong German forces held both banks of the long estuary that connected it to the sea, and their guns denied the use of the port to Allied shipping. Montgomery appeared more interested in gaining approval for his northern thrust than clearing the banks.

★

General Speidel visited Rommel at Herrlingen on 6 September. He had been Rommel's chief of staff but had been suspended the previous day and did not know why. He was *en route* to report to Berlin as ordered. According to Lucie, he warned that Rommel was being spoken of as 'a defeatist', but would say no more. The sense Rommel took from it (or which he preferred Lucie to take from it) was that the High Command wanted to apportion the blame for failings on the western front and he was their prime target.

Ruth Speidel telephoned the next day to say that her husband had been arrested and was in the Gestapo prison on Prinz Albrechtstrasse. Rommel wrote a letter of protest to Hitler, pleading for his release. He received no reply.

That afternoon two intruders were seen in the grounds behind the house but they ran off when challenged. The local innkeeper told Aldinger he had seen cars he did not recognize parked on the road alongside the house. Rommel now took the precaution of keeping a loaded pistol on his desk. He was not sure exactly what the threat was, but he was well enough now to go for walks and always took his pistol with him.

In Berlin Speidel was interrogated by the Gestapo. They wanted to know about Hofacker's visit to Rommel's headquarters on 9 July. Hofacker had confessed to talking to Speidel and telling him about the bomb plot before going in to speak privately with Rommel. Had Speidel reported this serious matter to Rommel?

Ike flew to see Monty in Brussels on 10 September. As Ike had wrenched his leg a few days before and was now walking on crutches, he preferred not to leave his plane and the meeting took place there. Monty still wanted his northern thrust to be given priority and he believed he had Ike 'over a barrel'. He pulled from his pocket Eisenhower's Directive 13889 Part II, issued on 5 September, and read aloud paragraph 4, 'I give priority to the Ruhr.' He looked up as he said, 'Repeat, Ruhr.'

It was not the way to make friends with the Supreme Commander. Ike replied coldly, 'I did not mean this to be at the expense of other operations.'

Monty was not going to take no for an answer: 'I then said that we would do no good by trying to sustain two thrusts; we must put everything into one selected thrust and give it priority.'

It sounded too much like a demand and Ike, placing a hand on Monty's knee, said, 'Steady, Monty, you can't talk to me like that. I'm your boss.' With his final words Ike had also put a metaphorical finger on the problem. He explained to Monty again that they must get across the Rhine first on a wide front, incorporating advances towards the Ruhr *and* the Saar.

After the meeting Monty wrote that 'Eisenhower is a nice chap; but he is on the wrong track.' Ike thought Monty was off the track altogether, commenting to staff officers that 'His suggestion is simple – give him everything. Which is crazy.'

With his single thrust again refused, Monty proposed a more limited plan. He would go only as far as the Rhine, get there in one 'leap', and take Arnhem. He would need extra supplies, but not 'everything'. The plan appealed to Eisenhower because it could deliver a bridgehead over the Rhine to the north, while allowing Patton to continue his advance towards the Rhine to the south. The British and Americans could then be 'released' into Germany at the same time, allowing no one to steal the show.

Monty's aim was to take the bridge over the Rhine at Arnhem in Holland. Two rivers had to be crossed before the Rhine, so two American airborne divisions would be dropped to take the Meuse and the Waal crossings, and 1st British Airborne would be dropped across the Rhine. The armoured divisions of Monty's Second Army would then advance along a corridor opened by the airborne troops, cross the Meuse and the Waal by the bridges taken, and link up with 1st Airborne to establish a bridgehead on the Rhine. In one bold leap over German troops, the operation would kick open the door to Germany. Operation Market Garden was scheduled for 17 September.

Enigma transcripts revealed German strength in the area of

Arnhem and Monty's chief of intelligence, Bill Williams, warned him that 9th and 10th SS Panzer Divisions were within easy reach of the bridge: 'No one could tell Monty not to go. I briefed him about the reported panzer strength at Arnhem. I could not get him to change his mind.' Eisenhower's intelligence officers brought the panzers to his notice on 15 September – their presence near Arnhem had by then been confirmed by the Dutch underground – and he sent Bedell Smith to talk to Monty about it: 'I tried to stop him but I got nowhere.' When General Stanislaw Sosabowski, commanding the Polish 1st Independent Parachute Brigade, complained that the plan was 'disastrous' and that Montgomery was 'recklessly over-confident', he was answered with flattery: 'The gallant Poles can do it.'

Montgomery's supreme self-confidence expressed itself in a certainty that his plans would work. At El Alamein during the first phase of Operation Lightfoot, when his senior commanders had wanted to pull out of the attack, it was only Monty's force of will that had kept them going and the battle was eventually won – he never accepted that his original plans had been adjusted meanwhile by ideas originating with those same commanders. For Montgomery, operations always went to plan and the plans were always his own. But El Alamein was a battle of attrition. In western Europe, while he was undoubtedly engaged in a campaign that would be won largely by attrition, the battles (of which Caen was one and Arnhem another) had to be won by tactical attack. His supreme confidence in his own plans blinded him to the clear advice of those who pointed out the difficulty of attaining a tactical victory at Arnhem.

Market Garden was a major operation and troops, fuel and ammunition were being transferred to Monty from all other units. Bradley heard a rumour that a good part of his US First Army was to be handed over. On 15 September Patton gave his opinion: 'Monty does what he pleases and Ike says, "Yes, Sir." Brad thinks I can and should push on. Brad told Ike that if Monty takes control of XII and VII Corps of the First Army as he wants to, he, Bradley, will ask to be relieved.' Patton told him that he, Patton, would resign too.

The next day Bradley rang Patton to tell him that Monty was asking Ike to halt all American forces and transfer their fuel and ammunition to him so that 21st Army Group could 'make a dagger-thrust at the heart of Germany'. Bradley quipped that Monty's move would as likely be 'more like a butterknife thrust'.

Nevertheless Patton was worried. He felt that if Monty's thrust was successful, Ike would concentrate all efforts on exploiting that breakthrough and there would be insufficient fuel to continue his own advance towards the Saar. He wrote in his diary: 'To Hell with Monty, I must get so involved in my own operations that they can't stop me.'

Orders normally came before an offensive but he decided to see what would happen if that was reversed. He had his own source of fuel that the higher command knew nothing about. Some of it came from the enemy: captured German fuel was kept and not reported to headquarters. Some of it came from his friends: First Army was convinced that their own missing fuel had been appropriated by Patton's men masquerading as First Army officers, and Patton denied it with convincing indignation, while admitting in a letter to his father-in-law that he had 'already stolen enough gas to put me in jail for life'. The German and First Army fuel combined was not enough to allow Third Army to reach the Rhine, but he was still advancing.

Completing this comic circle of deceit and misunderstanding, Monty was unable to work out how Patton could possibly still be moving forward, and suspected that Eisenhower (who claimed to be giving Market Garden priority) was in fact allowing Patton's advance to continue using fuel *he* should have had. Patton, of course, believed exactly the reverse. He told war correspondent Cornelius Ryan, 'If Ike stops holding Monty's hand and gives me the supplies, I'll go through the Siegfried Line like shit through a goose.'

★

Market Garden was launched on 17 September with an artillery bombardment followed by the largest airborne assault of the war, involving more than 5000 aircraft, 2000 gliders and 20,000 paratroops. It began with astounding success. The two US airborne divisions were dropped in their planned zones and the British armour moved forward. The bridges over the canal at Eindhoven, the river Meuse at Grave and the Waal at Nijmegen were taken.

Then it went wrong. The British 1st Airborne Division was dropped six miles from the planned zone, and their jeeps were lost, leaving them to make a four-hour trek on foot to the road and railway bridges at Arnhem. This delay meant that German reinforcements reached Arnhem before the British. Worse still, that reinforcement included the 9th and 10th SS Panzer Divisions, which had been refitting just outside the town. Ironically the panzers had just returned from a training exercise, which involved defeating a mock airborne assault. Bad weather delayed 1st Airborne being reinforced by Polish 1st Independent Parachute Brigade. The plan for Monty's armour to reach Arnhem in three days had been hopelessly optimistic and the bridges at Nijmegen were not secured until 20 September.

As at Caen, Monty's reports on the progress of the battle appeared to take no account of the facts. He announced that it was going 'according to plan'. The armour left Nijmegen on the morning of 21 September to link up with 1st Airborne, but by then it was too late. A decision was made to evacuate as many men as could be saved from Arnhem. The operation had failed to secure a bridgehead across the Rhine. The 1st Airborne Division alone had 1300 men killed and 6400 taken prisoner.

The phrase 'a bridge too far' was not used at the time, but if it had been it would not have referred to the bridge at Arnhem. The British armour had been allowed three days to reach Arnhem, but by the time it had reached its first objective, the bridge over the canal at Eindhoven, it was already sufficiently behind schedule to render the whole timetable impracticable. Eindhoven was a bridge too far from Second Army's starting point to reach in the time allowed.

Why did Monty, who already had a reputation for taking *no* risks, take the biggest risk of all? Bradley expressed surprise when the operation was first suggested: 'Had the pious, teetotal Montgomery wobbled into SHAEF with a hangover, I could not have been more astonished than I was by the daring adventure he proposed.' Soldiers prefer the word 'daring' because its connotations are so different from its synonym in war, which is 'risky'. 'Risky' is used before an operation that, if successful, is thereafter known as 'daring', but if it fails it is invariably labelled 'foolish'. Brian Urquhart, one of Monty's intelligence officers, called it 'an unrealistic, foolish plan that had been dictated by motives which should have played no part in a military operation'.

Those motives were thought at the time to include Monty's (and thereby Eisenhower's) concession to pressure from Churchill, who wanted the destruction of the *Vergeltungswaffe* sites given priority over everything else to reduce the suffering of Londoners, and taking Arnhem would have allowed that. Yet Monty had no difficulty in standing up to the Prime Minister on operational matters and he would certainly not have planned such a high-risk endeavour to achieve what he considered a political end.

Monty could deny Churchill's needs. He could not deny Bernard Montgomery's. Only personal desperation could have led him to act so wholly out of character. And that desperation arose from the long-running dispute with Eisenhower over two issues that now became joined – command and strategy. If Arnhem succeeded, the Allies would in all probability 'go with a winner' and throw everything into the Montgomery thrust into the Ruhr at the expense of all other operations. They would then be operating to Montgomery's single-thrust strategy and, as the army commander on the spot, he could expect that any 'request' for overall command would be granted. We must suspect that he took the risk at Arnhem because it was the only operation that would, in one stroke, allow him to get his way – in command and in strategy – and enable him to direct the war to the early end that he genuinely believed was possible.

Bill Williams saw through the supposed limited objectives of Market Garden to Monty's long-term strategy:

He thought that success would tilt the centre of gravity and give the British priority of supplies before the US armies. Probably Monty thought then it was just a question of who put in the final punch against a defeated enemy before a final victory. If this airborne drop succeeded in front of his Second Army drive, his punch not Patton's would be the triumphal road to final victory.

If Monty's motives in proposing a high-risk operation can be understood, that still fails to explain why Eisenhower went along with it. The Americans at SHAEF believed he did so as a sop to Montgomery – that having turned down his plan for a single thrust, he accepted Market Garden to appease him, and would have turned it down if considered on military grounds alone. (After the war when this was put to Eisenhower he strongly denied it.)

If he had intended Market Garden to keep Monty quiet, it did not. He continued to press for his 'two demands': his appointment as overall land-forces commander, and a single thrust through Germany to Berlin by his 21st Army Group (with the concomitant transfer of ammunition, fuel and American troops from the southern advance to Monty's own). Ike insisted they keep to the original plan (which had been part of the pre-invasion agreement) to advance on a broad front to clear all German forces west of the Rhine before entering Germany. Monty's refusal to accept that decision – clearly communicated to him by Ike on a growing number of occasions – now caused friction between them. Ike had been patient, he had allowed Market Garden, and now he had had enough.

Eisenhower could be thankful for one thing. Monty did not suggest that Arnhem had in fact been successful because it drew enemy forces away from Patton's southern thrust towards Metz and the Saar. This time he had a more ingenious explanation. The operation had been ninety per cent successful because ninety per cent of the target area had been taken. Unfortunately the remaining ten per cent included the bridge at Arnhem. Such an argument fooled no one and discredited him among his peers, many of whom considered its outcome required of him a discreet silence.

Churchill put on his best bulldog face and assured the British people that it was a 'decided victory'. Prince Bernhard, speaking for the people of the Netherlands, begged to differ: 'My country can never again afford the luxury of a Montgomery success.'

It is a commonplace to say that Monty never admitted to a mistake, but this once he did: 'The airborne forces at Arnhem were dropped too far away from the vital objective – the bridge. I take the blame for this mistake.' But he later explained who the real culprit was. There had been critical delays while extra fuel was brought in and he believed this could have been prevented if Eisenhower had kept Patton halted on the Meuse and had given full logistic support to Market Garden. 'If the operation had been properly backed, it would have succeeded in spite of my mistakes.' As Eisenhower had said, Monty wanted everything 'and that was crazy'.

Monty stepped up his campaign for a single thrust and to gain overall command with two signals sent on 21 September. The first was to Ike: 'I have always said stop the right [Bradley/Patton] and go with the left [Montgomery], but the right has been allowed to go on so far that it has outstripped its maintenance and we have lost flexibility.' He asked again that 'the right' be ordered to halt its advance. Monty did not know it, but Patton had already been stopped, not by Ike but by the strong German fortifications around Metz. The second signal was to Bedell Smith: 'I recommend that the Supreme Commander hands the job [overall land command] over to me and gives me powers of operational command over First US Army.'

That same day Patton flew to Paris for lunch with Eisenhower.

They talked about Montgomery and he noted that Ike was nearing the end of his patience: 'Things look much better today. Ike still insists the main effort must be thrown to the British. However he was more peevish with Montgomery than I have ever seen him. In fact, called him a "clever son-of-a-bitch", which was very encouraging.'

Monty had been halted short of Arnhem and he blamed Ike for not reining in Patton. Patton had been halted at Metz and in his frustration, according to Bradley, he 'raged at me, Ike and Monty'. In truth, in September 1944 both Monty and Patton were halted by the enemy. In general the Allies had been too certain that the war was already won, and had failed to appreciate the strength and resolution of surviving German forces.

Following the evacuation of British and Polish troops from Arnhem Monty wrote his own press release to be issued on 28 September: 'There can be few episodes more glorious than the epic of Arnhem. In years to come it will be a great thing for a man to say, I fought at Arnhem. All Britain will say, You did your best, you did your duty, and we are proud of you.'

It read like the proposal for a blockbuster film.

15. The Rommel Murder

On 1 October Rommel received a letter from Ruth Speidel pleading with him to find out what he could about her husband, who remained with the Gestapo. He wrote again to Hitler, detailing the outstanding service of his chief of staff in Normandy. Having no reason to connect Speidel with the bomb plot, he assumed there was some attempt to apportion blame for the collapse of the western front and *he* could hardly be accused without Rommel being next in line. As a precaution he drew up maps and accounts of his own handling of German forces following the *Grossinvasion* lest he need to produce them. He added two sentences to his letter to Hitler:

> You, *mein Führer*, know that I have done everything in my power in the western campaign of 1940, in Africa 1941–3, in Italy 1943, and in the west in 1944. Just one thought possessed me – to fight and win for your New Germany. *Heil mein Führer!* Rommel

It is not known whether the letter reached Hitler. But by now Martin Bormann's report of 28 September had certainly done so. Bormann was Hitler's secretary and had received information from Eugen Maier, one of the leaders of the Ulm Nazi Party, that Rommel had told him, 'The Führer's mental abilities have declined and he cannot be trusted.'

At the beginning of October 1944 Patton complained that he had 'too little gas and too many Germans, not enough ammo and more than enough rain'. He 'talked up' his situation and when he was asked at a press conference whether the Nazis would go underground when the Allies entered Germany, he replied, 'Yes. Six feet.' But that show of confidence belied his frustration. He was held at Metz by strong German fortifications and a determined defence, his operations were hindered by heavy rain and, in any case, his army was receiving insufficient fuel to permit any further advance. He was stalled in the Lorraine and that was not in the Patton script.

The halt affected much of the Allied force. It was wet, but the near-constant rain and the resultant mud fields the armies moved through and often enough lived in were not the only causes of what came to be called 'the October pause'. The Allies were waiting for the port of Antwerp to be opened up to shipping so that supplies – most urgently of fuel – could come in. For once Patton did not blame Montgomery for his own lack of fuel: 'We roll across France in less time than it takes Monty to say "Regroup" and here we are stuck in the mud of Lorraine. Why? Because some so-and-so who never heard a shot fired in anger believes in higher priorities for ping-pong sets than for ammunition and gas.' That was for him a useful insight. Having suspected that Montgomery was somehow behind every delay in fuel reaching Third Army he now recognized that there were genuine supply problems, although they had more to do with the failure to open Antwerp than with ping-pong sets.

The Gestapo had finished with Speidel by 4 October and he was put before a military court. He had admitted under interrogation that Hofacker had informed him of the plot to kill Hitler, but claimed he reported that properly to his superior, Field Marshal Rommel, and could not be blamed if Rommel had failed to pass that information on to Berlin. Speidel pleaded his innocence on that basis and in doing so he implicated Rommel. If he had told Rommel of the plot he had acted correctly; if Rommel knew of the plot and failed to tell Berlin, he had by his failure made himself a conspirator.

Although Rommel was not present and knew nothing of these

proceedings the court was in effect choosing to condemn either Speidel or Rommel, although of course only Speidel faced the hangman's rope if their decision went against him. It is possible that Hitler was still not ready to believe the worst of his field marshal and attempted to influence the court in his favour, because it was announced that 'The Führer has expressed the view that there can be no doubt Speidel is guilty.' A guilty verdict for Speidel would clear Rommel. However, Speidel's defence argued convincingly that the prosecution had produced no evidence that Speidel had not told Rommel about the plot, and the case against him was dismissed. That decision turned the focus of attention on the case against Rommel.

Monty attended the SHAEF conference held at the Hôtel Trianon Palace in Versailles on 5 October, which was meant to decide the priorities that should be assigned to present and future operations. He announced that in his opinion all effort should be put into the thrust towards the Ruhr and that this could be done without taking Antwerp first.

Admiral Bertram Ramsay, the naval Commander-in-Chief, noted later that 'This afforded me the cue I needed to lambast him for not having made the capture of Antwerp the immediate objective at highest priority.' Ramsay argued that the Allied armies had virtually come to a halt because of a lack of supplies that could have been brought in through Antwerp. Even Alan Brooke, Monty's most loyal supporter, agreed: 'I feel that Monty's strategy for once is at fault.'

Monty left the meeting with a clear understanding that SHAEF considered Antwerp his most urgent task. His response was to do as he had done for weeks: he ignored it. This is commonly explained by reference to his single-minded focus on the advance into the Ruhr, but he may have had an altogether more discreditable reason for giving the port such low priority. He later dictated to Dawnay this entry in the Second Army log: 'From a purely British point of view Antwerp had never been a vital necessity;

the Pas de Calais ports provided all that we required. But to the Americans it had become vital . . .'

Four days after the Versailles meeting, an exasperated Eisenhower informed Monty that if Antwerp was not open to shipping by the middle of November then 'all operations will come to a standstill'. He asked in writing that Monty give 'the most important' of their current endeavours 'his personal attention'.

Instead of doing as the Supreme Commander had, with surprising courtesy, indicated he must, Monty replied the next day with yet another call for a single land commander. He complained that 'Both British and American armies are involved in the capture of the Ruhr but the job is not handed over to one commander,' and that as a result operations had become 'untidy and ragged'. To be fair to Monty he did indicate the 'one commander' could be the present commander of either army, meaning Bradley or himself. However, he knew that by the unwritten rules of the alliance an American supreme commander could not give overall land control to a fellow countryman, just as earlier Monty's command of all land forces for Overlord ruled out the possibility of a British supreme commander. The suspicion must arise that, as in Italy (when he had halted his army on the road to Salerno and refused to move until Alexander gave in to his demands), he was now suggesting that if he was given overall command of the operation, he would *then* be able to take Antwerp.

By now Eisenhower was extremely angry. He replied that the matter was not about command but about Antwerp, and that the current supply problems left the US Army in 'a woeful state . . . by comparison you are rich!' However, if Monty wished to pursue the issue of command then he should 'refer the matter to higher authority for any action they may choose to take however drastic'. The threat was unmistakable. If this was referred to the politicians there could be only one outcome. The war effort was being driven by a preponderance of American troops and American matériel, and the American Supreme Commander would be supported rather than the British field marshal.

When Churchill stood up to Monty on the matter of visiting the front in Normandy and it was made known that his job was at risk, he hastily withdrew, and now he wrote to Ike: 'You will hear no more from me on the question of command. I will weigh in 100% to do what you want, and we will pull it through without a doubt. I have given Antwerp top priority in all operations in 21 Army Group and all energies and efforts will now be devoted towards opening up the place.' He signed the letter 'Your very devoted loyal subordinate'. Eisenhower was not to know that for Monty this represented a tactical withdrawal, not surrender.

Rommel was out when Field Marshal Keitel rang him at Herrlingen on 7 October. Aldinger took a message: Field Marshal Rommel was to attend an interview in Berlin on 10 October. A special train would be sent to collect him the previous evening.

When Rommel arrived home he rang Keitel back but was connected to General Burgdorf, head of Army Personnel, who explained that 'The Führer has ordered Field Marshal Keitel to discuss your next assignment with you.' Rommel said that his doctor had advised him not to travel – that was not the case but it was a convincing excuse – and asked that an officer visit him at home. Burgdorf agreed that he would talk to Keitel about it.

He told Lucie that he was to be offered a new appointment, and kept his fears to himself until Admiral Ruge (who had been his naval adviser and confidant in Normandy) came to dinner on 11 October and stayed overnight. The two men sat talking until midnight and Rommel mentioned the summons to Berlin. 'I am not going. I would never get there alive. They would kill me on the way and stage an accident.'

Two days later Keitel called from Hitler's headquarters. General Burgdorf would come to see him in Herrlingen at noon the next day, along with General Maisel. As both these men worked for the Personnel Department, Rommel was uncertain whether his fears were well founded or Hitler really did have a new role for him. He told Aldinger the generals were most likely coming to discuss a new appointment. Rommel was not aware that since 20 July both men had been assigned to the special unit investigating officers suspected of involvement in the bomb plot.

King George VI had expressed a wish to visit the Allied forces, and when among the troops he had a liking for (relatively) 'roughing it'. Monty suggested a bunk in his caravan at Eindhoven and the offer was accepted. He rejected suggestions that his caravan might not be a safe lodging for the King of England: 'This is utter nonsense. No enemy aircraft have been over Eindhoven since I was here . . . Motoring on the roads is perfectly safe.' The King arrived on 11 October for a six-day stay, and travelled out each day to visit units of the British troops, the US First Army and the Canadian Army. His Majesty was informed that it was Monty's habit to retire at 2130 each evening; his guests were expected to follow suit.

On 14 October the King visited the headquarters of General Hodges' First Army near Liège. Among the generals gathered there to take lunch with him were Bradley and Patton. Eisenhower made a loyal speech – 'If ever there is another war, pray God we have England as an ally. And long live King George VI!' – and Patton, for once, had nothing to say.

Meanwhile, Monty wrote a short letter home to David that ended: 'Rommel is not very well and he has never quite recovered from his attack of pneumonia.' He was referring to one of his dogs.

In the early hours of Saturday, 14 October, a dark green Opel with Berlin number-plates drove at speed along the *Autobahn* heading for Herrlingen. The driver wore the black uniform of the Waffen SS. On the rear seat sat Generals Wilhelm Burgdorf and Ernst Maisel.

Manfred Rommel was up early too. He had been given two days' leave from his anti-aircraft battery and by catching the night train he reached Herrlingen at dawn. He got home to find his father at breakfast and joined him.

43 A British soldier helps an elderly woman to safety through the ruins of Caen, 10 July 1944

44 French citizens welcome the first American tank to advance through their town

45 Patton at a press conference in August 1944 tells the world that he has arrived in Normandy

46 Rommel looks glum as he studie the rapid Allied build-up along the Normandy coast

47 Patton's Third Army advances through the gap in the German line near Avranches to begin his American blitzkrieg at the beginning of August 1944

48 Patton and Montgomery discuss the Normandy campaign, while General Omar Bradley looks on

49 Monty plays with the puppies he named 'Hitler' (a Jack Russell) and 'Rommel' (a King Charles Spaniel) at his Blay HQ in July 1944

50 Patton preparing to move forward at first light behind an armoured column

51 Churchill and Stuart tanks of Monty's 21st Army Group advance towards the Seine

52 Patton crossing the Seine at Melun, followed by a column of light tanks

53 Monty crossing the Seine via a pontoon bridge at Vernon, 1 September 1944

54 American troops march through the ruins of a German town

55 Rommel's coffin arrives at Ulm for his state funeral, 18 October 1944

56 Lucie and Manfred at the *Rathaus* (town hall) in Ulm, where Rommel's funeral service was held

57 After crossing the Rhine, Patton stops to thank his Third Army engineers who built the bridge and enabled him to beat Monty across

58 Admiral von Friedburg signs the surrender document drawn up by Monty at the latter's Lüneburg Heath HQ, 4 May 1945

59 Monty rides the white Arab stallion he named after its previous owner, Rommel, on which the German had intended to ride into Cairo after its capture

60 Patton rides 'Favory Afrika' in Salzburg. This Lipizzaner stallion had previously been selected for presentation to Emperor Hirohito of Japan

61 Patton's bull terrier, Willie, waits with his deceased master's trunk for return to the United States

'At twelve o'clock today two generals are coming to see me to discuss my future employment,' my father said. 'So today will decide what is planned for me; whether a People's Court or a new command in the East.'

'Would you accept such a command?' I asked.

He took me by the arm: 'My dear boy, our enemy in the East is so terrible that every other thought has to give way before it. If he succeeds in overrunning Europe, it will be the end of everything which has made life worth living. Of course I would go.'

At 1130 Rommel changed out of his civilian jacket and trousers, preferring to meet the generals in his Deutsches Afrika Korps uniform.

Generals Burgdorf and Maisel arrived promptly at noon. Rommel introduced them to Lucie and Manfred, then invited them into his study. Burgdorf did the talking; Maisel was there only as a witness. There were no preliminaries. He told Rommel that several officers arrested in connection with the 20 July plot to kill the Führer had, during their interrogation by the Gestapo, implicated the field marshal. On the basis of their evidence Hitler was charging him with high treason. He handed over a letter from Keitel confirming this.

Burgdorf then read out the most pertinent sentences from the written testimonies of Hofacker, Speidel and Stülpnagel. It was Hofacker's evidence that appeared most damning: that he had visited Rommel on Stülpnagel's instruction to recruit the field marshal, and that Rommel had said to him, 'Tell your gentlemen in Berlin that when the time comes they can count on me.' There was evidence that if the assassination attempt had succeeded, Rommel would have become the next head of state in place of the Führer.

According to Maisel, Rommel appeared surprised but remained calm. He told the generals that these statements meant nothing as they could only have been elicited under Gestapo torture. He studied the letter from Keitel and looked up at Burgdorf. 'Does this come from the Führer?' Burgdorf assured him that it did. At that point he asked Maisel to leave the room.

Immediately Burgdorf was alone with Rommel he told him what had not been put in writing by Keitel. The Führer offered him a choice: he could face trial for treason by the People's Court or he could avoid that by immediately committing suicide. If he chose the latter, his part in the plot would not be revealed to the German people, he would receive a state funeral and Lucie would receive a full field marshal's pension. Burgdorf had brought with him a poison that worked in three seconds.

Rommel hardly hesitated before choosing the poison. He was certain that if he chose the People's Court he would not reach Berlin alive. In any case he was aware that the conspirators had suffered from the *Sippenhaft* policy by which close family members were punished. His only request was that he be allowed to swallow the poison somewhere other than in the house. Burgdorf agreed that he would be given the capsule in the car on the road to Ulm. His body would then be taken to the town's Wagnerschule Hospital. The announcement of his death would indicate that he had died as a result of a brain seizure. Every detail had been planned.

Rommel went upstairs to find Lucie.

My husband came to see me in the bedroom. It is impossible for me to describe what I saw in his face. 'What's the matter?' I asked. 'In fifteen minutes I'll be dead,' he said absentmindedly. 'The Führer has given me the choice of taking poison or going on trial before the People's Court.' He was suspected of taking a leading part in the 20 July affair. He chose to take poison, he told me, because he was sure he would not reach the People's Court alive.

Manfred came into the bedroom, expecting to be told what new appointment the generals had offered his father.

'Come outside with me,' he said in a tight voice. We went into my room. 'I have just had to tell your mother,' he began slowly, 'that I shall be dead in a quarter of an hour.' He was calm as he continued: 'To die by the hand of one's own people is hard. But the house is surrounded and Hitler is charging me with high treason.'

'Can't we defend ourselves . . .' He cut me off short.

'It's better for one to die than for all of us to be killed in a shooting affray.'

Finally he called Aldinger upstairs and told him what was about to happen.

I said that he must make an attempt to escape. We could try to shoot our way out together. 'It's no good, my friend,' he said. 'The roads are blocked by SS cars and the Gestapo are all around the house.' I said we could at least shoot Burgdorf and Maisel. 'No,' said Rommel, 'they have their orders. Besides, I have my wife and Manfred to think of.' He had been promised that no harm should come to them if he took poison. 'I have made up my mind,' he said. 'I will not allow myself to be hanged by Hitler. I planned no murder. I only tried to serve my country, but now this is what I must do.'

Rommel walked downstairs and put on his leather coat. The dachshund that he had been given as a puppy a few months earlier in France, leaped up at his legs playfully. He asked Manfred to shut the dog in the study. Then Rommel, Manfred and Aldinger walked outside together.

The two generals were standing at the garden gate. We walked slowly down the path, the crunch of the gravel sounding unusually loud. As we approached the generals they raised their right hands in salute. 'Herr Field Marshal,' Burgdorf said. The SS driver swung the car door open and stood to attention. My father pushed his Marshal's baton under his left arm, and with his face calm, gave Aldinger and me his hand once more before getting into the car. The two generals climbed into their seats and the doors were slammed. My father did not turn as the car drove quickly off up the hill and disappeared round a bend in the road.

According to the driver, SS Corporal Heinrich Doose, they left Herrlingen on a quiet, secondary road that led uphill towards Ulm. General Burgdorf ordered him to stop the car and Doose pulled on to the roadside by the edge of a wood.

General Maisel took me away, further up the road. After about five minutes Burgdorf called us back to the vehicle. I saw Rommel sitting in the back of the car, obviously dying, unconscious, collapsed into himself, sobbing, not moaning or groaning but sobbing. His cap had fallen off. I sat him upright and put his cap back on his head.

They drove quickly to the hospital in Ulm. Dr Meyer, the senior doctor, rushed to attend to the VIP patient but saw that all efforts at resuscitation would fail: 'One look at the man and it was obvious he had not died a natural death.' Minutes later when Meyer officially certified the field marshal dead and suggested an autopsy, Burgdorf ordered him not to touch the body and said, 'Everything is being handled from Berlin.'

During the following days Lucie received many letters from generals, soldiers and friends expressing their sympathy, but first to arrive was a telegram from Berlin: 'Please accept my sincere condolences on the heavy loss you have suffered as a result of your husband's death. The name of Field Marshal Rommel will be forever linked to the heroic battles in North Africa. Adolf Hitler.'

Among the mourners at Rommel's state funeral on 18 October 1944 was Admiral Ruge. He arrived on a special train arranged for senior officers travelling from Berlin to Ulm and walked through the town to the *Rathaus* (town hall). The building had been hung with swastika banners both inside and out. Rommel's coffin, completely covered with a swastika, was carried in at precisely 1300 and a military band played the '*Trauermarsch*', the second movement of Beethoven's Third Symphony (written in honour of Napoleon). Field Marshal von Rundstedt addressed those present, speaking from the podium.

A pitiless destiny has snatched him from us just as the fighting has come to its crisis ... The German nation has loved and celebrated Field Marshal Rommel. With his passing, such a great soldierly leader has gone from us as is only rarely given to a nation. This tireless fighter in the cause of the Führer and the Reich was imbued with the National-Socialist spirit ... His heart belonged to the Führer.

Lucie sat listening: 'When von Rundstedt was speaking, I longed to call out that they were all acting a lie. But they would have hushed it up somehow or else my husband would have been publicly disgraced. And I had to think of Manfred. I did not care any more for myself but Manfred would have been killed.'

As von Rundstedt left the podium the band played 'Ich hatt' einen Kameraden' (I had a comrade), expressing the soldier's grief for a comrade fallen in battle, while outside the Rathaus a battery fired a nineteen-gun salute. The coffin was carried from the hall and lifted on to a motorized gun-mount, then driven through Ulm to the crematorium on a hill outside the town. The streets were lined with Hitler Youth. As the townspeople paid their last respects, many were in tears. Ruge felt that something was wrong:

In one of the crematory anterooms I had a chance to take Aldinger aside and, making certain that nobody heard us, asked him: 'What is actually the matter here?' Tears rushed to his eyes and he said: 'On Saturday they came.' Then his voice broke, but I knew enough. On our way back to Berlin in the special train it was clear to me that, with Rommel, Hitler had eliminated the only man who possessed enough respect inside Germany as well as outside to end the war.

The Allies had made little progress during the last month. It had been hoped that as they closed on Germany the German people would rise up against the Nazis and end the war for them. But that had not happened and Eisenhower decided on a fresh offensive in an effort to break through the Siegfried Line before the winter storms and snows. On 18 October he met Monty and Bradley in

349

Brussels. Bradley's US Army Group would make the attack in the second week of November, and in anticipation of that he was to push east and establish a bridgehead on the Rhine south of Cologne. Ike's promise to Monty that priority would later switch back to the thrust into the Ruhr meant nothing: if Bradley broke through he might, with Patton's drive, go all the way to Berlin.

While Patton waited for the November offensive he ensured that the troops were entertained and took a little time off himself. He made several visits to see Jean Gordon in Paris, and on 5 November he lunched with Marlene Dietrich, who was performing for the men. Some of her biographers claim that they began an affair which continued through the winter. There were certainly rumours, but no hard evidence. In her autobiography she called him 'a great man' and described him, most appropriately, as 'looking like a tank too big for the village square'. He referred to her only once in writing, describing her show for the troops as 'very low comedy, almost an insult to human intelligence'. He told Hughes that he found her 'terrifically charming' and that she had warned him the press were eager to catch them together in compromising circumstances and that they should be careful. That comment begs the question whether or not they had reason to be careful.

On 7 November Montgomery returned briefly to London to be presented with his field marshal's baton by the King and to have an urgent repair made to his dentures. He saw Alan Brooke and told him that Bradley's offensive was a bad idea. In his belief, the inexperienced American officers and men were not up to the job. Brooke tended to agree, although when he noted their discussion in his diary he seemed a little irritated that Montgomery continued to return to the issue of command:

Monty goes on harping over the system of command in France which prolongs the war. He had got this on his brain as it affects his own personal position and he cannot put up with not being in sole command of land operations. I agree the set-up is bad but it cannot easily be altered as USA are preponderant and consider they have a major share in the running of the war.

Monty told him that the war could be won early in 1945 if the Allies concentrated their strength (by transferring American forces to his own 21st Army Group) and appointed him as overall ground commander. Then Monty could win the war *his* way.

Hitler, too, believed that the war could be won by concentrating all available forces to make a single thrust through the extended enemy line, but he believed it could be done by *German* forces. He developed a plan predicated on another belief he shared with Monty, that the Americans would prove easier to overcome than the British. The veterans of Fifth and Sixth SS Panzer Armies would strike through the Ardennes, sweep through the weakest American sector held by two inexperienced divisions, take Antwerp to cut the Allies' supply line, and trap the British armies to the north behind German lines. He asked Field Marshals Model and von Rundstedt to work out the details for this last great offensive.

16. Bulging Ambitions – Monty and Patton 'Win' the War

Patton's greatest opportunity came to him at the worst of times. Eisenhower had authorized Bradley's Army Group to push forward from Lorraine in November, cross the Moselle and the Siegfried Line, and move into the Saar. To pull his officers out of the lethargy induced by the October pause, Patton quoted Grant: 'In every battle there comes a time when both generals think themselves licked. Then he who is fool enough to keep on fighting wins.' Montgomery had failed to take Arnhem, which would have offered an easy route into the Ruhr. It seemed to Patton that if he could now break into the Saar, Ike might abandon his broad-front strategy and add the full Allied strength to this southern thrust, with Third Army in its vanguard. There was still a chance that the war could be won by the end of the year.

It rained constantly across the flatlands of Lorraine and the water-logged terrain was unfit for tanks, but when his senior commanders requested a postponement he told them that such an order would come only after they had 'made recommendations as to your successors'. The operation went ahead as scheduled. The Germans were not as accommodating. They had anticipated an American attempt to break into the Saar and reinforced the Moselle line with Fifth Panzer Army.

The immediate capture of Metz would allow Third Army to cross the Moselle and close on the Siegfried Line, but when his attack was repulsed he laid siege to the city. It was the first time the enemy (as opposed to the Allies) had stopped him. It was also his first battle of attrition and he proved himself as ruthless as Monty in accepting heavy losses among his own men in order to 'crumble' the enemy, sending in attack after attack, each with little chance of success in its own right but further reducing German strength. Even Bradley said, 'For God's sake, George, lay off.' He kept on and finally entered Metz on 22 November, his infantry fighting street by street and house by house. It took several days to clear the enemy from every part of the city and the network of tunnels beneath it. By then he was far behind schedule. Just as Monty had praised his men after Arnhem, all the more because the operation had failed, so now, as Patton saw that he could not reach the Saar before the winter snows, he told the men of 5th Division, 'Your country is proud of you. Your deeds in the battle of Metz will fill the pages of history for a thousand years.'

The advance continued, and by mid-December he had reached the Siegfried Line. German forces that had readily withdrawn when challenged in central France now put up a determined resistance as Third Army edged closer to the German border. Patton experienced the slow, hard crawl that is more a part of war than blitzkrieg, and Bradley recognized how hard an education that was for him: 'He slugged his way forward some thirty-five miles to the Saar river, crossed it, but was stopped dead at the Siegfried Line. His failure to break through deeper into Germany infuriated him . . . In his frustration he raged at me, Ike and Monty.'

He directed his anger at them because they had handed him an opportunity to win the war but they gave it *too late*. He believed that if he had been allowed priority for fuel and ammunition in September (when it was given instead to Monty) he could have reached this point before the German defences were sufficiently organized to receive him, before the snow fell to slow him, and thus broken through. Now he stood within sight of Germany and cursed. To his north and south First and Ninth Armies came to a standstill too.

Montgomery had made an identical point after the failure of Market Garden: that if he had been given full logistical support his operation might have succeeded. Both men blamed Eisenhower's broad-front strategy by which he first put additional weight behind Monty's thrusts to Arnhem, then additional weight behind Patton's thrust through Lorraine, but not his *total* weight behind either. The campaign can be characterized as a conflict between Monty and Patton, but that was of lesser consequence compared with the true battle between the single-thrust plans advocated by both men, and the broad-front policy followed by Ike. Fear of upsetting British and American national sensitivities prevented Ike plumping decisively for Monty or Patton as the general whose powerful thrust might 'win the war'. It would be an exaggeration to say that maintaining the alliance threatened to cost the Allies the war because Germany had passed beyond the point at which it could recover, but it was almost certainly extending the war.

On 6 December Eisenhower and Bradley went to see Monty at his headquarters in Maastricht in Holland to sort out their differences. Monty argued yet again for a single attack, to move north of the Ardennes and strike into the Ruhr, and that by using fifty divisions they would 'concentrate such strength on this main selected thrust that success will be certain'. As his Army Group currently consisted of fifteen divisions the extra forces would have to be transferred from the south, giving him command of a predominantly American army and making a second attack to the south impossible. Bradley accused him of wanting to 'command the whole show' (which was hardly a stinging criticism, as that was precisely what Monty wanted).

Bradley put in a counter-claim and a threat. He wanted permission to make a further thrust south of the Ardennes into the Saar, and if Ike 'put the Twelfth Army Group under Marshal Montgomery, he would have to relieve me of command'. Eisenhower knew that if Bradley went Patton would go too, and Washington would not countenance the loss of his most successful generals. He told Monty that American forces had achieved a 'brilliant breakthrough' in Normandy, which 'almost carried us across the Rhine', and he had no intention of stopping their operations.

Ike compromised. General Simpson's US Ninth Army would be transferred to Monty, giving him an additional ten divisions for a thrust north of the Ardennes. A second thrust would be made by Bradley's Army Group, spearheaded by Patton's Third Army, to the south of the Ardennes. It was more of the same: a broad advance with two points of concentration.

Because Monty had not got everything he wanted, he now decided that both Eisenhower and Bradley should go. He wrote to Brooke to explain why: 'We shall split our forces and we shall fail. I think now that if we want the war to end within any reasonable period you will have to get Eisenhower's hand taken off the land battle. In my opinion he just doesn't know what he is doing. And you will have to see that Bradley's influence is curbed.'

Hitler was unknowingly in full agreement with the strategy both Monty and Patton advocated. One single powerful thrust might win the war. While maintaining a broad front, the Allies had concentrated their forces each side of the Ardennes, with the British to the north and the Americans to the south. A strike at the centre of the line might divide them.

He launched Herbstnebel (Autumn Fog) at 0530 on 16 December. Following an artillery bombardment, seven panzer divisions of Model's Army Group B attacked along a seventy-mile front in the Ardennes and Luxembourg, defended by five infantry divisions of General Hodges' US First Army. The Allies had considered this thickly wooded area to be impenetrable by armoured columns, especially once the winter snows had fallen, and classified it as a 'quiet zone'. Hodges' infantry had been exhausted by prolonged engagement and were there primarily to rest.

The attack caught the Allies and particularly their commanders at rest. Eisenhower and Bradley were enjoying a game of bridge at Versailles when the first report reached them. Montgomery was playing golf in Eindhoven. All thought it likely this was a 'spoiling operation' to disrupt preparation for their own coming offensive. The Germans had broken through the First Army front at five separate points and Eisenhower ordered 7th and 10th Armored Divisions to move immediately into the area. The 10th was part of Patton's Third Army and when Bradley rang to say it was being taken from him to help defend the Ardennes sector, he shouted over the telephone that any fool could see this was not a serious offensive, and that the Germans were attempting to weaken his own attack into the Saar to begin in three days' time. Bradley held out the receiver for Eisenhower to hear. 'Tell him,' Ike said, 'that Ike is running this damned war,' and returned to his cards.

The German 'bulge' in the Allied line became bigger throughout the next day and Berlin received reports of ground easily taken, panzers destroying columns of American trucks crammed with troops, and enemy prisoners being marched to the rear. On 18 December Hodges and his staff were forced to evacuate their headquarters at Spa. Montgomery stopped massing his forces for the planned thrust into the Ruhr and sent four divisions south towards an area between the Meuse and Brussels, into which the enemy was moving.

The next morning the 'bulge' was expanding still and the Germans were closing rapidly on Bastogne. The six roads that met there were the key to control of the whole area, and from there the panzers could strike out towards Antwerp. For the first time some thought the unthinkable: that the Allies, on the verge of victory in Europe, had relaxed their guard and the Führer was about to topple them back into the English Channel.

Eisenhower drove to Verdun to confer urgently with Bradley and Patton. Patton's first suggestion – 'Hell, let's have the guts to let the sons-of-bitches go all the way to Paris, then we'll really cut 'em up and chew 'em up' – was ignored. Ike ordered him to change the direction of his advance east towards the Saar and move rapidly north to Bastogne; the town had to be held. He asked Patton how quickly he could attack and was told 'in forty-eight hours'. Some among Eisenhower's staff thought that too ambitious. Third Army had to turn through ninety degrees and move fifty miles to the battle area in severe winter conditions, and immediately engage the best panzer divisions that the Germans had left. Patton lit a cigar as he reassured them in his own inimitable style: 'The Kraut has stuck his head in a meat grinder and I've got the handle.'

By 20 December enemy forces had surrounded Bastogne and the small American force in the town was besieged. Patton was moving up from the south to relieve them and Monty's divisions were heading down from the north.

The German thrust had divided the American front. Two-thirds of Bradley's Army Group was north of the enemy salient, while he was 'cut off' at his headquarters in Luxembourg, far from the troops he would be expected to command. Eisenhower felt that it made sense to co-ordinate the two forces north of the salient by placing all US troops there under Monty's command. Although he was more aware than anyone of the antagonism between the British and the Americans that this might exacerbate, he believed the necessity of a co-ordinated response to the German offensive left him no alternative. Bradley was not happy with the decision but there was nothing he could do.

Monty was, of course, delighted, and misinterpreted it as a halfway step to giving him what he had long pressed for – overall land command – although for the moment Bradley retained control of American troops not involved around Bastogne. At 1030 on 20 December he took command of 'the American armies north of the German salient'. Before midday he arrived at the First Army headquarters near Liège in a Rolls-Royce with a Union Flag tied to the bonnet and flanked by eight motorcycle riders. He 'took over' rather brusquely and in his own words began to 'tidy up' the battlefield. The BBC correspondent Chester Wilmot believed that his approach was deliberate:

The Americans had spurned his leadership. Now they had turned to him to extricate them from a predicament which, he believed, would never have developed if he had been left in command of all ground forces.

That afternoon Montgomery did not endear himself to his American audience, for his confident tone seemed to carry a note of censure.

Almost his first act was to order First Army to withdraw from what he considered to be exposed positions, to establish a firm front further west. As his British divisions were at that time moving forward, this decision would later be criticized as fed by a pro-British bias, but in fact it had the agreement of all the American generals concerned and meant that First Army could consolidate its strength for a counter-attack.

When no immediate counter-attack came, Bradley flew to meet Monty at Zonhoven on Christmas Day. Monty said his forces were not yet present in sufficient strength to counter-attack, and took the opportunity to tell the American why at a time when the Allies had supposedly won the war, the German offensive had been possible at all: 'I was frank with Bradley. I said it was our fault; we had tried to develop two thrusts at the same time, and neither had been strong enough to gain a decisive result. The enemy saw his chance and took it. Now we were in a proper muddle.'

By the next day the 'bulge' was sixty miles deep and thirty miles wide, and its 'arrowhead' at the Meuse had a width of five miles. The Germans had destroyed 300 Allied tanks and taken 25,000 prisoners, but had suffered heavy losses too and the advance was losing momentum. The time was right for a decisive counter-attack. Monty was still not ready.

Eisenhower travelled to Hasselt on 28 December to confront him, and Kay Summersby described the meeting: 'E. and Monty had a long talk. Monty tried to convince E. that there should be one commander of the entire battle front [not only of the 'bulge' area], and left no doubt of who should be that commander. He was not very co-operative.' The suspicion must arise that Montgomery was again 'negotiating' with the Supreme Commander and that he might suddenly find himself ready to counter-attack if he got what he wanted. He told Ike that he could not attack until the New Year, and that the German offensive proved that his argument for a single land-forces commander had been correct. Afterwards he thought he had convinced the American to see things his way: 'Eisenhower was in a somewhat humble frame of mind and clearly realizes that the present trouble would not have occurred if he had accepted British advice and not that of American generals.'

The next day, presuming Ike was ready to concede and needed only one final push, Monty wrote to him: 'Co-ordination will not work. One commander must have powers to direct and control the operations; you cannot possibly do it yourself. I put this matter up to you again only because I am anxious not to have another failure.' Then, in an unparalleled display of sheer cheek, he provided Ike with the order he must issue to make things right: 'From now onwards full operational direction, control, and co-ordination is vested in C-in-C 21 Army Group [i.e. Montgomery].'

Any man weaker than Ike might have given in. But Monty's arrogance riled him. Even Brooke, when the letter came to his notice, advised Monty as a friend that this was 'too much of "I told you so" to assist in producing the required friendly relations'. For Ike there were no relations left. Telling his staff 'I am tired of the whole business,' he drafted a signal to the combined chiefs of staff stating that with Montgomery still pressing for a land-forces commander 'it was impossible for the two of them to carry on working in harness together'. Either Monty must be sacked or *he* would resign.

De Guingand, Monty's chief of staff and not blind to his master's failings, was tipped off and persuaded Ike to hold on to the signal for twenty-four hours. He flew to Monty's headquarters, explained what was afoot and pointed out that if forced to choose the CCS would, for the sake of the alliance, retain Eisenhower and dispense with Montgomery. De Guingand described him as 'looking completely nonplussed ... it was as if a cloak of loneliness had descended on him'. A letter was quickly written (based almost entirely on a draft de Guingand had prepared) in which Monty declared himself 'very distressed that my previous letter may have upset you', withdrew his comments and signed himself 'Your very devoted subordinate'. Ike destroyed the signal to the CCS.

As Patton advanced against the south flank of the German salient, neither Eisenhower nor Bradley considered his attack with three divisions to be strong enough against an enemy fighting with unexpected determination. His answer to that concern highlights the difference between him and Monty. The latter would have agreed (for by the book it was undoubtedly the case) and halted to build up his strength. However, Patton argued that his success would come from taking the enemy by surprise, and that by waiting he would lose as much as he gained. Ike explained that he was worried because the veteran von Rundstedt was in command of the panzer divisions Patton would face, and as Monty was holding the Germans to the north, he wanted to be assured that Patton could hold them to the south. 'Hold von Rundstedt?' Patton said. 'I'll take von Rundstedt and shove him up Montgomery's ass!'

By the morning of 22 December he had three divisions facing Bastogne and began the final advance along a twenty-mile front in heavy snow. Every man in Third Army had received a small card with Patton's Christmas greeting printed on one side and a short prayer on the other. The first told the men that they would 'march in our might to full victory', while the second asked the Almighty 'to restrain these immoderate rains with which we have had to contend' and 'grant us fair weather for battle'.

After several days of fierce fighting, Patton's troops made contact with the besieged American force in the town just before dawn on 26 December, and the siege was lifted. He was first to praise his own action, telling Beatrice that 'the relief of Bastogne is the most brilliant operation thus far and is in my opinion the outstanding achievement of this war'.

Although that technically ended the enemy offensive, it was as Third Army attempted to reduce the size of the enemy salient east of the city that some of the heaviest fighting of the battle took place. Although by then Patton had seven divisions and the enemy was falling back, his advance from Bastogne to Houffalize was particularly bloody.

By 3 January Monty's divisions, including those of First Army, were moving south, but there was little indication of a counter-attack in force. When war correspondent Leland Stowe was granted a private interview with Patton, he first asked about the pistol belt and ivory-handled pistol. 'I wear that gun because I killed my first man with it,' the general said. Then Stowe asked the question he hoped would provide a headline: 'Why have the British not attacked?' Patton gave him what he wanted: 'Why in hell don't they attack? They're just being true to form. More afraid of losing a battle than anxious to win one. Well, that's Montgomery for you.' To make matters worse, the British papers and the BBC were reporting that Monty was rushing to the relief of the Americans who, it was said, had failed to halt the German offensive and were now hard pressed.

It was not until 16 January that the two Allied armies linked up and by the end of the month the 'bulge' had been eliminated. Patton's Third Army had suffered 27,860 casualties. While German losses were high too, once again a large number of troops had escaped eastward. Some at SHAEF blamed the late arrival of Monty's divisions for delaying the link-up of Allied forces and allowing so many of the enemy to escape.

The worst of the battle with the Germans was over. The worst of the battle between Monty and Patton was about to begin. At first all was calm. Patton held a press conference at which he praised his American troops and ebulliently celebrated what was in truth the largest battle the US Army had ever fought: 'The purpose of this operation as far as the Third Army is concerned is to hit this German son-of-a-bitch – pardon me – in the flank, and we did, with the result that he is damn well stopped and going back . . . To me it is a marvel what our soldiers can do. I know of no equal to it in military history. I take off my hat to them.'

Montgomery held *his* press conference at Zonhoven. Bill Williams, his chief of intelligence, could see trouble looming when Monty chose to wear a purple Airborne beret: 'I begged him not to hold the press conference. The journalists liked him because he was crystal clear and they could understand exactly what had happened, or rather what Monty wanted them to understand, because he was capable of telling whopping lies.' But he genuinely believed that he had personally won a great victory and nothing could stop him telling the world about it.

He began by ensuring the reporters understood the command structure. 'General Eisenhower placed me in command of the whole northern front.' His narrative of the battle made much use of the personal pronoun. 'As soon as I saw what was happening, I took steps to ensure that the Germans would not get over the Meuse. And I carried out certain movements to meet the threatened danger. I employed the whole power of the British group of armies.'

'The British group of armies' amounted to one British division. All other divisions involved in the battle were American. What came next was a master-class in condescension towards General Collins, commanding the US VII Corps, and the whole of Hodges' First Army of which it was part: 'VII Corps took a knock. I said, "Dear me, this can't go on. It's being swallowed up in the battle." I set to work and managed to form the corps again. Once more the pressure was such that it began to disappear in a defensive battle. I said, "Come, come," and formed it again and it was put in the offensive by General Hodges.'

Monty did mention the enemy in conclusion – 'he was first "headed off" from vital spots, then "seen off", and he is now being "written off"' – but the most vivid image left in the minds of the war correspondents was one he painted for them: 'You have thus the picture of British troops fighting on both sides of American forces, who had suffered a hard blow.'

Bill Williams had been right to worry: 'Monty's attitude did much harm to Anglo-American relations. It was crazy of him to say that a battle which had cost so many American lives and produced the crisis of the campaign was "one of the most interesting battles I have handled". This was absurdly patronizing.'

He exaggerated the British part in the battle to the extent that it appeared as if the Americans were in severe trouble and were saved by the British, or at the very least by Montgomery. The opposite was the case. The battle had been primarily an American victory. But Monty had handed the British pressmen a good story and they were not going to argue with that. It was written up as if he had intervened personally to save the floundering Americans, and while some papers used words that were not Monty's, the sentiment clearly was.

Eisenhower complained that American commanders were 'deeply resentful – they believed he had belittled them'. In fact they were furious. Bradley said Monty was 'all-out, right-down-to-his-toes mad'. Both he and Patton told Eisenhower they could no longer serve under Monty and threatened to resign. General Collins had most cause for complaint:

Monty really got under my skin by downgrading the American troops. This is what irritated the hell out of me and Brad. He suggested that now the British were masters, everything would be all right. Only one British division participated in the fighting. That press conference so irritated Bradley and Patton, and many of us who fought on the northern front of the Bulge, that it left a sour note.

Monty tried to extricate himself, telling war correspondents that 'I salute the brave fighting men of America' and 'I am absolutely devoted to Ike', but it was too late. His comments had triggered a crisis that threatened the Anglo-American alliance and Churchill acted quickly to repair the damage. Speaking in the House of Commons, he announced (in what amounted to a rebuff to Monty) that US forces had done 'almost all the fighting'. He pointed out that 'The Americans engaged thirty or forty men for every one we engaged, and they lost sixty to eighty men for every one of ours.'

One sentence was surely meant for Monty himself: 'Care must be taken in telling our proud tale not to claim for the British Army an undue share of what undoubtedly is the greatest American battle of the war.'

Despite this, Anglo-American relationships remained fragile. General Hastings Ismay, Churchill's chief military adviser, hoped that someone would 'muzzle or better still chloroform Monty. I have come to the conclusion that his love of publicity is a disease, like alcoholism or taking drugs, and that it sends him equally mad.' The man himself was not repentant: 'So great was the feeling against me on the part of the American generals, whatever I said was bound to be wrong . . . In contradistinction to the rather crestfallen American command, I appeared, to the sensitive, to be triumphant – not over the Germans but over the Americans.'

That, of course, was precisely the point.

The German Army may have lost the battle but German intelligence sensed a propaganda coup. They had been following the increasing friction between the Allies with interest and knew that, for the Americans, this antagonism found its focus in General Montgomery. They now attempted the propaganda equivalent of the Ardennes offensive by making a quite different thrust between the British and Americans.

On the evening of Monty's conference Chester Wilmot had sent his report to the BBC in London by radio signal. The Germans picked it up and skilfully redrafted it to tone up the sense of British superiority – already in its original form enough to irritate the Americans – to the point that it could only be interpreted as anti-American. It was then rebroadcast in perfect 'BBC' English:

Montgomery tackled the German Ardennes offensive and transformed it into a headache for Rundstedt. He found no defence lines, the Americans somewhat bewildered, few reserves on hand and supply lines cut. He quickly studied maps and started to tidy up the front. He took over scattered American forces, planned his action and stopped the German drive. The battle of the Ardennes can now be written off, thanks to Field Marshal Montgomery.

Several British newspapers picked up the transmission, mistook it for a genuine BBC report and used the 'information' in their columns. The *Daily Telegraph* reproduced it in full. American troops heard it on their radios and read it in their British papers, and were incensed. Anti-British feelings soared.

Eisenhower confirmed that 'Both the Prime Minister and I tried every device at hand to counteract this but it seemed to cause lasting resentment due largely to the fact that American troops here have their principal sources of information through London newspapers or the BBC.' He slightly missed the point. American troops had reacted almost as badly to newspaper and radio reports of what Monty actually said as they had to the doctored version.

German intelligence felt that this was a propaganda battle they had won. It is nearer the truth to say that Monty won it for them. The fake report had the effect it did because its words and sentiments were recognized as Montgomery's own, even though in this case they had been scripted for him in Berlin. The writers did not have a hard task – from 'unintentionally insulting' to 'deliberately insulting' is not a great literary step.

During the early weeks of 1945 the viewpoint of the Allies changed dramatically. After the failure of the Ardennes offensive, German forces had no strength left for anything other than defence. In that sense Germany had already lost the war (although Hitler and much of his army insisted on playing it out to the bitter end). It did not follow that Britain, America and Russia would win on an equal footing.

In the east the Soviet winter offensive rolled forward faster than the British or Americans had anticipated, and by February the Red Army was within 160 miles of Berlin. In the west the Siegfried Line had not been seriously breached and the Allies had no bridge-head on the Rhine. It seemed likely that Stalin might occupy the whole of Germany including the strategically important Baltic and North Sea ports. The western Allies saw an urgent need to cross the Rhine into Germany, not because that might affect the outcome of the war but because it would affect the political and military makeup of post-war Europe.

As this posturing for position began between the western Allies and the Soviets, a similar process began between the British and the Americans. Eisenhower felt that the British were doing everything possible to 'build up' Montgomery's achievements so that it would appear to be their national hero who 'won the war'. He wrote to tell the US Army chief of staff, General Marshall, on 24 March: 'When operations carried out under Montgomery's direction are of considerably less magnitude than those in other parts of the front, there is some influence at work that insists on giving him credit that belong to other commanders.' That was mostly the partisan British war correspondents. But their chosen hero was not unwilling, as General Whiteley pointed out: 'Monty wants to ride into Berlin on a white charger.' Ike told Cornelius Ryan that 'Montgomery had become so personal in his efforts to make sure the Americans and me in particular got no credit that in fact we hardly had anything to do with the war!'

At the same time Washington pressed Ike to ensure that the war ended with the plaudits and the material gains going to the US. Americans felt that, however the final battle went, they had won the war of attrition by supplying vast matériel resources to sustain the Allies while German matériel was reduced and could not be replaced (particularly after the bombing of the Ruhr and the destruction of the steel industry).

As Soviet tanks surged towards Berlin all available German forces were retransferred to the eastern front. This German weakening of the defence in the west, combined with the Allied build-up of men and matériel, should have allowed the next Anglo-American objective – crossing the Rhine – to be easily attained. But bickering over strategy handicapped the effort. Ike still favoured a broad-front advance. The British felt this would not carry them across the Rhine quickly enough to reach 'the main prize, Berlin, before or at the same time as the Red Army'. Churchill and Brooke pressured Eisenhower to allow Monty's 'single, full-blooded thrust' towards Berlin.

A difference of opinion that had previously only arisen at SHAEF and in the various army headquarters, now occupied both London and Washington. The British chiefs of staff wrote to Marshall implying that Eisenhower was to blame for the German Ardennes offensive, because instead of giving full logistical support to Montgomery's thrust into the Ruhr he had allowed Patton's thrust towards the Saar to continue and thereby use up scarce resources. Marshall countered that as a majority of Allied land troops were American there could only be an American in overall command of them, and suggested that unless London agreed to Ike's broad-front strategy he would resign as Supreme Commander, triggering a major crisis in the alliance.

It was not simply a national divide. Many of the British officers at SHAEF were fiercely loyal to Ike, while Patton sided with the British. His dislike of the British in general and Monty in particular took a back seat to the military situation. He and Monty had always agreed that a single thrust was needed, although each believed it should be his own.

Roosevelt, Churchill and Stalin had met at the Yalta summit conference in November 1944. It proved easier for the politicians to agree the post-war division of Europe than for the chiefs of staff, in a concurrent meeting, to agree Anglo-American strategy for the final phase of the fighting. The British again attempted to have Montgomery named as overall land-forces commander. US secretary of war Henry Stimson noted that 'Marshall, who is always very tolerant with the British, finally "lit out". Montgomery has won the reputation of being a good deal of a self-seeker among our commanders at the front. He wants everything in the way of help and command and then is rather over-cautious in his advances.'

Eisenhower was caught in the middle. For purely military reasons – 21st Army Group was simply not strong enough on its own – he allowed Monty to retain US First Army under his command, and now arranged to transfer a further six divisions to him from Bradley in the south. That was perfectly in line with his broad front strategy, taking from the stronger force to give to the weaker with the intention that both reach the Rhine at about the same time. To Bradley and Patton it appeared that he was again putting more weight behind Monty's northern thrust to the Rhine than Patton's move in the south. Their reaction was almost hysterical. Bradley felt that the Americans left in the south would have to 'sit on our ass until hell freezes', while Patton suggested that he 'tell them to go to hell and we will resign'. He felt that 'This is another case of giving up a going attack in order to start one that has no promise of success except to exalt Monty, who has never won a battle since he left Africa and only El Alamein there. I won Mareth for him.'

Patton went on leave to Paris (and Bradley thought for a time that he had 'given up' on the war). He stayed at the Hôtel George V and saw Jean Gordon. When he attended the Folies Bergère, 'which is perfectly naked', he was given a standing ovation by the audience. He visited the Roman amphitheatre at Trier and could 'smell the seat of the legions'. By the time he returned to his army he had calmed down, and he and Bradley resolved to press on at all speed with what they had. Patton pointed out that 'Monty is so slow and timid that he will find a German build-up in front of him and will stall. I will be the first on the Rhine yet.'

The advance eastward was slow but relentless. The supply situation was now better than at any previous time in the campaign, although whenever a day's full allowance of fuel failed to reach Third Army, Patton instinctively blamed Ike for still allowing Monty priority with supplies, and thought this was 'a foolish and ignoble way for the Americans to end the war'. He was eager to reach the Rhine before Monty but of greater importance to him now was reaching Berlin before the Red Army. Every report from the eastern front indicated that the Soviets were closing on the German capital faster than he was, and sometimes his patience snapped. When a column was held up by a motorized gun stuck in a tunnel and the colonel in charge asked what he should do, Patton barked, 'You can blow up the Goddamn tunnel, or you can blow up the Goddamn gun, or you can blow out your Goddamn brains, I don't care which.'

Monty began his big push towards the Rhine on 7 February, preceded by an air and artillery bombardment. It took him four weeks and a great deal of hard fighting to clear the west back area, particularly the Reichswald forest and the northern end of the Siegfried Line, during which British and Canadian casualties exceeded 15,000 and in the Ninth US Army more than 7000.

To the south Bradley moved towards the Rhine south of Düsseldorf and found less resistance as German forces pulled back across the Rhine. Those that were left were divided between a number of small towns instead of being concentrated in one mass, and easily overcome. On 6 March Third Army broke through a weak enemy force in the Eifel area and dashed towards the Rhine. It advanced twenty-five miles in three days and reconnaissance patrols reached the river north of Coblenz to find all the bridges demolished. As his divisions came up Patton looked for likely crossing sites. On the night of 22 March he put patrols across at Nierstein, and when they found the far bank unoccupied he sent a whole division over on pontoon bridges. He called Bradley the next morning to say, 'Don't tell anyone, but I'm across. There are so few Krauts around here they don't know it yet. So don't make any announcement.'

The Germans knew it later that day. As the *Luftwaffe* attacked the bridges Patton's anti-aircraft guns shot down every plane. The secret was out and he called Bradley again: 'Brad, for God's sake tell the world we're across. I want the world to know Third Army made it before Monty.'

On 24 March he crossed the Rhine at Oppenheim. He walked over a pontoon bridge and halfway across he stopped and un-buttoned his fly 'to take a piss in the Rhine. I didn't even piss this morning when I got up so I would have a full load.' He was proud to have beaten Churchill in polluting the river – the Prime Minister had made a point of urinating on the Siegfried Line and looked forward to doing so in the Rhine – but he took the greatest pleasure in beating Montgomery to the far bank. Eisenhower wrote to tell him that 'You have made your Army a fighting force that is not excelled by any other of equal size in the world.'

Just before Monty's assault on the Rhine near Wesel began on 23 March, Churchill, Brooke and General Simpson (director of Military Operations in the War Office) arrived to observe. Monty had prepared as only he could. The town of Wesel and the sur-rounding area had been laid waste by 50,000 tons of high explosive. He had assembled thirty divisions: 250,000 men and their equip-ment, with the addition of assault boats to make high-speed cross-ings. The Prime Minister watched from an armoured car as 6th and 17th Airborne Divisions – a total of 17,000 men – landed beyond the river. He was impressed by the spectacle, but more so by how Monty kept himself informed of what was happening: 'A succession of young officers presented themselves. Each had come back from a different sector of the front . . . As in turn they made their reports and were searchingly questioned by their chief the whole story of the battle was unfolded. This gave Monty a complete account . . . by highly competent men.'

The following day Churchill said that he would like to go across the Rhine, and he was prepared to argue it out if the general attempted to stop him, but Monty said, 'Why not?' Churchill, Monty, Brooke and Simpson managed to fit themselves into a tank landing craft and the captain motored steadily across the river. The four men walked for thirty minutes on the far bank. The Prime Minister was clearly enjoying himself and Monty suggested they return to the landing craft and motor down-river to the town of Wesel. When the captain pointed out that the river was mined between their present position and the town, Monty had the solution: he had a Rolls-Royce and driver at hand, and they drove there. Churchill and Monty were soon clambering over the twisted remains of the iron railway bridge like young boys, when German shells began to fall nearby. Simpson felt he had to act: 'Prime Minister, they are shelling both sides of the bridge and now they have started shelling the road behind you. I must ask you to come away.'

Now that Monty was across the Rhine he decided to strike immediately for Berlin and signalled Eisenhower on 27 March: 'My intention is to drive hard using the ninth [American] and second [British] armies . . . my Tac HQ will move to Wesel, Münster, Widenbruck, Herford, Hanover – thence by *Autobahn* to Berlin I hope.'

Ike was furious. First, he had not agreed that Monty could retain Ninth Army under his command, yet he acted as if it was 'his'. Second, it was up to the Supreme Commander to decide how the advance east of the Rhine would be conducted and with which armies, but Monty simply announced the route of his triumphal dash to Berlin, and added that he was taking Ninth Army to assist him. Third, although Monty did not know it yet, the Allies would not be going to Berlin.

Eisenhower replied the next morning and, with one powerful thrust, he won what had become an increasingly acrimonious tussle with Montgomery. Perhaps his anger finally allowed him to exercise the firm control he should have exerted much earlier, but it is as likely that the certainty the war was won freed him from his role of protector of the alliance. His curtly worded signal simply gave Monty his orders:

US Ninth Army will revert to Bradley . . . Bradley will be responsible for occupying the Ruhr and will deliver his main thrust on the axis Erfurt–Leipzig–Dresden to join up hands with the Russians. The mission of your army will be to protect Bradley's northern flank.

Monty had attempted more than once to relegate an American army to protecting his flank and the final sentence was meant to hurt.

17. What to Do with Nazis and Communists

The frantic transfer of German forces to hold back the Red Army in March 1945 left only sixty-five divisions in the west, all of them short of tanks and ammunition, and none of them at full strength, to face the Anglo-American armies surging across the Rhine. Eisenhower commanded ninety-two full-strength divisions. He had declared the previous September that 'we should concentrate all our energies on a rapid thrust to Berlin' and now his commanders on the east bank of the Rhine were in position to do exactly that.

They discovered only now the 'agreement' that had been made with Stalin. At the Yalta conference Roosevelt, Churchill and Stalin had divided Germany into occupation zones. As Berlin was in the Soviet sector it would be left to the Red Army to take the city. Any Anglo-American advance risked bumping into Soviet forces and an unprecedented Blue-on-Blue incident.

Churchill had disagreed vehemently, but at Yalta he had been the junior partner with no veto over the decision of the two heavyweights. His concerns had deepened when Roosevelt told Stalin that all American troops would be withdrawn from Europe within two years of victory. Now at the point of that victory he still felt that Berlin should have been 'the prime and true objective of the Allied armies'. He believed that the Soviet Union would become 'a mortal danger to the world'. In London the celebrations began before the official surrender of German forces and the Prime Minister admitted that he 'moved amid cheering crowds with an aching heart and a mind oppressed by forebodings'.

To the generals who had done the fighting, making a gift of

Berlin to the Soviets appeared nonsensical. Monty called it 'dirty work' and wondered what the point was in winning the war militarily if they then lost it politically. Patton confronted Ike and asked why it had been done:

EISENHOWER: Berlin has no strategic or tactical value.
PATTON: I don't see how you figure that. We had better take Berlin and quick.
EISENHOWER: George, why would anyone want it?
PATTON: I think history will answer that for you.

The German Army had not yet surrendered although for the Allied armies a series of 'mopping-up' operations had replaced major actions. Third Army found the going easy during the first two weeks of April as it took to the *Autobahnen* and motored past the city of Erfurt to Ohrdurf, where Patton established his headquarters. On 12 April he visited the Ohrdurf concentration camp along with Eisenhower and Bradley. No one could tell them exactly what had happened there – the SS guards had killed all of their prisoners to prevent that – but the sights were enough. Patton was physically sick. He later ordered that as many men of Third Army as possible must visit the camp.

By 25 April the Red Army had encircled Berlin and five days later the Soviet flag was raised over the Reichstag. Hitler poisoned and shot himself. Goebbels killed himself and his six children. Soviet troops took the Reich Chancellery and Marshal Zhukov claimed Berlin on behalf of Stalin and the Soviet Union. Swastika banners were torn down and replaced by red flags bearing the hammer and sickle.

When the Russian and American armies linked up at Torgau, an improvised Red Army choir sang 'The Star Spangled Banner'. Bradley presented Marshal Konev with an American jeep. In return Konev gave him a horse, for which Bradley had no use and an ADC was told to 'see that Patton gets the animal'.

★

Admiral von Friedeburg led the four-man German delegation that arrived unexpectedly at Montgomery's Tac HQ on Lüneburg Heath on 3 May. A Union Flag was quickly run up outside the command caravan and Monty came out to demand in what he called 'a very sharp, austere voice, "Who are you and what do you want? I have never heard of you."' Friedeburg offered to surrender all German forces in the north of the country, including those facing the Russians. After telling the delegation off for the mass murder of the Jews, the destruction of Coventry and other sins, Monty demanded the unconditional surrender of all German forces within twenty-four hours or he would convince them 'with 10,000 bombers'.

They returned to sign the next day. A tent had been erected for the occasion and Monty described the actual moment of surrender:

The German delegation went across to the tent watched by groups of soldiers and war correspondents, all very excited. They knew it was the end of the war. The Germans stood up as I entered ... They were clearly nervous and one of them took out a cigarette; he wanted to smoke to calm his nerves. I looked at him, and he put the cigarette away. I read out in English the Instrument of Surrender. I then called on each member of the German delegation by name to sign the document.

He then ordered all offensive action to cease at 0600 the following day, pending the signing of a formal surrender by General Jodl on behalf of the German High Command. Hitler's successor, Admiral Karl Dönitz, delayed this as long as he could to allow as many German troops as possible to surrender to British or American armies and thereby escape the Red Army. Under pressure from Eisenhower, he finally gave Jodl permission to sign on 7 May.

Monty signalled Brooke to say that he had been 'persuaded to drink some champagne'. Despite the celebrations his mood remained sombre. It was his belief that 'the oncoming Russians were more dangerous than the stricken German'. His last task of the war had been the capture of the German port of Lübeck on the Baltic, not because it was of any great use to the enemy, but to prevent the Russians using it as a stepping-stone for the occupation of Denmark. He had taken Lübeck on 2 May.

It was left to Patton, who saw the situation precisely as Monty did, to spell that out more forcefully in a press conference held on the day Jodl signed the surrender document:

What the tin-soldier politicians in Washington and Paris have managed to do today is to kick hell out of one bastard and at the same time forced us to help establish a second one as evil or more evil than the first . . . We'll need Almighty God's constant help if we're to live in the same world with Stalin and his murdering cutthroats.

That was not bluster; he meant every word of it. The same day he met Robert Patterson, the US under-secretary of war, and told him that the army must 'keep our boots polished, bayonets sharpened and present a picture of force to the Russians – this is the only language they understand'. Because Patton always expressed himself colourfully it was not always appreciated that his opinions were well reasoned. He explained to Patterson that the US Army had come to Europe to help nations reclaim the right to govern themselves, not to hand them on from Hitler's rule to Stalin's, and it should stay in Europe until that job was done.

His antipathy towards the British had often expressed itself in his descriptions of individuals, and his dislike of the Russians now did so too. He met Marshal Zhukov and wrote to tell Beatrice that 'he was in full dress uniform like comic opera and covered in medals. He is short, fat, and has a prehensile [sic] chin like an ape.' Marshal Tolbukhin was 'a very inferior man and sweated profusely'. His impression of Russian officers as a whole was that they 'give the appearance of recently civilized Mongolian bandits'. He told Cornelius Ryan that 'I just can't stand being around and taking any lip from those sons-of-bitches.'

Unfortunately he could not avoid a number of engagements with his Russian equivalents to give or receive an award or medal, or to entertain to lunch or dinner. Whenever possible he eased his passage by drinking heavily and found that the Soviet officers did so too. On 13 May when he lunched with the commander of the Soviet Fourth Guards Army, the Russian drank so freely that he fell into a stupor while Patton bragged that he himself 'walked out under my own steam', which apparently proved that 'we could beat the hell out of them'.

He was stone-cold sober on the day that he scared the hell out of them. A Russian general visited his HQ to demand that a number of riverboats on the Danube requisitioned by Third Army be handed over to him. Without asking why the Russian thought they might be *his* boats, Patton pulled out a pistol, slammed it hard on to his desk, and shouted to his ADC, 'Goddammit, get this son-of-a-bitch out of here! Who the hell let him in? I don't want any more Russian bastards in my headquarters.' The general left hastily. Patton then told the ADC that 'Sometimes you have to put on an act. That's the last we'll hear from those bastards.'

Monty shocked the Russians in a quite different way. He had lunch with Marshal Konstantin Rokossovsky on 7 May and found him to be 'tall, good-looking and well dressed'. For his part Rokossovsky could not quite make out his short, wiry host, and when told that the field marshal neither smoked, drank nor entertained women in his command caravan, he asked, through a bemused interpreter, 'What the devil does he do all day?'

Montgomery was appointed military governor of the British Zone of occupied Germany. Accurate figures were never established but there were approximately two million German prisoners of war, one million wounded German soldiers and civilians, and about two million civilian refugees who had fled the advance of the Red Army. The railway network had been destroyed, there were few undamaged hospitals and the public utilities were not operating. Monty's job was to restore order and sort out the mess.

He did so by issuing orders to the civilian population as he had previously done to his army. The first, on 30 May, announced that

'The German people will work under my orders to provide the necessities of life for the community. The population will be told what to do. I shall expect it to be done efficiently.' He was the first (but perhaps not the last) to describe himself as behaving like 'a military dictator'. In the circumstances it was the only way.

He believed, as did Patton, that civilians who had been members of the Nazi Party should be put back to work in their former civilian jobs, and ordered that German troops were to be 'demobilized and directed back to their civil vocations, as and when they were needed'. In fact, he issued many of his instructions 'through the German command organization' and specifically via Field Marshal Busch. The crisis was so acute that any command structure that worked, and any worker who could be returned to his job, was preferred to the strict policy of denazification sponsored by Eisenhower.

Like Churchill, and Patton too, he now viewed the Red Army as the real threat: 'The Russians, though a fine fighting race, were in fact barbarous Asiatics who had never enjoyed a civilization comparable to the rest of Europe. Their behaviour, especially in their treatment of women, was abhorrent.' There is a startling sentence in his *Memoirs* in which he notes that at this time he was 'given a "stand still" order regarding the destruction of German weapons and equipment, in case they might be needed by the Western Allies for any reason'. As no such order was received by American commanders it did not come from Eisenhower, and the only other person from whom Monty would accept an order was the Prime Minister. Churchill certainly believed that Stalin and the Red Army were a threat to the West, and as both the British and American armies had sufficient weapons of their own (there being many in the logistical 'queue' that had not yet reached their end users), this order could only refer to the potential return of German weapons to German soldiers.

Patton was appointed military governor of Bavaria and established his headquarters at Bad Tölz thirty miles south of Munich. He lived in a country house that had been owned previously by Max Amann, publisher of *Mein Kampf*, and overlooked Lake Tegernsee.

Patton had shown a surprising talent for diplomacy in French Morocco, but it was thought he might struggle in Bavaria and Eisenhower sent Robert Murphy, his diplomatic adviser, to offer any help or advice he might require. Murphy was shocked by the only question Patton wanted answering: 'He inquired with a gleam in his eye whether there was any chance of going on to Moscow, which he said he could reach in thirty days, instead of waiting for the Russians to attack the United States when we were weak and had reduced the army.'

In June he was required to return to the US on a thirty-day bond-raising tour. He told Hughes he was 'scared to death of going home' because he knew Beatrice was waiting to question him about Jean Gordon. Hughes was present at his farewell dinner the evening before he left; so was Jean.

He arrived in Boston on 7 June to a ticker-tape welcome as he was driven through streets lined with cheering crowds. It was estimated that more than a million people turned out to see him. It was the same in Denver, Los Angeles and Pasadena. His speeches attracted huge audiences – 100,000 at the LA Coliseum – and his visits were carried off with aplomb; he quipped to reporters who accompanied him around the Walter Reed Hospital in Washington, 'I'll bet you're just following me to see if I'll slap another soldier.' It was an indication of his total reinstatement as an American hero that he could make light of an incident that had almost cost him his military career. The press reported the crowds and the speeches, but also Patton's 'other side'. When he visited the amputee ward at Walter Reed he openly cried as he told the soldiers that 'If I had been a better general, most of you would not be here.' During and immediately following the tour the sales of war bonds soared. Not everything went so well. Back in Bavaria in early July he told Hughes that Beatrice had 'given him hell' over Jean Gordon.

His solution to 'the mess' in Bavaria was to do just as Monty did, despite Eisenhower's denazification policy requiring him to dismiss proven Nazis: 'If I dismiss the sewage cleaners and clerks my soldiers will have to take over their jobs. They'd have to run the telephone exchanges, the power facilities, the street cars, and that's not what soldiers are for.' He reinstated civilians and ex-soldiers to their previous jobs regardless of whether or not they had been members of the Nazi Party. He broke Ike's 'guide-lines' and expected to get away with it, especially when secretary of war Henry Stimson found Bavaria to be the best-governed of all US military areas. He saw the Red Army as the only threat now and explained that to Beatrice on 16 July: 'We have destroyed what could have been a good race and are about to replace them with Mongolian savages. I wish I were young enough to fight in the next big war. It would be fun killing Mongols . . . It's hell to be old.' That for Patton was the problem – not his age as such, but that it ruled out any further service as a soldier. He wrote in his diary that all he had left to do was 'to sit around and await the arrival of the undertaker and posthumous immortality'.

He still had the Russians to spar with. He attended a Soviet military review in Berlin and Marshal Zhukov proudly brought his attention to the new Stalin IS-3 tank. When Zhukov boasted that its gun had a range of seven kilometres, Patton replied: 'My dear Marshal Zhukov, if any of my gunners started firing at your people before they had closed to less than seven hundred yards, I'd have them court-martialled for cowardice.'

During a visit to the Garmisch-Partenkirchen prison camp he discovered that some civilians held there had previously worked in the civil service and for utility companies, and were imprisoned because they had also held high posts in the Nazi Party. His comment that it was 'sheer madness to intern these people' was reported by an American officer. Eisenhower went personally to Bavaria to investigate and found that far from following the denazification policy, the 'new administration' he had established to rule Bavaria included a large number of former Nazis, including some former SS men. Ike demanded that he 'stop mollycoddling the Goddamn Nazis' and keep to his guide-lines. Patton confided to his diary that 'If it's a choice between the Germans and the Russians, I prefer the Germans.'

Colonel Harkins was in Patton's office during a telephone conversation he had with General Joseph McNarney, who was Eisenhower's deputy and acted in his place whenever Ike was not in Germany. McNarney had rung to tell Patton the Soviets were complaining that German Army units in his area were not being disbanded quickly enough.

PATTON: Hell, why do we care what those Goddamn Russians think? We are going to have to fight them sooner or later. Why not do it now while our Army is intact and the damn Russians can have their ass kicked back to Russia in three months? We can do it easily with the help of the German troops we have, if we just arm them and take them with us. They hate the bastards.

MCNARNEY: Shut up, George. This line could be tapped and you'll be starting a war with the Russians with your talking.

PATTON: I want to get it started some way. That's the best thing we can do now. You don't have to get mixed up in it if you're so damn scared of your rank. Let me handle it from here. In ten days I can have enough incidents happen to have us at war with those sons-of-bitches and make it look like their fault, so we'll be justified in attacking them and running them out.

McNarney hung up. Patton had said too much. Everyone knew that he spoke from the hip and allowed for that, but McNarney feared Patton might really 'do something'. Colonel Harkins had overheard the call and Patton turned to him, apparently eager to convince him too. 'I really believe that we are going to fight them, and if this country does not do it now, it will be taking them on years later when the Russians are ready for it and we will have an awful time whipping them. We will need these Germans.'

Patton's pronouncements would be proven correct but at

the time they seemed indiscreet and naïve to an equal extent. Eisenhower's concern was to ensure that Germany could not re-emerge from the second war as it had from the first to rearm and begin the cycle again. His eyes were not on the Russians and what they might do.

It was later suggested that, following this telephone conversation with McNarney, Ike decided that Patton might be mentally unbalanced and an 'undercover psychiatrist' was posted to his headquarters as a supply officer to examine his behaviour without his knowledge, and his phone was tapped. However no report from the supposed psychiatrist and no transcripts or tapes of his telephone calls have ever been identified. In any case, Patton could always be relied on to condemn himself, and he soon did.

At a press conference at Bad Tölz on 22 September the gathered correspondents knew enough to ask Patton how his employment of ex-Nazis in high administrative posts in Bavaria squared with Eisenhower's denazification policy. He replied that he hated the Nazis as much as anyone and pointed out that he had just spent three years killing as many of them as possible, but that to get things done in Bavaria he had to 'compromise with the devil'. Then one asked a carefully loaded question: 'After all, General, didn't most ordinary Nazis join their party in about the same way that Americans become Republicans or Democrats?' Patton failed to see the headline this man already had in mind. 'Yes, that's about it.' The next morning it appeared in the US: 'AMERICAN GENERAL SAYS NAZIS ARE JUST LIKE REPUBLICANS AND DEMOCRATS'.

Two days later while Patton was sitting for the portrait painter Boleslaw Czedekowski, Ike's chief of staff Bedell Smith rang to say that he must retract the statement or lose command of Third Army. The image of Patton caught by Czedekowski at this critical time was duplicated in ink by the war correspondent Bill Mauldin. Patton had threatened to ban the *Stars and Stripes* newspaper from his Third Army if it continued to print Mauldin's 'Willie and Joe' cartoons. At a meeting between the two men Patton harangued him for drawing 'anti-officer cartoons'. Mauldin's ordeal was worth it for the pen portrait he came away with:

His hair was silver, his face was pink, his collar and shoulders glittered with more stars than I could count, his fingers sparkled with rings. His face was rugged, his eyes were pale with a choleric bulge, his mouth sharply downturned at the corners. Beside him, lying in a big chair, was Willie the bull terrier. If ever a dog was suited to its master this one was. Willie had his boss's expression . . . I stood staring at the four meanest eyes I'd ever seen.

On 28 September Ike decided that Patton's withdrawal of his statement was not enough, given the evidence of his employment of so many ex-Nazis. He was relieved of command of Third Army and given command of Fifteenth Army at Bad Nauheim, which was no army at all but an administrative term for the unit detailed to compile a history of the war in Europe. He was ordered to say nothing more about the whole affair, and he let off steam in his diary: 'I was unwilling to be party to the destruction of Germany under the pretence of de-Nazification. I believe Germany should not be destroyed, but rather rebuilt as a buffer against the real danger which is Bolshevism from Russia.' He added later that 'My chief interest in establishing order was to prevent Germany going Communistic.'

It is tempting to characterize Patton's deep hostility to the Russians as a symptom of mental instability, but all his life he had expressed his opinions using forceful, colourful (often blue) words, and usually there were sober reasons behind them. On 17 November he wrote a long analysis of 'the Russian problem' that included this observation:

One result of the Bolshevik conquest of half of Europe is that they have reduced the scale of living in those countries to the Russian scale, which is very low, and have prohibited the United States and England from selling to about a third of their former markets. This is bound to upset the political economy of England and America, and throw large numbers of men out of work and make them ready victims of Communism.

Soviet economists predicted that too, and in the US the Marshall Plan would later be developed to forestall precisely that effect.

At first light on the morning of Sunday, 9 December, Patton and Colonel Hobart Gay left Bad Nauheim at first light for a pheasant shoot by the Rhine west of Speyer, an area rich in game. It was to be Patton's last day in Europe and he wanted to enjoy it. The next day he was to fly to London, then sail from Southampton to New York aboard the USS *New York*.

He liked to move fast and he approved of his new nineteen-year-old driver, Horace Woodring: 'Woodring is the fastest. He's better than the best Piper Cub to get you there ahead of time.' The car, a 1939 Model 75 Cadillac, was sturdy and powerful, and Woodring accelerated along the *Autobahn*. Patton ordered a detour so that he could visit the ruins of a Roman military post, but by 1145 they were back on the road and driving through Mannheim where Woodring slowed down to about thirty miles per hour.

At the same time Robert Thompson was driving a US Army two-ton truck through the town and heading back to the quartermaster depot. He had spent the previous night in a beer hall packed with GIs and *fräuleins*, and was enjoying life in post-war Germany. He slowed to ten miles per hour as he approached his left turn into the depot.

Patton noticed burned out vehicles lining the road and said, 'Just look at that.' He was speaking to Gay but Woodring instinctively took a glance too. When he looked back to the road Thompson's truck was directly in front of the Cadillac and only twenty feet away. He stamped on the brakes and swung the car to the left but there was no avoiding a collision. Gay remembered shouting, 'Sit tight.'

Then we crashed. Patton was thrown forward and then backward. At my next recollection of the accident (I was unhurt except for bruises) his head was to the left and I was supporting him on my right shoulder in a semi-upright position. He was bleeding profusely from wounds of the forehead and scalp. He was conscious.

Both Gay and Woodring had seen the collision coming and braced themselves for the impact. Patton, his eyes on the burned-out vehicles, had not. He was thrown forward head first and struck the glass partition between the forward and rear compartments. A cut had opened his skull to the bone. Patton knew instantly that his injury was serious. He said, 'I think I'm paralysed.' He asked Gay to rub his fingers and could feel nothing when he did. 'This is a hell of a way to die.'

An ambulance rushed him to the US Army Station Hospital in Heidelberg, where Lieutenant Hill, the chief surgeon, examined him: 'He had lost quite a bit of blood. He had neither sensory nor motor function below his neck.' A blood transfusion was set up. Patton told the surgeon and his staff, 'If there's any doubt in any of your Goddamn minds that I'm going to be paralysed the rest of my life, let's cut out all this horse-shit right now and let me die.' X-rays proved that he had a broken neck and was paralysed from that point down. While Hill stitched up his wound where, on a balding scalp, it would leave a visible scar, Patton remarked that 'The only memento of my historic service in the First World War is the Goddamn scar on my ass.'

Eleven generals visited him during his first twenty-four hours in hospital. Beatrice was flown to Germany and arrived in Heidelberg on 11 December. Top British and American neurosurgeons were flown in, too, and all agreed that an operation to repair his spinal cord was impossible. Thirty reporters left the Nuremberg war-crimes trials to wait outside the hospital for news. When Colonel Harkins visited and told him he 'looked fine', Patton smiled: 'Paul, Goddammit, you're a lying son-of-a-bitch.' As his condition deteriorated he lamented, 'What an ironical thing to have happen, after the best of the Germans have shot at me, to get hurt in an automobile accident going pheasant hunting.' Beatrice sat at his bedside and read to him each evening from an English translation of the memoirs of General Armand de Caulaincourt, confidant of Napoleon.

He died in his sleep at 1800 on 21 December: a pulmonary embolism in his right lung had deprived his brain of oxygen. The hospital had formerly been a German cavalry school and Patton's body was at first moved to the basement stables, where it lay wrapped in his personal flag in one of the stalls. Later it was moved to the Villa Reiner, high on the mountain that overlooked Heidelberg, to lie in state in a flag-draped casket. MPs mounted a twenty-four-hour honour guard. A constant stream of soldiers visited to say goodbye to their general. In the US the newspapers framed the headline in black: 'GENERAL PATTON DEAD'. At the Virginia Military Academy the flag was lowered to half mast.

On 23 December his casket was carried on a half-track escorted by a platoon of 15th Cavalry in jeeps and armoured cars to the Protestant church in Heidelberg, followed by a cortège of generals. An honour guard of six thousand GIs lined the streets. Two US Army chaplains conducted the twenty-minute service. Several divisional bands played as the casket left the church for Heidelberg station, where the artillery of 1st Armoured Division fired a seventeen-gun salute as a special train left to carry the casket and mourners to Luxembourg City. Before he died he had told Beatrice there must be no question of his body being returned to the US for burial. 'If I should conk, it would be far more pleasant for my ghostly future to lie among my soldiers than to rest in the sanctimonious precincts of a civilian cemetery.'

Patton was buried the following morning in the American military cemetery at Hamm near Luxembourg in the Ardennes. Sergeant Meeks drove the half-track carrying the casket and the roads were lined with troops. He was buried next to a Third Army soldier killed in the battle of the Bulge one year earlier. A twelve-man squad fired a three-round volley over his grave. The *New York Times* wrote: 'History has reached out and embraced General George Patton. He will be ranked in the forefront of America's great military leaders. Long before the war ended, Patton was a legend. Spectacular, swaggering, pistol-packing, deeply religious and violently profane, easily moved to anger because he was first of all a fighting man.'

When Patton's baggage was gathered together to be flown back to the US, an uncomprehending Willie sat waiting among it for his master to return. Two weeks later Jean Gordon committed suicide.

At the time of Rommel's death the German people had been told that the field marshal had died of an embolism as a result of his injuries. But as the war came to an end, the speculation began. A number of high-ranking officers who had been linked with the von Stauffenberg plot chose to kill themselves rather than face what the Gestapo called 'rigorous interrogation'. Rommel's death fell within the same time-frame as these others, and some concluded that he had been involved and killed himself to escape justice.

Manfred Rommel put the record straight as soon as he was free to do so safely. In the first post-war edition of the *Südkurier* (published on 8 September 1945) he revealed that his father had been ordered by Hitler to swallow poison:

His former chief of general staff, Lieutenant General Speidel, who had been arrested a few weeks earlier, was supposed to have testified that my father participated in 20.07.1944 in a leading capacity and was only prevented from taking part directly as a result of being wounded. General Stülpnagel had made the same statements. The Führer did not want to debase his memory in the eyes of the German nation and gave him the option of suicide by poison capsule.

Manfred's statement revealed what had happened on 14 October, but in failing to reveal whether or not the accusations were true it merely inflamed the speculation. In the immediate post-war period the fiercely nationalist German people still viewed the conspirators as traitors. The first reaction to defeat, before a full realization of the extent of Nazi crimes had sunk in, was to embrace the idea of 'the nation' and the fallen Führer at its head. The conspirators had betrayed them all.

Aware of this feeling Lucie Rommel acted quickly to save her husband's reputation. On 9 September, the day after Manfred's account appeared, she wrote a letter as if to a friend but which was published in the 16 October edition of the *Südkurier:*

I want to correct the many rumours, in order to keep the name of Rommel pure and to uphold the Field Marshal's honour. I want to make it clear that my husband took no part whatever in the planning or execution of the July plot. As a soldier, he rejected such a measure. My husband always expressed his opinions and intentions openly to the very highest authorities, even when he knew they would not like it. He was always a soldier, never a politician.

Speidel had survived the war and he was just as eager not to be thought a conspirator. It must be assumed that he and Lucie agreed to combine their efforts, because in that same edition of the *Südkurier* he rushed to the defence of Rommel. In doing so he also cleared his own name, because if he *had* testified to the Gestapo about the field marshal's involvement in the plot then he himself must have been involved: 'I never made such a statement, not even in spirit, in front of the Gestapo; to the contrary, I described participation by the Field Marshal in 20.07.1944 as totally groundless.' Lucie's letter corroborated that: Rommel himself did not believe the supposed sources of the accusation against him: 'My husband was informed by the two generals that General von Stülpnagel, Lieutenant General Speidel and Lieutenant Colonel von Hofacker had made implicating statements. He answered the generals that he did not believe it because it was a fabrication.'

Lucie Rommel and Hans Speidel had underestimated the problem. It was not just that many among the German people believed her husband to have been one of the plotters. The victorious Allies *needed* Rommel to have been one of the plotters.

The Allies realized that eventually they would need to rebuild the German Army to strengthen the west European powers against the Soviet Union. Patton had said as much before his death, but he had said it too soon and too loudly. While the war was still fresh in the minds of the British and American public, and the dead still being mourned, such an opinion could only be spoken of behind closed doors. But at some point the German Army would have to be rehabilitated in the minds of its former enemies, and that process would include a focus on those within it who had themselves struggled against Hitler. There was no better figurehead than 'the good German', Field Marshal Rommel, plotting to assassinate Hitler and make peace.

Speidel had testified in 1945, in association with Lucie Rommel, that the field marshal had not been a conspirator against Hitler. Because he had served as Rommel's chief of staff, most Germans took his word for it. But in 1949 in his book *Invasion 1944: a Contribution to the Fate of Rommel and the Reich* he claimed precisely the opposite. Rommel had been a leading conspirator and was prepared to offer himself as the next head of state after Hitler's 'removal'. That same year Ernst Jünger, who had been a captain on Stülpnagel's staff and had drafted the statement the conspirators intended to issue after Hitler's departure, claimed in *Strahlungen* (Rays) that Rommel was a conspirator, and he believed that 'the blow that felled Rommel on the Livarot Road deprived our plan of the only man strong enough to bear the terrible weight of war and civil war simultaneously'. Ten years later Friedrich Ruge, who had been Rommel's naval adviser in Normandy and spent more time with him than anyone else during that crucial period, described in *Rommel und die Invasion* how the field marshal had struggled with his conscience about acting in any way that was disloyal to Hitler, but was about to open negotiations with Eisenhower when he was injured and hospitalized.

It might be thought that the evidence of these senior German officers who survived the war and who identified Rommel as a conspirator must be considered conclusive. However, we must take into account the post-war context in which they made their claims. At an indeterminate point – but approximately when the Allied administration became fully effective throughout Germany – those who had formerly distanced themselves from the conspirators now readily offered evidence of their involvement. In post-war Germany those who had held a high military rank under the Nazi regime and sought a similar office under the Allied administration could best do so by proving themselves to be anti-Nazi now *and to have been at the very least anti-Hitler then*. Almost the only convincing evidence was a provable connection with the conspirators. They may have been telling the truth or merely going along with the 'good German' myth that benefited both the Allies and themselves. Speidel rose to command the rebuilt army of West Germany, and Ruge to command its rebuilt navy.

There can be no certain conclusion. But the personal accounts of the two men closest to him at the time, Friedrich Ruge and Hellmuth Lang, agree in their depiction of a man devoted to the Führer, who even in the midst of a military dilemma and a personal struggle kept him informed by letter and in person of his changing beliefs, and who finally agreed that following Hitler's removal he would negotiate with Eisenhower and perhaps become head of state. There is no evidence that he agreed with the plot to kill Hitler, and he expressly stated afterwards that he had not.

The idea of 'the good German' served the Allies well, but by the end of the century another transformation had taken place in the way Germans viewed Erwin Rommel. This 'change of mind' was described in 2000 by Ralph Giordana in *The Falsehood of Tradition*. Giordana argued that Rommel, as a senior army officer taking orders from the High Command, acted as a representative of a criminal regime and must himself be considered a war criminal. A detailed discussion of what Rommel knew or did not know then becomes irrelevant because he must share in the collective guilt.

This change in the German perception of Rommel is symbolized by the plaque commemorating him, unveiled in the former officers' mess at Goslar (where he had served as a battalion commander) in

November 1961. General Friedrich Foertsch, inspector general of the *Bundeswehr* (the post-war army), told those attending that Rommel was 'the most magnificent soldier' and 'a role model for young soldiers'. In May 2001 when the plaque was removed, the reason given was that Rommel represented a criminal regime and to honour him was to honour that regime.

Montgomery flew home to spend Christmas 1945 with David and the Reynoldses. He was to be made a viscount and had already decided to take the title 'Montgomery of Alamein', although that had to be kept a secret until the New Year's Honours were published on 1 January. In fact, his staff in Germany already knew. Bill Williams, appropriately his chief of intelligence, said that 'We'd seen him practising his signature for weeks on the blotting pad.'

When Monty travelled to Newport in Wales to attend a banquet given in his honour, the press had already received copies of the speech he would make praising the conduct of 53rd Welsh Division in the war. It was not until he arrived that he discovered his mother had been invited too. He gave the mayor of Newport an angry ultimatum: 'I won't have her here. If she comes, I go.' She was already *en route* and a compromise was hastily negotiated, by which she was moved from the top table to a distant corner of the hall. He refused to approach or speak to her throughout the evening.

While visiting Saanenmöser in 1946 he met a twelve-year-old Swiss boy, Lucien Trueb, and continued to write to him and take him on holidays. Lucien later revealed that Monty would bathe and dry him, but confirmed that nothing improper occurred and that his parents were happy about the relationship. Lord Chalfont (in a biography written after Monty's death) noted his 'predilection for the company of young men'. A later biographer, Nigel Hamilton, described him as a repressed homosexual who had 'quasi love affairs' with boys and men. Neither produced evidence of any physical relationship, although at eleven Hamilton had been befriended by the field marshal and had received more than a hundred 'loving letters' from him.

In April 1946 he became chief of the Imperial General Staff, the top post in the British Army, and arrived at the War Office in London wearing his famous black beret. He was unexceptionable in the post, and when he left two years later *Newsweek* magazine estimated that the British Army had never been 'so incapable of going to war'. Appointed deputy Supreme Commander of the newly formed NATO, he found himself serving once again under Eisenhower.

Monty retired in 1958 and spent much of his time touring the world to tell, among others, Khrushchev, Mao and de Gaulle how to run their countries. He died on 24 March 1976 aged eighty-eight. When his safe was opened it was found to contain nothing but his army revolver, and in all probability the army was the only thing he truly cherished.

Part Three Appendix
Original Monty: Memoirs At War

Eisenhower's memoirs, *Crusade in Europe*, were published in November 1948 for an advance payment of half a million dollars. He knew that his publisher was paying for more than a narrative of the war from the perspective of its Supreme Commander. He wanted the low-down on the major personalities, and the more controversial the comments the better for sales. The best-known personality Ike could disparage without any political fallout in his own land was Montgomery.

American readers lapped up his criticism of the British. Great Britain had lost an empire and the United States had gained one, and the American people expected this to be represented in the narrative of the war as much as by its outcome. Monty wrote to tell Ike that it was 'a pity you should have thought it necessary to criticize me and my ways ... I feel sad when an officer I have tried to serve loyally criticizes me publicly.' To de Guingand he complained that 'Ike's book conveys the impression that America won the war. Why has he done this? It is not good taste.'

British papers serialized the book and editors knew what would most interest their readers. The *Sunday Dispatch* headline was typical: 'HOW MONTGOMERY UPSET AMERICANS by GENERAL EISENHOWER'. Monty responded with a letter-writing campaign to newspapers giving his impression of the book and its author: 'It is a mediocre work. It shows clearly that he wasn't a great soldier. Had he kept silent, he might have passed for one. A man who can't control historical facts in narrative certainly couldn't control an army in battle.' He told the editor of the

Sunday Times that he would 'one day write my own story of the war and give my account of certain matters'.

Meanwhile Bradley's memoirs, *A Soldier's Story*, were published in 1951. British newspaper headlines included 'BRADLEY CRITICIZES MONTY'. Patton was dead but Bradley had included many of *his* views and comments. Monty wrote to him to complain:

I cannot see what good will be done to Anglo-American relations by quoting Patton on saying he would like to push the British into the sea and give them another Dunkirk. Nor do I see how such a statement by Patton can help you in your analysis of the campaign in Normandy . . . Whatever you say about me I must accept in silence; this is no time for making more trouble.

That time eventually came in November 1958 with the publication of *The Memoirs of Field Marshal Montgomery*. Bill Williams had read the draft manuscript and felt that his own memory of what happened did not square easily with Monty's account: 'I could check certain things and doubt certain things and ask for proof of certain things. But in the end if he wanted to say something, that was it – even if one thought: I don't think that is provable.' Monty had also shown the manuscript to his friend Colonel Trumbell Warren: 'I said, "Sir, you can't publish this. You can do it after Eisenhower dies. You can't do it before." He said, "I'm going to." I said, "You'll be awfully unpopular with the Americans. He's the President of the United States." He said, "I don't care who he is."' The first UK edition of 140,000 copies sold out on the day after publication, and another 100,000 during the following six weeks.

It was serialized in the US in *Life* magazine, and after seeing that, President Eisenhower declared that the book would be 'a waste of time to read if I was looking for anything constructive'. He was so angry that he considered calling together all the American wartime generals for a ten-day conference at Camp David to draw up a memorandum rebutting Montgomery's *Memoirs*. That sounds like an overreaction, but Monty wrote to tell Warren he had heard that 'Ike considers my book made a definite contribution to the heavy defeat of the Republican Party in the Congressional Elections. I am told that Ike is more vehement about the book than anything else.' Monty was not repentant: he was boasting.

He could be relied on to make matters even worse. In an interview broadcast by CBS in April 1959 he declared that Eisenhower had failed to understand the battle plan for the Normandy campaign, and failed to understand the necessity of concentrating the Allied armies for a single thrust into Germany. This latter mistake, he claimed, allowed the German Ardennes offensive and prolonged the war. As if intent on insulting his American audience he went on: 'My observations would be that your leaders are not awfully well. Your President has had a heart attack and a stroke. The head of your State Department walks about on two crutches . . .'

Eisenhower called the interview 'a deliberate affront'. When de Guingand attempted, as in times past, to act as peacemaker between the two, Ike told him that any attempt by Monty to visit Washington 'would be bad judgement' and that 'any correspondence between us could not be helpful'.

In these passages from his *Memoirs*, Monty discusses the Caen controversy, and his plan to make a single, powerful thrust into the Ruhr to win the war.

Heavy battles in the Caen area

I never once had any cause or reason to alter my master plan . . . As soon as the armoured advance came to a standstill because of determined enemy resistance, and also because heavy rain turned the whole area into a sea of mud, I decided to abandon that thrust. Many people thought that . . . the battle had been a failure. But there was never at any time any intention of making the break-out from the bridgehead on the eastern front. Misunderstandings about this simple and basic conception were responsible for much trouble between British and American personalities.

The impression is left that the British had failed in the east (in the Caen sector) and that therefore the Americans had taken on the job of breaking out in the west. This is a clear indication that Eisenhower failed to comprehend the basic plan to which he himself cheerfully agreed . . . The misconception led to much controversy and those at Supreme Headquarters who were not very fond of me took advantage of it to create trouble as the campaign developed.

All this time the British forces were steadily playing their part on the eastern flank. By hard and continuous fighting they kept the main enemy strength occupied in the Caen sector. The greater the delay on the American front, the more I ordered the British forces to intensify their operations.

On 26 July, Eisenhower had lunch in London with the Prime Minister . . . and complained to the Prime Minister that I did not understand what I was doing. I heard later he had told the Prime Minister he was worried at the outlook taken by the American Press that the British were not taking their share of the fighting and the casualties. He gave the Prime Minister to understand that in his view the British forces on the eastern flank could and should be more offensive.

In a few days' time we were to gain a victory which has to be acclaimed as the greatest achievement in military history. The British had had the unspectacular role in the battle, and in the end it would be made to appear in the American Press as an American victory. But we all knew that if it had not been for the part played by the British Second Army on the eastern flank, the Americans could never have broken out on the western flank . . .

From that time onwards there were always 'feelings' between the British and American forces till the war ended. Patton's remarks from time to time did not help. When stopped by Bradley at Argentan he said: 'Let me go on to Falaise and we'll drive the British back into the sea for another Dunkirk.'

I gained the impression that the senior officers at Supreme Headquarters did not understand the doctrine of 'balance' in the conduct of operations. I had learnt it in battle fighting since 1940, and I knew from that experience how it helped to save men's lives. Eisenhower's creed appeared to me to be that there must be aggressive action on the part of everyone at all times. Everybody must attack all the time. I remember Bedell Smith once likened Eisenhower to a football coach; he was up and down the line all the time, encouraging everyone to get on with the game. This philosophy was expensive in life.

21 Army Group's thrust into the Ruhr

Eisenhower agreed that 21 Army Group was not strong enough to carry out the northern thrust alone. I said I wanted an American army of at least twelve divisions . . . He said that public opinion in the States would object. I asked him why public opinion should make us take military decisions which were definitely unsound. To adopt my plan he must stop the man with the ball: Patton, and his Third American Army . . . But my arguments were of no avail.

General Marshall had come to my headquarters on 8th October. I told him that since Eisenhower had himself taken personal command of the land battle . . . there was a lack of grip, and operational direction and control was lacking. Our operations had become ragged and disjointed, and we had now got ourselves into a real mess. Marshall listened, but said little. It was clear that he entirely disagreed.

The trouble was that Eisenhower wanted the Saar, the Frankfurt area, the Ruhr, Antwerp, and the line of the Rhine. To get all in one forward movement was impossible. If Eisenhower had adopted my plan he could at least have got Antwerp and the Ruhr, with bridgeheads over the Rhine in the north, and would have been well placed . . . I was of course greatly disappointed. I had hoped that we might end the German war quickly, save tens of thousands of lives, and bring relief to the people of Britain. But it was not to be. I went to see the Prime Minister to tell him that he must now expect the war to go on all through the winter and well into 1945.

It should be noted that Monty in his *Memoirs* makes a great many positive comments about Eisenhower, but these refer to his qualities as a man and are interspersed between unfailing criticisms of his abilities as a commander. Monty felt that Ike was a likeable man but a poor commander. Many, both American and British, believed the opposite of Monty: that he was a great commander but not a likeable man. It is arguable that the latter is to be preferred to command armies in time of war.

Monty's *Memoirs* was translated into fifteen languages and earned him a six-figure sum – he made much more than Rommel had from *Infanterie greift an*. Patton's campaign notes were edited by Beatrice and published as *War As I Knew It* in 1947. He condemned neither Eisenhower nor Montgomery, and wrote only about the war with Germany.

Epilogue: The Will to Fight

Churchill wrote that before the US entered the war he was not sure how it could be won. After Pearl Harbor he knew that victory would come from 'the proper application of overwhelming force'. From that moment it became a war of attrition. German manpower and resources would inevitably be exhausted before those of the Allies.

But Hitler had read Carl von Clausewitz too. He knew there was another, intangible factor that could still win the war for him: the will to fight. The will of British and American political and military leaders to continue the war might break as their troop losses mounted and public unease grew, and in that case the will of a Führer who would spend the last of his nation's blood to buy victory must prevail.

According to von Clausewitz the outcome of any conflict was determined by the ratio of 'means' and the 'will to fight' on each side. The means included troops, weapons and supplies. The will to fight resided in the politicians and the troops, but primarily in the military leaders who carried out the wishes of the politicians above them and inspired the troops below them. An enemy with a vast superiority of means could be beaten by an opponent with a stronger will to fight. In the Second World War the US provided the larger part of the means for the Allies. Hitler provided for the Axis forces an unprecedented will to fight, and inspired as much in his troops. Despite the matériel superiority of the Allies, the war might have gone against them without military leaders who demonstrated a sufficiently strong will to fight and who were able to transfer that will to their hastily recruited civilian armies.

★

Von Clausewitz was writing in 1812 when vast ranks stood in line and exchanged fire by cannon and musket, and victory went to the side with most men left standing. He made it a primary principle of war to attack strength, because that was the point at which the greater number of enemy troops could be destroyed.

Montgomery fought as von Clausewitz said he should, and in the great tradition of British generals. At El Alamein the opposing forces faced one another along an extended line and exchanged artillery and tank fire. It was a battle of attrition that Monty knew he must win because of his superior forces and supplies, *if* he and his commanders had sufficient will to fight.

Rommel's problem was not, as he thought, that Hitler refused to see the situation as it really was. The Führer knew that the Afrika Korps was inferior in numbers and supplies (and the reinforcements he could have sent went to the eastern front instead). His thinking was revealed by his *Sieg oder Todt* order: 'It would not be the first time in history that willpower has triumphed over the stronger battalions.' He allowed for the German inferiority of means, but in assuming an even greater British inferiority in willpower he had not factored in Montgomery.

Monty attacked the strongest sector of the German line because that was where the most panzer and infantry divisions could be 'crumbled'. When his most senior officers asked that they be allowed to withdraw their divisions, having suffered such heavy losses that lead regiments had been all but annihilated, he ordered them to press on. This, too, was a *Sieg oder Todt* order, issued several days before Hitler's own. Montgomery's superior will to fight – and it was his alone, all of his commanders being willing to pull back – brought the British a victory that might otherwise (despite their numerical and supply superiority) not have been won. They had failed before when stronger than the enemy. The difference at El Alamein was Montgomery.

He is usually judged by how well he managed the 'means'. His plans and his battle management were undoubtedly superb and he has been named the master of *Materialschlacht*, supervising the build-up of superior manpower and supplies before engagement to ensure the mathematics of attrition could equal only a British victory. But his greatest strength of all at El Alamein – and the resource that made the greatest difference – was his 'will to fight', with the concomitant acceptance of huge losses for Eighth Army. Instead of recognizing that Montgomery had the stronger will, Hitler assumed some weakening of Rommel's will, and it was this that he called 'defeatism', which led to a rift between the Führer and his favourite general.

Although the battle could only have been won Monty's way, de Guingand voiced an alternative possibility when he suggested that the point of attack be swapped to the weakest point of the enemy line. Monty adopted that as 'his plan', partly to appease officers he felt might 'break' if continuously driven against the enemy's strength, but he maintained the overall principle of attrition.

In the 1940s Basil Liddell Hart in Britain and Heinz Guderian in Germany challenged the principle established by von Clausewitz that it was necessary always to attack the enemy at his point of strength. Writing in the age of mobile armoured forces they argued that victory could be won more easily by attacking weakness. If during a battle a point of weakness developed in the enemy line, a concentrated attack should be made at that point, a gap opened up, then mobile forces passed through into the enemy's rear. Liddell Hart criticized von Clausewitz as 'the Mahdi of mass and mutual massacre' for attacking only strength, and advised instead 'the exploitation of the point of least resistance'.

Attacking weakness required a different kind of offensive. Liddell Hart and Guderian developed the concept of blitzkrieg independently, although both claimed to have done so first and to have influenced the other. By moving rapidly through a gap opened up at the point of weakness, and avoiding concentrations of enemy forces to keep up the momentum of the drive, armoured columns could take a distant objective, leaving it to the infantry coming up behind to 'mop up' pockets of resistance. That such an armoured mass moving at speed could overcome a numerically superior enemy was proven by the invasion of Poland, conducted according to Guderian's principles. Rommel applied them with even greater *élan* in the invasion of France and during his early conquests in North Africa.

The von Clausewitz and Liddell Hart/Guderian methods are illustrated by two images from the desert. Monty is often described before the battle at El Alamein standing with his chief of staff in his command caravan, stabbing at a map with his finger and announcing that Rommel would attack *there*, and that the enemy could be defeated in advance if plans and defences were properly prepared. The most common image of Rommel from that same period shows him standing upright in his command vehicle, his map discarded and pointing the way to his driver. He led from the front and made much use of his *Fingerspitzengefühl* (literally 'feeling in the fingertip'), an instinctive response to battle situations, the sense of what is possible that takes precedence over anything that has been planned.

Winning by blitzkrieg advance instead of by attrition required a different kind of military leader. The means were now armour and speed, and because of that the will to fight had to take on a different form. Rommel expressed what was required in a letter to Lucie on 3 April 1941, early in the North Africa campaign when he was making quick and unexpected gains: 'I took the risk against all orders and instructions because the opportunity seemed favourable.' He sensed the opportunity 'at his fingertips' but only his willpower overcame the risk he faced from the enemy and the inertia imposed upon him by the orders of the High Command. Patton saw it too: 'Leadership is the thing that wins battles. I have it, but I'll be damned if I can define it. Probably it consists in knowing what you want to do and doing it and getting mad if anyone steps in the way.'

Von Clausewitz had said that no military leader ever became great without audacity, but he assumed that such boldness would be exercised within the context of attacking strength. Boldness defined in that way was demonstrated by Montgomery at El Alamein. But when Patton quoted Frederick the Great, '*L'audace, l'audace, l'audace – tout jour [sic] l'audace*', he referred to the dash of the cavalry, and the taking of risks the outcome of which could not be calculated in advance.

Some operations had of necessity to exclude all risk. If Overlord had not been so meticulously prepared in advance, using Monty's genius for the master plan, the complexities of delivering the Allied armies, their equipment and supplies to the beaches might have defeated the invasion with only a helping hand from the Germans. The subsequent breakout and blitzkrieg advance through France was an operation of a quite different kind and required Patton's own *Fingerspitzengefühl*, audacity and willingness to take risks. After the war Bedell Smith admitted in a letter to Eisenhower that the Allies only succeeded because of both men: 'I am no Montgomery lover, but I give him his full due and believe that for certain types of operation he is without an equal. On the other hand, he would not have been worth a damn for the wide sweeping manoeuvre where Patton shone so brilliantly.'

Von Clausewitz expected that the expanded ego of a great military leader would inspire him to take risks: 'The greater the commander's self-confidence the more he will seek to break loose from the tendency to be strong everywhere at once in order to give one point a preponderant importance, even if it should be possible by running greater risks'.

There, summed up in 1812, was the strategic situation of the Allies on the western front in 1944, when such confidence was demonstrated by both Monty and Patton, but tragically not by Eisenhower.

The will to fight was a function of the ego and it is not surprising that such excessive self-confidence expressed itself in other ways too. Patton was brusque and eloquently profane in speaking to Eisenhower, but he knew when to stop pressing his own case and do as he was ordered. Monty would never have used 'language' or raised his voice to Ike, yet he was vastly more disrespectful in continuing to demand a 'single thrust' and a 'single command' even after agreeing he would do so no more. Rommel addressed Hitler in person and in writing with a forcefulness that to the very end flowed from his loyalty to the man as much as from his disenchantment with the Supreme Commander.

All three were accused of showmanship by their peers, implying that it was a selfish luxury irrelevant to their leadership. On the contrary von Clausewitz believed that a leader's will to fight had to be instilled in his men and that this was not done by issuing orders. Both Monty and Patton spent the first part of the war training their men, making soldiers out of civilians. Patton's warrior face and the barracks language of his speeches to the men, Monty's silly hats and his posture – 'there will be no retreat' – were an open display of the will to fight and its transfer to those thus enamoured of their generals. What appeared to be showmanship was an adjunct of motivation and had not been sought out, but the celebrity status it gave them among the men and the public was not unwelcome. When Rommel wrote home to ask Lucie to cut out and keep the newspaper reports about him, that was not merely to assist with the book about the campaign that he planned to write.

Swollen egos are easily bruised. Monty never recovered from the injury he incurred when Eisenhower assumed direct command of land forces. Rommel took it as a personal affront whenever Hitler chose a plan or operation suggested by von Rundstedt over his own. Patton took every reverse inflicted on him by the Allies (and he felt these were far more common than those inflicted by the enemy) as a personal attack made by Eisenhower, Bradley or Montgomery, and if he was uncertain which then he blamed Montgomery.

The darkest consequence of the will to fight is the high losses suffered by the troops. Rommel at Tobruk and Patton at Metz proved themselves as ruthless as Monty at El Alamein in 'spending' troops to break through strong enemy defences. But their reaction to it differed. During a live television interview with Montgomery, the writer Bernard Levin asked how he felt before a battle knowing that, however well he had prepared, his orders were sending many of his men to their deaths. Incredibly Monty replied that he had never thought of it like that. Patton is best remembered for slapping a man suffering from battle fatigue, but there are many reports of visits to field hospitals during which he knelt beside GIs to ask their forgiveness. On coming across a fatally injured man on the battlefield, he cradled him and prayed with him until he died. Rommel's oath of allegiance to the Führer meant everything to him and he obeyed orders that he knew to be strategically wrong, but when ordered to sacrifice his beloved Deutsches Afrika Korps to the last man at El Alamein, he refused.

Montgomery was an easy man to dislike, and on a number of occasions his insistence that an apparent defeat was in fact a victory, or that an operation had gone according to plan when it clearly had not, acted against him. Liddell Hart wrote to him on 17 March 1946 to point out that 'Your manner has always been your worst handicap.' No one defeated Montgomery more conclusively than Montgomery himself.

Patton had all the qualities of the cavalry he loved. Field Marshal Brooke described him as 'a dashing, courageous, wild and unbalanced leader, good for operations requiring thrust and push but at a loss in any operation requiring skill and judgement'. According to war correspondent Leland Stowe, he called himself 'an obstreperous, fighting, cantankerous bastard, and proud of it'.

Rommel's men simply said of him, '*Er hat die Strapazen mitgemacht*' which roughly translates as 'He shared the shit'. Liddell Hart pointed out that 'His successes were achieved with inferiority of resources and without any command of the air. No other generals on either side won battles under these handicaps.'

Hitler was a great general. His madness was not in his strategy but in his unreasonable perseverance with it. He combined in one what no Allied general could, by applying the theories of both von Clausewitz (impressed upon his army by the force of his own will) and Guderian (via Rommel's genius for blitzkrieg) to the war. His 'von Clausewitz offensive' was defeated by Montgomery, who bettered him in *Materialschlacht*, attrition and the will to fight expressed as perseverance. His 'Guderian' offensive was beaten by Patton, who bettered him in the application of blitzkrieg and the will to fight expressed as audaciousness. Of course, it was not that simple and only a complex interplay of matériel and mobile warfare finally won the campaign for the Allies. But if neither Montgomery nor Patton had been available to the Allies, and Hitler's mental decline had not broken his trust in Rommel, the Wüstenfuchs might well have won the war for Germany on the sandy beaches of Normandy.

Acknowledgements & Sources

Grateful thanks to the following:
Bundesarchiv Militärarchiv, Freiburg; Churchill Archives Centre, Churchill College, Cambridge; Citadel Archives, Charleston, South Carolina; Dwight D. Eisenhower Library, Abilene, Kansas; EP Microfilm Ltd, Wakefield; Imperial War Museum, London; Institut für Zeitgeschichte, Munich; Library of Congress (Manuscript Division), Washington, DC; Liddell Hart Centre for Military Archives, King's College, London; George C. Marshall Library, Virginia Military Academy, Lexington; Military Academy Library, West Point; Sterling Library, Yale University, New Haven, Connecticut; US Army Military History Institute, Carlisle, Pennsylvania; UK National Archives, London; US National Archives, Washington, DC.

Archival sources (papers, correspondence and diaries)

Alexander, Harold: UK National Archives, London

American War Diaries: US National Archives, Washington, DC

Armbruster, Wilfried: EP Microfilm Ltd, Wakefield

Bedell Smith, Walter: Eisenhower Library, Abilene, Kansas

Bradley, Omar: US Army Military History Institute, Carlisle, Pennsylvania; Military Academy Library, West Point

British War Diaries: UK National Archives, London

Brooke, Alan: Liddell Hart Centre for Military Archives, King's College, London

Butcher, Harry: Eisenhower Library, Abilene, Kansas

Cabinet Papers: UK National Archives, London

Churchill, Winston: UK National Archives, London

Clark, Mark: Citadel Archives, Charleston, South Carolina

Coningham, Arthur: UK National Archives, London

Cunningham, Andrew: Churchill Archives Centre, Churchill College, Cambridge

De Guingand, Francis: UK National Archives, London; Imperial War Museum; Liddell Hart Centre for Military Archives, King's College, London

Dempsey, Miles: UK National Archives, London; Liddell Hart Centre for Military Archives, King's College, London

Eisenhower, Dwight: Eisenhower Library, Abilene, Kansas

Gay, Hobart: US Army Military History Institute, Carlisle, Pennsylvania

German War Diaries: Imperial War Museum, London

Grigg, James: Churchill Archives Centre, Churchill College, Cambridge

Hodges, Courtney: Eisenhower Library, Abilene, Kansas

Hughes, Everett: Library of Congress (Manuscript Division), Washington, DC

Kesselring, Albert: US National Archives, Washington, DC

Lang, Hellmuth: Institut für Zeitgeschichte, Munich

Lattmann, Hans: EP Microfilm Ltd, Wakefield

Leese, Oliver: Imperial War Museum, London

Liddell Hart, Basil: Liddell Hart Centre for Military Archives, King's College, London

Lucas, John: US Army Military History Institute, Carlisle, Pennsylvania

Marshall, George: Marshall Library, Virginia Military Academy, Lexington

Mellenthin, Friedrich von: US National Archives, Washington, DC

Mims, John: US Army Military History Institute, Carlisle, Pennsylvania

Montgomery, Bernard: Imperial War Museum, London; UK National Archives, London; Royal Warwickshire Regimental Museum, Warwick

Overlord Papers: UK National Archives, London

Patton, George: Library of Congress (Manuscript Division), Washington, DC; US Army Military History Institute, Carlisle, Pennsylvania;

Military Academy Library, West Point; George C. Marshall Library, Virginia Military Academy, Lexington

Reynolds, Tom and Phyllis: Imperial War Museum, London

Rommel, Erwin: Bundesarchiv Militärarchiv, Freiburg; Institut für Zeitgeschichte, Munich; EP Microfilm Ltd, Wakefield; Imperial War Museum, London; US National Archives, Washington, DC

Rommel, Lucie: Bundesarchiv Militärarchiv, Freiburg; Institut für Zeitgeschichte, Munich

Rundstedt, Gerd von: US National Archives, Washington, DC

SHAEF papers: UK National Archives, London

Simpson, Frank: Imperial War Museum, London

Speidel, Hans: US National Archives, Washington, DC

Stimson, Henry: Sterling Memorial Library, Yale University, New Haven, Connecticut

Summersby, Kay: Eisenhower Library, Abilene, Kansas

Tedder, Arthur: UK National Archives, London; Eisenhower Library, Abilene, Kansas

Warning, Elmar: EP Microfilm Ltd, Wakefield

Wilmot, Chester: Liddell Hart Centre for Military Archives, King's College, London

Bibliography

Alexander, Harold, *The Alexander Memoirs 1940–1945*, Cassell, London, 1962

Barnett, Corelli, *The Desert Generals*, William Kimber, London, 1960

Bayerlein, Fritz, 'El Alamein' in *The Fatal Decisions*, Michael Joseph, London, 1956

Blumenson, Martin (ed.), *The Patton Papers*, Houghton Mifflin, Boston, 1974

— *Patton: The Man Behind The Legend*, William Morrow, New York, 1985

Blumentritt, Günther, *Von Rundstedt*, Odhams, London, 1952

Bradley, Omar, *A Soldier's Story*, Henry Holt, New York, 1951

Butcher, Harry, *Three Years with Eisenhower*, Heinemann, London, 1946

Bryant, Arthur, *The Turn of the Tide*, Collins, London, 1959

Carver, Michael, *El Alamein*, Batsford, London, 1962

Chalfont, Alun, *Montgomery of Alamein*, Weidenfeld & Nicolson, London, 1976

Churchill, Winston, *The Hinge of Fate*, Cassell, London, 1951

Clark, Mark, *Calculated Risk*, Harrap, London, 1951

Connell, John, *Auchinleck*, Cassell, London, 1959

Dawnay, Kit, 'Inside Monty's Headquarters' in *Monty At Close Quarters* (ed. T. Howarth), Leo Cooper, London, 1985

D'Este, Carlo, *Decision in Normandy*, Collins, London, 1983

— *A Genius For War*, Harper Collins, New York, 1995

De Guingand, Francis, *Operation Victory*, Hodder & Stoughton, London, 1947

— *Generals at War*, Hodder & Stoughton, London, 1964

Eisenhower, David, *Eisenhower at War*, Random House, New York, 1986

Eisenhower, Dwight D., *Crusade in Europe*, Heinemann, London, 1949

Eisenhower, John, *General Ike*, Simon & Schuster, New York, 2003

Farago, Ladislas, *Patton: Ordeal and Triumph*, Westholme, Yardley, Pennsylvania, 2005

Fraser, David, *Knight's Cross*, Harper Collins, London, 1993

Gisevius, Hans, *To The Bitter End*, Cape, London, 1948

Görlitz, Walter, *The German General Staff*, Hollis and Carter, London, 1953

Guderian, Heinz, *Panzer Leader*, Michael Joseph, London, 1952

Halder, Franz, *Private War Journals of Franz Halder*, Westview, Boulder, Colorado, 1976

Hamilton, Nigel, *Monty: The Making of a General 1887–1942*, Hamish Hamilton, London, 1981

— *Monty: Master of the Battlefield 1942–1944*, Hamish Hamilton, London, 1983

— *Monty: The Field Marshal 1944–1976*, Hamish Hamilton, London, 1986

Heckmann, Wolf, *Rommel's War In Africa*, Granada, London, 1981

Henderson, Johnny, *Watching Monty*, Sutton, Stroud, 2005

Hirshson, Stanley, *General Patton*, Harper Collins, New York, 2002

Horne, Alastair, and Montgomery, David, *The Lonely Leader*, Macmillan, London, 1994

Horrocks, Brian, *A Full Life*, Collins, London, 1960

Irving, David, *Rommel: The Trail of the Fox*, Weidenfeld and Nicolson, London, 1977

Irving, David, *The War Between the Generals*, Viking, London, 1981

Kesselring, Albert, *A Soldier's Diary*, William Morrow, New York, 1954

Lamb, Richard, *Montgomery in Europe 1943–45*, Buchan & Enright, London, 1983

Lewin, Ronald, *Rommel as Military Commander*, Batsford, London, 1968

— *Montgomery as Military Commander*, Batsford, London, 1971

Liddell Hart, Basil, *Memoirs*, Cassell, London, 1965

Luck, Hans von, *Panzer Commander*, Cassell, London, 1989

Manstein, Erich von, *Lost Victories*, Methuen, London, 1958

Marshall, Charles, *The Rommel Murder*, Stackpole, Mechanicsburg, Pennsylvania, 2002

Mauldin, Bill, *The Brass Ring*, Norton, New York, 1971

Mellenthin, Friedrich von, *Panzer Battles 1939–1945*, Cassell, London, 1955

Montgomery, Bernard, *Normandy to the Baltic*, Hutchinson, London, 1946

— *El Alamein to the River Sangro*, Hutchinson, London, 1948

— *Memoirs of Field Marshal Montgomery*, Collins, London, 1958

Montgomery, Brian, *A Field Marshal in the Family*, Constable, London, 1973

Moorehead, Alan, *Montgomery*, Hamish Hamilton, London, 1946

Morgan, Frederick, *Peace and War*, Hodder and Stoughton, London, 1961

Patton, George, *War As I Knew It*, Bantam, New York, 1980

Patton, Robert, *The Pattons*, Brassey's, Washington, DC, 1994

Reuth, Ralf, *Rommel, The End of a Legend*, Haus, London, 2005

Reynolds, Michael, *Monty and Patton*, Spellmount, Staplehurst, 2005

Rommel, Erwin, *The Rommel Papers* (ed. Basil Liddell Hart), Collins, London, 1953

— *Infantry Attacks*, Greenhill, London, 1990

Ruge, Friedrich, *Rommel in Normandy*, Presidio Press, London, 1979

Schmidt, Heinz, *With Rommel in the Desert*, Harrap, London, 1951

Schweppenburg, Geyr von, *The Critical Years*, Allen Wingate, London, 1952

Showalter, Dennis, *Patton and Rommel*, Berkley Caliber, New York, 2005

Speidel, Hans, *Invasion 1944*, Henry Regnery, Chicago, 1950

Summersby, Kay, *Eisenhower Was My Boss*, Prentice Hall, New York, 1948

Tedder, Arthur, *With Prejudice*, Cassell, London, 1966

Thompson, R. W., *The Montgomery Legend*, Allen & Unwin, London, 1967

Warlimont, Walter, *Inside Hitler's Headquarters*, Praeger, New York, 1964

Westphal, Siegfried, *The German Army in the West*, Cassell, London, 1951

Wilmot, Chester, *The Struggle for Europe*, Collins, London, 1952

Young, David, *Rommel*, Collins, London, 1950

Index

French Army Tank School, Chamlieu, 42

French refugees, 297

French Resistance, 295

Freyberg, General Sir Bernard, 156-7

friendly fire incidents (Blue-on-Blue), 312, 319

Gatehouse, General A. H., 149, 165

Gause, General, 130, 136-7

Gay, Colonel Hobart R. (Hap), 209, 211, 232, 384-5

George V, 40

George VI, 2, 194, 258, 259, 327: Montgomery and, 329; visit to France, 16 June 1944, 278; visit to Eindhoven, October 1944, 244

German Army: Allied need to rebuild, post-1945, 378, 388-9; career/class structure, 12, 13, 71; expansion, pre-1939, 56-7; Hitler, oath of allegiance to, 56, 405; as Imperial German Army, 12-13, 29, 43, 51; and *Dolchstass* belief, 51; infantry, 321; post-1945 (West German), 390-91; as *Reichswehr* (interwar) *see Reichswehr;* Army Group B, in Ardennes under Model 355-9;7th Army 317; 276th Infantry Division, 296; 277th Infantry Division, 296; 352nd Infantry Division, 272; Württemburg (124th) Infantry Regiment *see* Württemburg ; Deutsches Afrika Korps *see* Deutsches Afrika Korps;Panzerarmee *see* Panzerarmee; Propaganda Company, 90-91

German propaganda 10, 11: Army propaganda, 90-91; based on Rommel, 90-92, 104, 120-21, 138-9, 181, 316; on his command of Atlantic Wall 255-5: exploiting rift in Anglo-American relations, January 1945, 364-5

Das Reich (propaganda newspaper), 91

Victory in the West (film) 91-2, 104, 167

Germany: anti-Semitism 52, 53, 67; Austria, annexation of, 66; Belgium, invasion of, 74, 79-81, 83; Communism/anti-Communism 51-2, 55-6; Czechoslovakia, occupation of, 67; in First World War *see* First World War; France, invasion of *see* France, German invasion; inter-war period, 51-2, 55-7; Italy and, 65, 103, 178, 200; *see also* Italy, German occupation; Poland, invasion of, 69, 70-72, 76; post-war reconstruction, Rommel on, 277, 293; proposed political settlement with Allies 1944/5, 290; Rommel on need for 277, 282-3, 290, 291, 292-4, 312; RAF bombing 230-31, 245, 251; Russia, invasion of *see* Russia, German invasion; Sudetenland, occupation of, 66, 69; *see also* Hitler, Adolf

Germany, Allied invasion, 1944/5, 353: Baltic ports 366, 375-6; Berlin *see* Berlin; Lùbeck 375-6; Rhine crossing *see* Rhine crossing; Ruhr valley 325, 329, 330-31, 335-6, 341-2, 350, 352, 354-5, 367, 371, 397; and bombing of, 366; Russian Army in, 365-6, 371, 373-4; Saar valley 325, 331, 333, 352, 353, 354-5, 356; German surrender, 4 May 1945, 375, 376

Germany, Allied occupation, 1945, 377-8: denazification policy 22, 378, 380-81, 382, 390; German prisoners of war, 377, 378; German weapons/ equipment, policy on, 378; Montgomery as military governor, British Zone, 377-8, 379-80; Patton as military governor, Bavaria 22, 378-83; Yalta agreement on, 373; on occupation zones 373

Gibraltar, 172

Gillard, Frank, 275

Giordana, Ralph: *The Falsehood of Tradition,* 390

Patton, General George Smith:
alcohol consumption, 65, 377; birth/
childhood 2, 23; on bond-raising
tour of US, June 1945, 379; *Cavalry
Journal*, articles in, 66; character: 1,
22, 23, 25–6, 36, 38, 63, 64, 87,
141–2, 182–3, 192, 405: arrogant/
egotistical 1, 63, 142, 172, 184,
186, 234, 326, 352–72, 403, 404;
bigoted 24, 25, 88, 232;
insubordinate 326–7, 333, 380–84;
tactless 22, 192, 256–8, 280–83;
violent/brutal 66, 202, 216–17,
220–21, 232, 234–5; decorations
248; his diary 40, 49, 192, 198, 215,
220, 232, 239, 333, 380; as a
diplomat 22, 170, 379; his dog
(Willie) 251, 258, 383, 387;
education: 23–4; military 4, 24–6,
64, 319–20; as a slow learner 23, 24,
25; family 23, 25, 63, 333:
correspondence with, 28, 35–6; *see
also* individual family members; in
First World War 39–40, 43–6;
health, 246; hobbies, 384; in
interwar years, 52–3, 63–6, 67–8,
76–7, 87–90, 112–13; language, use
of obscenities/profanities, 2, 22, 25,
65, 90, 132, 133, 182, 206, 259,
263–4, 383, 403; as a leader *see*
leadership qualities/style; marriage,
26, *see also* Patton, Beatrice; mental
health, concern over, 282;
Montgomery and, 48–9, 211, 220,
247–8, 258–9, 261, 268; on
Montgomery 180, 232, 239, 291; *see
also* Anglo-American relations;
motto: *L'audace, l'audace* , 4, 63, 313;
nickname: Old Blood and Guts, 2,
113; on his old age, 380; personal
appearance, 2, 24, 25, 27, 112–13,
133, 251, 252, 263, 350, 382–3: Colt
revolvers, 2, 38, 141, 205, 361; his
poetry 30, 42–3, 49, 65; post-war
career 22, 378–83; portrait of, by
Boleslaw Czedekowski 21, 22–3,

361, 382; propaganda based on *see*
American propaganda; reincarnation,
belief in, 42, 63, 67; reputation 22,
52, 77, 112–13, 171–2, 186, 187–8,
218, 253, 313, 320, 379, 386, 405: in
Britain, 188; *see also* Anglo-
American relations; in Germany,
253; **Rommel and**, 3, 4, 48–9, 186,
192, 315; Rommel on, 3; on
Rommel, 3, 188; in Second World
War *see* individual battles/operations;
sexual activities, 65, 88–9, 284, 318,
350; speeches: 2, 89, 90, 221, 251–2,
256–8, 288, 315, 324–5, 352, 353,
379; invasion speech to US Third
Army, 263–7; as a sportsman, 25, 26,
27, 65: in Stockholm Olympics,
1912, 27; tactics *see* tactics/tactical
ability; *War As I Knew It* (ed.
Beatrice Patton), 321–2, 398; death/
funeral, 384–7
Patton, George (Sr) (father of George
Smith Patton), 23, 24
Patton, Robert (brother of George Smith
Patton), 23–4
Patton, Ruth Ellen (daughter of George
Smith Patton), 65
Patton, William Tazewell (great-uncle of
George Smith Patton), 23
Patton sabre (US Sabre Mark 1913), 27,
36, 40
Pearl Harbor, Japanese attack,
7 December 1941, 115: Patton's
1935 report on feasibility of, 65; *see
also* Japan
Pearson, Drew, 234, 239
Percival, Major A. E., 54
Pershing, General John (Black Jack) 49,
191: in First World War 39–40, 43;
in Mexico, 1916 36–7, 117; Patton
and, 141; criticism of Patton 235;
Patton, correspondence with 53, 76,
235; as Patton's role model 36–7
Pétain, Henri, 168
Poett, General Nigel, 274
Poland: Danzig, 4, 13, 70, 72